Interactive
Microcomputer-Aided

Structural
Steel Design

Interactive Microcomputer-Aided

Structural Steel Design

HOJJAT ADELI

The Ohio State University

PRENTICE HALL, *Englewood Cliffs, New Jersey 07632*

Library of Congress Cataloging-in-Publication Data

ADELI, HOJJAT (date)
  Interactive microcomputer-aided structural steel design.

  Bibliography: p.
  Includes index.
  1. Building, Iron and steel—Data processing. 2. Structural
design—Data processing. 3. Computer-aided design. I. Title.
TA684.A23 1988   624.1′821′0285   87-25796
ISBN 0-13-469982-3

Editorial/production supervision: Merrill Peterson
Interior design: Joan Stone
Cover design: Ben Santora
Manufacturing buyer: Rhett Conklin

Printed in the United States of America

10 9 8 7 6 5 4 3 2 1

ISBN 0-13-469982-3 01

Prentice-Hall International (UK) Limited, *London*
Prentice-Hall of Australia Pty. Limited, *Sydney*
Prentice-Hall Canada Inc., *Toronto*
Prentice-Hall Hispanoamericana, S.A., *Mexico*
Prentice-Hall of India Private Limited, *New Delhi*
Prentice-Hall of Japan, Inc., *Tokyo*
Simon & Schuster Pte. Ltd., *Singapore*
Editora Prentice-Hall do Brasil, Ltda., *Rio de Janeiro*

To my wife
Nahid

Contents

Preface

This book presents a new approach to structural steel design called interactive microcomputer-aided design. Many steel designers are still designing structures with the aid of design tables, charts, and graphs. The last edition of the authoritative *Manual of Steel Construction* published by the American Institute of Steel Construction (AISC) (47) as well as a number of other recently published manuals is evidence of this fact. These traditional design aids often cannot be used directly, and the designer must resort to the wearisome process of interpolation. A survey of recent design textbooks also indicates that this is the way future structural engineers are taught design courses in most engineering schools.

The advent of inexpensive microprocessor-based computers is changing the way structures are designed, and a growing number of engineers and engineering programs are responding to the change. Computer-aided analysis of structures is now performed routinely in most engineering schools and engineering firms. Introduction of computers into the design process, however, has been relatively slow. While the process of analysis can easily be automated, total automation of the design process without the interaction of the human designer is at present impractical except for very limited cases.

This book presents the fundamental subjects of steel design on microcomputers through the approach of interactive design. In this approach, the designer or user of the system is in charge, and the system works as an assistant to him or her. The philosophy and principles behind this approach are described in Chapter Three. Computer-aided design (CAD) programs presented in this book are all developed in Advanced BASIC. BASIC language lends itself to such interactive CAD systems very effectively.

The present volume includes nine chapters and is intended for use in an introductory senior level course on structural steel design. The second volume of the book, to be published later, is intended for use in an intermediate senior or first-year graduate level course on design of steel structures. In the last chapter of this volume, an introduction to the potential applications of artificial intelligence in computer-aided design of structures will be presented. In the near future, we will be able to develop computer "expert systems" that will emulate an experienced human designer by extracting, articulating, and implementing a human expert's knowledge and/or knowledge acquired through the use of numerical experimentation and computers. A knowledge-based expert system may be viewed as an extension of the approach of interactive design presented in this book (5).

In an attempt to keep the book self-contained as much as possible, the fundamentals of design are covered at the beginning of each chapter. The design basis is the 1980 AISC specification (47) as well as the recently proposed Load and Resistance Factor Design (LRFD) specification (48). Even though this book emphasizes computer applications, no designer can achieve a rational and meaningful design without a good understanding of the mechanics involved. Without indulging in lengthy theoretical developments, I have tried to present succinctly the reason behind each design equation as well as its limitations. Realizing the necessity of solving a few problems by hand for students or novice designers, I have included in every chapter a few manually solved examples before presenting the interactive programs.

I am thankful to Professors Robert Hanson of the University of Michigan, Ronald Harichandran of the Michigan State University, and Ravindra Vyas of the University of Utah for reviewing the manuscript and providing suggestions and comments. The research presented in this book has been partially sponsored by grants from Bethlehem Steel Corporation and Ohio State University College of Engineering. Several graduate students have worked on the interactive design of various types of structures which have been published in several journal articles (references 6, 10–20, and 22–23). James Fiedorek and Khing Phan have been especially helpful in developing some of the computer programs.

It should be emphasized that the primary objective of this book is to present a new approach to structural design. As the design is an open-ended problem, the development of CAD programs should also be considered a continuing effort. While the computer programs presented in this book have been used by some 50 students in my Structural Steel Design course offered in the fall of 1986, they may still contain "bugs" and certainly have room for improvement. The present book should be considered as a first attempt. I am simply anxious to share it with other students and colleagues. In fact, these programs can be easily modified or extended by students. I will appreciate receiving comments and suggestions by readers.

Hojjat Adeli
Columbus, Ohio

Note: BASIC codes for the programs presented in sections 4.7, 5.10, 6.6, 7.7, 8.4, and 9.12 can be obtained from the author for a nominal fee.

Notations
and Abbreviations

A	Cross-sectional area (in.^{2})
A_b	Nominal body area of a fastener (in.^{2})
A_{bs}	Cross-sectional area of bearing stiffener (in.^{2})
A_e	Effective net area of an axially loaded tension member
A_{eff}	Effective area (in.^{2})
A_f	Area of flange (in.^{2})
A_g	Gross area
A_n	Net area of an axially loaded tension member
A_{sc}	Cross-sectional area of a steel cable
A_{st}	Cross-sectional area of a stiffener or a pair of stiffeners (in.^{2})
A_w	Area of web (in.^{2})
AI	Artificial intelligence
AISC	American Institute of Steel Construction
AISCM	AISC Manual
AISCS	AISC Specification
C	Structural earthquake response factor
C_b	Bending coefficient dependent upon moment gradient = $1.75 + 1.05(M_1/M_2) + 0.3(M_1/M_2)^2$ (less than or equal to 2.3)
C_c	Column slenderness ratio separating elastic and inelastic buckling

C_e	Combined height, exposure, and gust factor
C_m	Coefficient applied to bending term in the beam-column interaction formula dependent upon curvature caused by applied moments
C_q	Wind pressure coefficient
C_t	Reduction coefficient in computing the effective net area of an axially loaded tension member
CAD	Computer-aided design
D	Dimension of the structure in a direction parallel to the applied earthquake forces (ft)
	Diameter of a hole
	Outside diameter of a tubular member
D_n	Dead load
E	Modulus of elasticity of steel (29,000 ksi)
E_n	Earthquake load
F_a	Allowable axial compressive stress in the absence of bending (ksi)
F_{as}	Allowable axial compressive stress in the absence of bending moment for secondary members such as bracings (ksi)
F_b	Allowable bending stress in the absence of axial force (ksi)
F_b'	Allowable bending stress in the compression flange of plate girders as reduced for hybrid girders or because of large web depth-to-thickness ratio (ksi)
F_{cr}	Critical compression flange stress (ksi)
F_e'	Euler stress for a prismatic member divided by factor of safety (ksi)
F_i	Lateral earthquake force applied to level i
F_t	Allowable axial tensile stress (ksi)
F_u	Specified minimum tensile strength of steel (ksi)
F_v	Allowable shearing stress (ksi)
F_x	Lateral earthquake force applied to level x
F_y	Specified yield stress of steel (ksi)
F_{yf}	Yield stress of the flange steel (ksi)
F_{ys}	Yield stress of the stiffener steel (ksi)
F_{yt}	Yield stress of the tension flange (ksi)
F_{yw}	Yield stress of the web steel (ksi)
F.S.	Factor of safety
G	Shear modulus of elasticity (11,200 ksi)
	Nomograph designation of end condition used in column design to determine the effective length
I	Importance Factor
	Moment of inertia of a section (in.^{4})

I_{st} Moment of inertia of a single stiffener or a pair of intermediate stiffeners with respect to an axis in the plane of the web and perpendicular to the plane of stiffeners (in.^{4})

I_x Moment of inertia of a section about the x-axis (in.^{4})

I_y Moment of inertia of a section about the y-axis (in.^{4})

J Torsional constant of a cross section (in.^{4})

K A coefficient that depends on the type of the lateral force resisting system

          Effective length factor for a column or beam-column

          Curvature

L Length of a member (ft)

          Span of a beam (ft)

L_1 Length of a portion of intermittent fillet weld (in.)

L_n Floor live load

L_r Roof live load

L_u Unbraced length (length between points which are braced against lateral displacement of compression flange or twist of the cross section) (in.)

LRFDS   Load and Resistance Factor Design Specification

M Bending moment (Kip-ft)

M_1 Smaller moment at the ends of the unbraced length of beams (Kip-ft)

M_2 Larger moment at the ends of the unbraced length of beams (Kip-ft)

M_D Moment produced by dead load (Kip-ft)

M_L Moment produced by live load (Kip-ft)

M_n Nominal flexural strength (Kip-ft)

M_p Plastic moment capacity of a section (Kip-ft)

M_u Required bending moment based on the factored loads (Kip-ft)

N Length of base plate (in.)

          Bearing length (in.)

P_{cr} Maximum strength of an axially loaded compression member or beam (Kip)

P_e Euler buckling load (Kip)

P_n Nominal axial strength (Kip)

P_u Required axial strength (Kip)

P_y Plastic axial load = AF_y (Kip)

Q First moment of area (in.^{3})

R Reduction coefficient for live loads (percent)

R_e Hybrid girder factor

S Site-structure resonance factor for earthquake loading

S_n Snow load

S_x Elastic section modulus about the x-axis (in.^{3})

S_{xc} Section modulus corresponding to the compression flange (in.^{3})

S_{xt}	Section modulus corresponding to the tension flange (in.^{3})
S_y	Elastic section modulus about the y-axis (in.^{3})
T	Fundamental elastic period of vibration of the structure in the direction under consideration
T_s	Characteristic site period
UBC	Uniform Building Code
V	Total lateral seismic force
	Shear force (Kip)
V_n	Nominal shear strength (Kip)
V_u	Required shear force based on the factored loads (Kip)
W	Total dead load of the structure
W_n	Wind load
Z	Seismic zone coefficient
	Plastic section modulus (in.^{3})
Z_x	Plastic section modulus with respect to x-axis (in.^{3})
Z_y	Plastic section modulus with respect to y-axis (in.^{3})
a	Clear distance between transverse stiffeners (in.)
a_r	Ratio of web area to compression flange area
a_1	Spacing of intermittent fillet weld (in.)
b_{bs}	Width of bearing stiffener (in.)
b_f	Flange width of a rolled beam or plate girder (in.)
b_s	Width of intermediate stiffener (in.)
d	Depth of beam, girder, or column (in.)
	Diameter of a fastener (in.)
d_h	Diameter of hole (in.)
f	Stress
f_a	Computed axial stress (ksi)
f_b	Computed bending stress (ksi)
f'_c	Compressive strength of concrete
f_{cb}	Compressive stress in the bearing stiffener (ksi)
f_{cw}	Bearing compressive stress in the web plate (ksi)
f_t	Computed tensile stress (ksi)
f_v	Computed shear stress (ksi)
f_{vs}	Shear between girder web and transverse stiffeners (Kips per linear inch of single stiffener or pair of stiffeners)
g	Center-to-center spacing of two consecutive holes in the transverse direction (normal to the direction of stress) (gage) (in.)
h	Clear distance between flanges of a beam or column (in.) ($h = d - 2t_f$)

h_c	Twice the distance from the neutral axis of the girder cross section to the inside face of the compression flange minus the fillet or corner radius (in.)
h_i	Height of level i above the base
h_x	Height of level x above the base
h_n	Height of the structure (ft)
h_1	Depth of intermediate stiffener (in.)
k	Distance from outer face of flange to web toe of fillet of rolled sections or equivalent distance for welded sections (in.)
m	Ratio of web to flange yield stress in hybrid girders
q	Force per unit length of fillet weld (Kip/in.)
q_a	Allowable strength of fillet weld (Kip/in.)
q_s	Wind stagnation pressure at the standard height of 30 ft (psf)
r	Radius of gyration (in.)
r_T	Radius of gyration of a section comprising the compression flange plus $\frac{1}{3}$ of the compression web area, taken about an axis in the plane of the web (in.)
r_x	Radius of gyration with respect to x-axis (in.)
r_y	Radius of gyration with respect to y-axis (in.)
s	Center-to-center spacing of the two consecutive holes in the direction of stress (pitch) (in.)
	Sag of a cable
s'	Spacing of cables around the circumference of a circular suspension cable roof
t	Thickness (in.)
t_{bs}	Thickness of bearing stiffener (in.)
t_f	Thickness of flange (in.)
t_s	Thickness of intermediate stiffener (in.)
t_w	Thickness of web
w	Intensity of distributed load (Kip/in.)
w_D	Intensity of distributed dead load (Kip/ft)
w_g	Gross width
w_L	Intensity of distributed live load (Kip/ft)
w_n	Net width
w_w	Size of the fillet weld (in.)
γ	Specific gravity of steel $= 0.490$ Kip/ft^3
Δ	Beam deflection
ε	Strain
λ_c	Column slenderness parameter
λ_p	Limiting slenderness parameter for compact elements

λ_r	Limiting slenderness parameter for slender elements
ν	Poisson's ratio ($= 0.3$ for steel)
ρ	Radius of curvature
ϕ	Resistance factor
ϕ_b	Resistance factor for flexure
ϕ_c	Resistance factor for compression
ϕ_t	Resistance factor for tension
ϕ_v	Resistance factor for shear

Interactive
Microcomputer-Aided

Structural
Steel Design

1

Steel
and Steel Structures

1.1 PROPERTIES OF STEEL

The stress-strain relationship for mild structural steel is shown graphically in Fig. 1.1. This curve is obtained from a tensile test in which a prismatic bar of cross-sectional area A and length L is subjected to two equal but opposite forces P at its two ends. In Fig. 1.1(a), elongation, strain, and stress are shown by δ, ε, and f, respectively. Several important points and stress limits can be identified on this figure.

Point A is the proportional limit, the largest stress for which Hooke's law applies. Point B is the elastic limit (no permanent deformation). Point C is the yield point with yield stress F_y. Point E is the ultimate limit with ultimate stress F_u. The three points A, B, and C are usually very close to each other. Plastic strains are usually 10 to 15 times the elastic strain. The strain-hardening portion of the stress-strain diagram is not used in structural steel design. Yielding of steel with practically no stress increase is a significant property of the material for resisting dynamic loads. A mild structural steel undergoes substantial deformation before failure by fracture. In contrast, failure of a brittle material such as concrete or glass is sudden (Fig. 1.2) without any advance warning. The strain at failure [point F in Fig. 1.1(b)] for mild steels is 150 to 200 times the elastic strain.

When the normal stress f is found by dividing the axial force P by original cross-sectional area A, the stress-strain diagram $OABCDEF$ will be obtained. Beyond approximately point D, however, a lateral contraction or *necking* occurs,

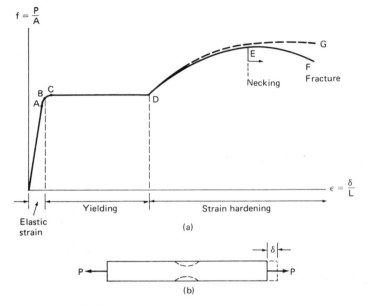

Figure 1.1 Stress-strain relationship for mild steel.

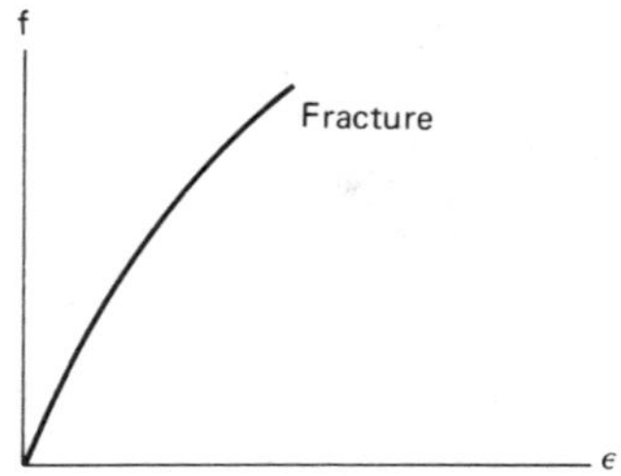

Figure 1.2 Stress-strain relationship for a brittle material.

as shown in Fig. 1.1(b). If the narrow cross section of the bar at the neck is used in calculating the normal stress, the dashed curve *DG* will be obtained instead of *DF*.

There is an inverse relationship between ductility and strength of steel. In other words, high-strength steels have low ductility. As a result, the designer may have to compromise between these two properties.

1.2 ADVANTAGES AND DISADVANTAGES OF STEEL AS A STRUCTURAL DESIGN MATERIAL

The following advantages in general may be credited to steel as a structural design material:

1. High strength/weight ratio. Steel has a high strength/weight ratio. Thus, the dead weight of steel structures is relatively small. This property makes steel a very attractive structural material for
   a. High-rise buildings
   b. Long-span bridges
   c. Structures located on soft ground
   d. Structures located in highly seismic areas where forces acting on the structure due to an earthquake are in general proportional to the weight of the structure
2. Ductility. As discussed in the previous section, steel can undergo large plastic deformations before failure, thus providing a large reserve strength. This

property is referred to as *ductility*. Properly designed steel structures can have high ductility, which is an important characteristic for resisting shock loading such as blast or earthquakes. A ductile structure has energy-absorbing capacity and will not incur sudden failure. It usually shows large visible deflections before failure or collapse.

3. Predictable material properties. Properties of steel can be predicted with a high degree of certainty. Steel in fact shows elastic behavior up to a relatively high and usually well-defined stress level. Also, in contrast to reinforced concrete, steel properties do not change considerably with time.

4. Speed of erection. Steel structures can be erected quite rapidly. This normally results in quicker economic payoff.

5. Quality of construction. Steel structures can be built with high-quality workmanship and narrow tolerances.

6. Ease of repair. Steel structures in general can be repaired quickly and easily.

7. Adaption to prefabrication. Steel is highly suitable for prefabrication and mass production.

8. Repetitive use. Steel can be reused after a structure is disassembled.

9. Expanding existing structures. Steel buildings can be easily expanded by adding new bays or wings. Steel bridges may be widened.

10. Fatigue strength. Steel structures have relatively good fatigue strength.

The following may be considered as disadvantages of steel in certain cases:

1. General cost. Steel structures may be more costly than other types of structures.

2. Fireproofing. The strength of steel is reduced substantially when heated at temperatures commonly observed in building fires. Also, steel conducts and transmits heat from a burning portion of the building quite fast. Consequently, steel frames in buildings must have adequate fireproofing.

3. Maintenance. Steel structures exposed to air and water, such as bridges, are susceptible to corrosion and should be painted regularly. Application of weathering- and corrosion-resistant steels may eliminate this problem.

4. Susceptibility to buckling. Due to high strength/weight ratio, steel compression members are in general more slender and consequently more susceptible to buckling than, say, reinforced concrete compression members. As a result, considerable materials may have to be used just to improve the buckling resistance of slender steel compression members.

1.3 TYPES OF STEEL STRUCTURES

1.3.1 Framed Structures

These are the most common types of steel structures. This book concentrates on the design of framed structures. Framed structures in general consist of tension members, beams, columns, beam-columns, and members acted upon by combined bending and torsion. Framed structures may be divided into six categories: beams, plane trusses, space trusses, plane frames, space frames, and grids. Many three-dimensional structures such as buildings may be considered as consisting of two-dimensional or planar structures in two perpendicular directions.

Common planar steel structures are simple and continuous beams (such as the plate girder of Fig. 9.1), single-story rigid frames (Fig. 1.3), multistory rigid frames (Fig. 1.4), braced frames (Fig. 1.5), frames with shear walls (Fig. 1.6), trusses (Fig. 1.7), and Vierendeel frames (Fig. 1.8).

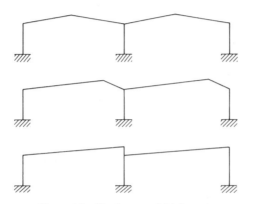

Figure 1.3 Single-story rigid frames.

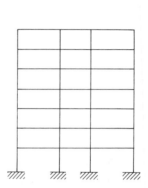

Figure 1.4 Multistory rigid frames.

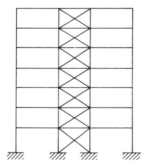

Figure 1.5 Braced frame.

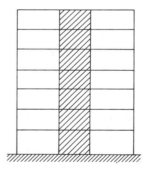

Figure 1.6 Frame with shear wall.

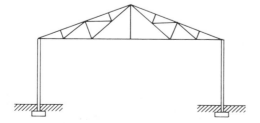

Figure 1.7 Truss on vertical walls.

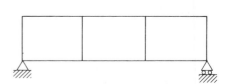

Figure 1.8 Vierendeel frame.

Figure 1.9 North and South Newburgh-Beacon Bridges over the Hudson River in New York. (From Ref. 49, printed by permission of the AISC.)

Steel truss bridges are often used for 400–1200-ft spans. Figure 1.9 shows the north and south Newburgh-Beacon bridges over the Hudson River in the state of New York, completed in 1963 and 1980, respectively. Both of them have a three-span cantilever through-truss and a number of continuous deck trusses.

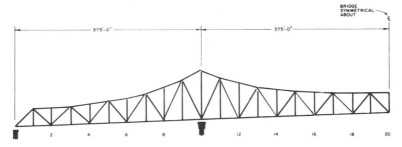

Figure 1.10 Sewickley Bridge over the Ohio River near Pittsburgh. (From Ref. 49, printed by permission of the AISC.)

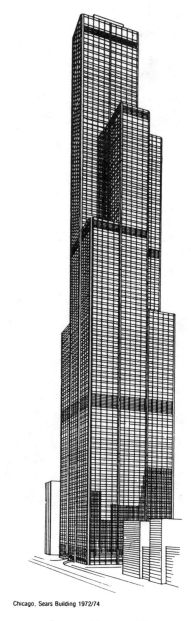

Chicago, Sears Building 1972/74

Figure 1.11 One-hundred-nine-story, 445-m-high Sears Building in Chicago. (Reprinted from Ref. 32 by permission of Collins Professional and Technical Books.)

Figure 1.10 shows the prize-winning Sewickley bridge over the Ohio river near Pittsburgh. This bridge is a three-span continuous Warren truss with a main span of 750 ft.

Figure 1.11 shows the 109-story, 445-m-high Sears building in Chicago, which is the tallest building in the world. The amount of steel used in this building (total weight divided by total floor area) is 33 psf.

1.3.2 Tensile Structures

These structures are also called *cable structures* or *suspension-type structures.* Tension cables play an important role in design of these structures. Tension is in general the most efficient means of supporting loads. Very light structures can be built by using high-strength cables. Providing sufficient stiffness for tensile structures is, however, a problem particularly for buildings located in high-wind or seismic areas.

Examples of bridge tensile structures are suspension bridges and cable-stayed bridges (Fig. 1.12). Figure 1.12 shows the Weirton-Steubenville asymmetrical cable-stayed bridge with a main span of 820 ft and a back span of 688 ft in West Virginia.

Figure 1.12 Suspension and cable-stayed bridges. (From Ref. 49, printed by permission of the AISC.)

Cable structures are also used as roof structures. Figure 1.13 shows a suspension roof built in Raleigh, North Carolina, in 1953 with a span of 325 ft (99 m). The main cables are suspended from two intersecting arches. In order to increase the stiffness of the roof, prestressed secondary cables are used at right angles to the main cables. Figure 1.14 shows a stadium with a span of 308 ft (94 m) built in Uruguay. The roof consists of a single layer of cables, an outer reinforced concrete compression ring, and an inner steel tension ring. The roof is covered by precast concrete slabs. The Oakland-Alameda County Auditorium, built with a span of 420 ft (128 m) in 1967 has a similar construction. In this structure the rainwater is collected off the roof through pumping.

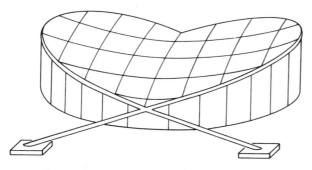

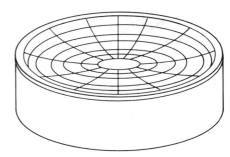

Figure 1.13 Suspension roof built in Raleigh, North Carolina.

Figure 1.14 Suspension roof built in Uruguay.

Figure 1.15 shows the cable-suspended elliptical roof of the Pan-American arrival building at John F. Kennedy International Airport in New York. This roof consists of radial beams supporting reinforced concrete slabs. The beams cantilever 150 ft over an outer ring and are supported by cables anchored at an inner ring. The outer ring is supported by columns. The inner ring is anchored to tension members connected to a massive concrete foundation in the ground.

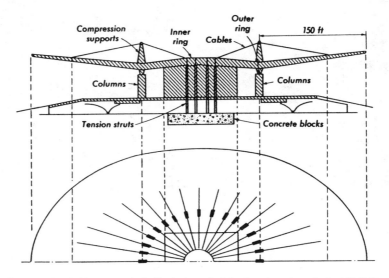

Figure 1.15 Cable-suspended elliptical roof of the Pan-American Arrival Building at Kennedy Airport in New York. (Mario Salvadori and Robert Heller, *Structure in Architecture: The Building of Buildings*, © 1986 3rd ed., p. 121. Reprinted by permission of Prentice-Hall, Inc., Englewood Cliffs, New Jersey.)

1.3.3 Thin-Plate Structures

Examples of this type of structure are liquid storage tanks (such as elevated water tanks, storage bins, and silos) and shell roofs.

1.4 STRUCTURAL SHAPES

Common structural shapes available in the American Institute of Steel Construction Manual (AISCM) [47] are shown in Fig. 1.16. The most widely used type of section is wide-flange shape or W shape [Fig. 1.16(a)]. The inner surface of the flange of a W shape has very little slope (from 0 to 5 percent). In contrast, the inner flange surfaces of American Standard or S shapes [Fig. 1.16(b)] and American Standard channels or C shapes [Fig. 1.16(c)] have a slope of roughly $16\frac{2}{3}$ percent (or 2 in 12 inches).

Shapes W, S, and C are designated by two numbers such as $W14 \times 132$, $S24 \times 121$, and $C15 \times 50$. The first number in W and S designations indicates the "nominal" (not the actual) depth of cross section. The actual depth of $W14 \times 132$ is $d = 14.66$ in., and the actual depth of $S24 \times 121$ is 24.50 in. On the other hand, the first number in the C designation indicates the actual depth of the cross section. The second number in W, S, and C designations indicates the weight (in pounds) per unit length (in feet) of the shape.

The web thickness of W and S shapes in the AISCM is always smaller than the flange thickness. The HP (bearing pile) shapes are similar to W shapes, but their web thickness is the same as their flange thickness. The increased web thickness

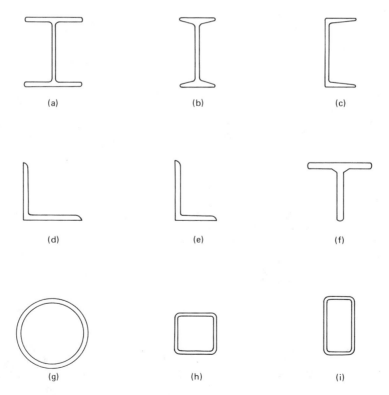

Figure 1.16   Structural shapes.

of the HP shapes is necessary to resist the impact of pile driving. The M shapes are miscellaneous shapes that cannot be classified as W, S, or HP shapes. Similarly, MC designates miscellaneous channel. In the AISCM, one finds 187 W shapes, 31 S shapes, 15 HP shapes, 8 M shapes, 29 C shapes, and 40 MC shapes.

Figure 1.16(d) shows an equal-legs angle, and Fig. 1.16(e) shows an unequal-legs angle. An angle is designated by three numbers. For example, L9 × 4 × $\frac{5}{8}$ indicates an angle with leg sizes of 9 and 4 inches and thickness of $\frac{5}{8}$ in. There are 131 angles in the AISCM.

Figure 1.16(f) shows a Tee (or simply T) shape. The T shapes are made by splitting the webs of W, S, and M shapes. For example, a WT7 × 66 is a T shape obtained by splitting a W14 × 132 shape.

Other shapes sometimes used in steel structures are pipes [Fig. 1.16(g)], square tubes [Fig. 1.16(h)], and rectangular tubes [Fig. 1.16(i)]. Dimensions and geometric properties of different structural shapes have been tabulated in Part One of the AISCM. The flange thickness in these tables for S, C, M, and MC shapes is the average flange thickness.

1.5 DESIGN APPROACHES

1.5.1 Allowable Stress Design

The majority of steel structures are designed according to the elastic design philosophy. In this approach, known as *allowable stress design* or *working stress design* method, the designer estimates the working loads that the structure must safely carry during its lifetime. Then, the structure subjected to the working loads is proportioned so that nowhere in the structure the stress exceeds the allowable stress. The allowable stress for steel is usually defined in terms of the yield stress F_y as follows:

$$F_{all} = \frac{F_y}{\text{F.S.}} \tag{1.1}$$

where F.S. is the factor of safety; this factor is always greater than one.

In this book, we follow the American Institute of Steel Construction Specification (AISCS). This specification consists of two parts and is given in Part Five of the AISCM. Part One of the AISCS is fundamentally based on the allowable stress design philosophy. Unless otherwise noted throughout this book, it is assumed that the design is based on the AISCS.

1.5.2 Plastic Design

In this approach, also known as *limit design*, *collapse design*, or *ultimate strength design*, the working loads are multiplied by factors greater than unity called *load factors* (L.F.). Then the structure is proportioned so that its ultimate load capacity is at least equal to the factored working loads. Thus, in this approach it is necessary

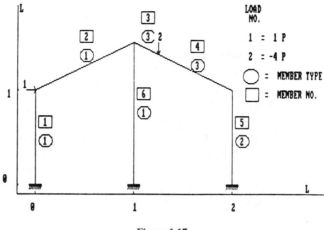

Figure 1.17

to perform a limit or collapse analysis based on postulated failure mechanisms. For example, the failure mechanism of the frame shown in Fig. 1.17 is shown in Fig. 1.18. Note that these two figures have been plotted by a microcomputer [11].

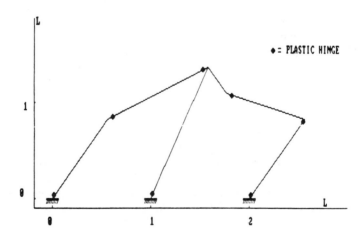

Figure 1.18   Failure mechanism for frame of Fig. 1.17.

Although this approach has not been used extensively in the past, it is only logical to observe its more widespread use in the future. This approach takes into account the ductility and plastic reserve strength of steel structures. For certain types of structures, it results in a more efficient structure. Part Two of the AISCS is based on the plastic design philosophy.

The factor of safety in the allowable stress design approach and load factor in the plastic design approach, sometimes called "factor of ignorance," take into account different uncertainties involved in load and strength evaluations. Some of these uncertainties are as follows:

1. In order to make structures amenable to analysis, idealized conditions are usually assumed.
2. There are tolerances in the evaluation of material strength. Also, material strength may change during the lifetime of a structure due to corrosion or fatigue, the exact prediction of which is not possible.
3. Loading on the structure can only roughly be estimated. This is especially true for geophysical loading such as earthquakes and winds.
4. Stresses developed in the structure during fabrication and erection are often ignored.
5. Residual stresses are developed in the structural shape during the manufacturing process. They may result from [37]:
   a. Uneven cooling after hot-rolling of structural shapes. (In W shapes, for example, after hot-rolling, the flange tips and the mid-portion of the web are thinner and consequently cool faster than the other portions. This results in compressive residual stresses at flange tips and at mid-depth of the web, and tensile residual stresses at the intersections of the flange and web.)
   b. Cold bending or cambering during fabrication.
   c. Welding.
   d. Punching of holes.
   e. Cutting operations during fabrication.
6. There are tolerances in dimensions of the structural members.

1.5.3 Load and Resistance Factor Design

A new specification has been published recently by AISC for structural steel buildings which is based on limit state philosophy and called Load and Resistance Factor Design (LRFD) specification (LRFDS) [48]. It is based on the limit states of strength and serviceability combined with a first-order probability analysis for determination of load and resistance factors. The advantage of such a probability-based design is a more uniform and consistent approach to the load and strength evaluation.

LRFD is a method of designing structural components so that no applicable limit state is exceeded when the component is subjected to all the appropriate load combinations. In this approach equations of the following type should be satisfied [27, 43].

Factored strength of a component $\geq$ factored nominal load effect

or

$$\phi R_n \geq \lambda_A \sum_{i=1}^{n_L} \lambda_i Q_i \tag{1.2}$$

where R_n is a nominal resistance and Q_i is the load effect i. Greek letter ϕ is a resistance factor for taking into account the uncertainties in the calculation of resistance, λ_i is the load factor for the load effect i for taking into account the uncertainties in determining the load effect i, λ_A is an analysis factor for taking into account the uncertainties of the structural analysis, and n_L is the number of load combinations.

It should be noted that for design according to LRFDS, the designer does not have to be involved in the application of the probability theory and statistics for determining the load and resistance factors. These factors have already been estimated for different conditions and are included in the LRFDS.

Loads

2.1 DEAD LOAD

Dead load is defined as the vertical gravity load due to the weight of the fixed elements in the structure including

- structural or load-bearing elements of the structure such as beams, columns, bracings, walls, and floor and roof slabs
- fixed partitions
- floor and roof covering and suspended ceilings
- mechanical and electrical systems such as pipes, air-conditioning and heating ducts and fixtures, and electrical conduits
- facade cladding, ornamental attachments, cornices, etc.
- storage tanks

The dead load of a structure is usually estimated by the designer. This estimate may be up to 20 percent in error.

2.2 LIVE LOAD

Live load, also known as *occupancy load*, is defined as the vertical gravity load that is variable during the lifetime of the structure. Examples of live loads are weights of people, movable partitions, automobiles, mechanical equipments (for example, computers), furniture, books, and safes. Determination of the exact condition and magnitude of the live load is an impossible task. Also, the actual distribution of the live load on a structure is irregular and complicated, so it may not be used easily in structural analysis and design.

Considering the unknown nature of the magnitude and location of the live loads and in order to protect the public safety, minimum live loads are usually recommended by building codes based on surveys of loads and experience.

TABLE 2.1   UNIFORM LIVE LOADS [50]

Use of occupancy		Uniform load (psf)
Category	Description	
1. Assembly areas and auditoriums and balconies therewith	Fixed seating areas	50
	Movable seating and other areas	100
	Stage areas and enclosed platforms	125
2. Cornices, marquees, and residential balconies		60
3. Garages	General storage and/or repair	100
	Private or pleasure car storage	50
4. Hospitals	Wards and rooms	40
5. Libraries	Reading rooms	60
	Stack rooms	125
6. Manufacturing	Light	75
	Heavy	125
7. Offices		50
8. Residential		40
9. Schools	Classrooms	40
10. Sidewalks and driveways	Public access	250
11. Storage	Light	125
	Heavy	250
12. Stores	Retail	75
	Wholesale	100

Minimum uniform live loads as recommended by the Uniform Building Code (UBC) [50] are presented in Table 2.1. It should be noted that these loads are generally conservative. The probability of having these uniform live loads everywhere in a structure is practically nil. Realizing this fact, building codes permit reduction of live loads under certain conditions.

According to UBC, the design live loads recommended in Table 2.1 for floors and roofs may be reduced for any structural member supporting an area more than 150 square feet according to the following formula:

$$R = r(A - 150) \leqslant 23.1(1 + D/L) \qquad (2.1)$$

where

R = Reduction in percent

r = Rate of reduction equal to 0.08 percent for floors
(for roofs, see the UBC)

D = Dead load per square foot of area supported by the member

L = Live load per square foot of area supported by the member

The reduction factor obtained from Eq. (2.1) shall not be greater than 40 percent for members carrying loads from one floor and 60 percent for other members.

Note.   No reduction of the live load is allowed for the following cases:

- For floors in places of public assembly
- For live loads greater than 100 psf

2.3 SNOW LOAD

The U.S. Weather Bureau has estimated the snow load in different parts of the U.S. on the basis of a 50-year mean recurrence interval. Figure 2.1 shows an isolines map provided by the American National Standards Institute (ANSI) for use in structural design. The snow load varies from 5 psf in the south to 80 psf in the northeastern United States.

It is evident that the snow accumulation over a roof decreases with its slope. Thus, according to UBC, snow loads larger than 20 psf may be reduced for each degree of slope over 20 degrees by R_S, given by

$$R_S = \frac{S}{40} - \frac{1}{2} \qquad (2.2)$$

where S is the total snow load in psf. For example, if $S = 60$ psf and the slope of the roof is 25 degrees,

$$R_S = \tfrac{60}{40} - \tfrac{1}{2} = 1$$

and the design snow load will be

$$60 - (25 - 20)(1) = 55 \text{ psf}$$

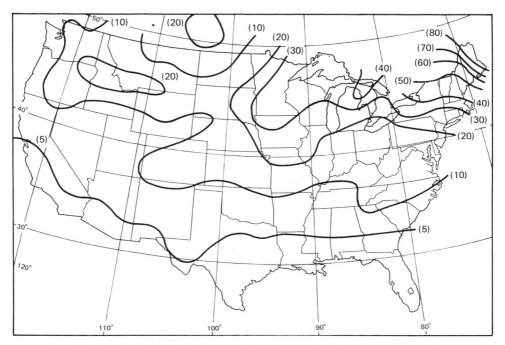

Figure 2.1 Minimum snow loads in different areas of the United States according to ANSI [44], in psf.

2.4 RAIN AND ICE LOADS

Water loads are not ordinarily considered in design of structures. Flat roofs, however, may be subjected to considerable water loads, especially when the drains are clogged. Accumulation of water on flat roofs produces deflection, which in turn leads to more deflection. This phenomenon, referred to as *ponding*, may cause the failure of the roof.

Ice may accumulate on overhangs and protruding elements. Icicles formed on certain structures such as bridge trusses may increase not only the weight but also the area exposed to wind, resulting in increased wind pressure.

2.5 WIND LOAD

An ideal fluid striking an object exerts a pressure on the object equal to

$$p = CV^2 \tag{2.3}$$

where V is the wind velocity and C is a constant. According to UBC, the design wind pressure acting on a structure at any height is equal to [50]

$$p = IC_eC_qq_s \tag{2.4}$$

In this expression, q_s is the wind stagnation pressure at the standard height of 30 ft

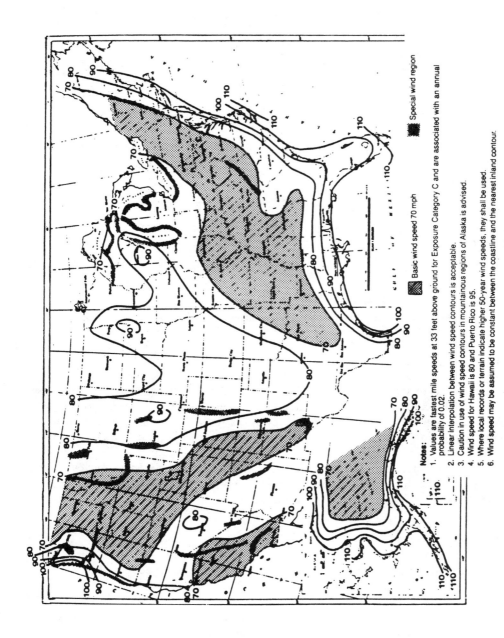

Figure 2.2 Minimum basic wind speed for different areas in the United States. (Reprinted from Ref. 50 by permission.)

Notes:

1. Values are fastest mile speeds at 33 feet above ground for Exposure Category C and are associated with an annual probability of 0.02.
2. Linear interpolation between wind speed contours is acceptable.
3. Caution in use of wind speed contours in mountainous regions of Alaska is advised.
4. Wind speed for Hawaii is 80 and Puerto Rico is 95.
5. Where local records or terrain indicate higher 50-year wind speeds, they shall be used.
6. Wind speed may be assumed to be constant between the coastline and the nearest inland contour.

Basic wind speed 70 mph

Special wind region

TABLE 2.2 WIND STAGNATION PRESSURE (q_s) AT STANDARD HEIGHT OF 30 FT [50]

Basic wind speed (mph)	70	80	90	100	110	120	130
Pressure q_s (psf)	13	17	21	26	31	37	44

given in terms of the basic wind speed in Table 2.2. The minimum basic wind speed for different locations in the United States is given in Fig. 2.2.

Coefficient C_e is the combined height, exposure, and gust factor coefficient given in Table 2.3. UBC considers two different exposure types for structure site. Exposure B is for terrains with buildings, forest, or other surface irregularities at least 20 ft high and covering at least 20 percent of the area extending one mile from the structure site. Exposure C is for generally open and flat terrains extending one-half mile or more from the structure site.

Coefficient C_q is the pressure coefficient for the structure given in Table 2.4. UBC describes two different methods for calculation of C_q and wind loading. Method one, called the Normal Force Method, can be used for design of any type of structure, including gable frames. In this approach, it is assumed that wind pressures act normal to all exterior surfaces simultaneously. Method two, called the Projected Area Method, can be used for any structure less than 200 ft high except gable frames. In this approach it is assumed that the horizontal pressures act upon the full vertical projected area of the structure and the vertical pressures act simultaneously upon the full horizontal projected area.

Finally, coefficient I in Eq. (2.4) is an importance factor. This factor in general is taken as 1.0 except for essential facilities, for which a value of 1.15 is used. Essential facilities are those structures which must remain safe and operational after a windstorm for emergency purposes. They include

1. Hospitals and medical centers with surgery or emergency treatment facilities
2. Fire and police stations
3. Government disaster operation and communication centers
4. Buildings with primary occupancy as assembly hall for more than 300 people.

TABLE 2.3 COMBINED HEIGHT, EXPOSURE, AND GUST FACTOR COEFFICIENT (C_e) [50]

Height above average level of adjoining ground (ft)	Exposure B	Exposure C
0–20	0.7	1.2
20–40	0.8	1.3
40–60	1.0	1.5
60–100	1.1	1.6
100–150	1.3	1.8
150–200	1.4	1.9
200–300	1.6	2.1
300–400	1.8	2.2

TABLE 2.4 WIND PRESSURE COEFFICIENT (C_q) FOR PRIMARY FRAMES AND SYSTEMS [50]

	Description	C_q
Method 1 Normal Force Method	Windward wall	0.8 inward
	Leeward wall	0.5 outward
	Leeward roof or flat roof	0.7 outward
	Windward roof	
	Slope < 9:12	0.7 outward
	Slope 9:12 to 12:12	0.4 inward
	Slope > 12:12	0.7 inward
Method 2 Projected Area Method	On vertical projected area	
	Structures 40 ft high or shorter	1.3 horizontal any direction
	Structures over 40 ft high	1.4 horizontal any direction
	On horizontal projected area	
	Enclosed structure	0.7 upward
	Open structure (with more than 30% of any one side open)	1.2 upward

2.6 EARTHQUAKE LOAD

In this section we cover the seismic provisions of the Uniform Building Code [50], which is the most widely used seismic code in the United States. According to UBC, any structure shall be designed to resist the following total lateral seismic force in the direction of each of the two main axes of the structure nonconcurrently:

$$V = ZIKCSW \tag{2.5}$$

where W is the total dead load of the structure and Z is the seismic zone coefficient. The country is divided into five different zones (Fig. 2.3) as follows:

1. Zone 0 with no damage, $Z = 0$
2. Zone 1 with minor damage, $Z = \frac{3}{16}$
3. Zone 2 with moderate damage, $Z = \frac{3}{8}$
4. Zone 3 with major damage, $Z = \frac{3}{4}$
5. Zone 4: areas within zone 3 but close to certain major fault systems, $Z = 1.0$.

Coefficient C in Eq. (2.5) is a structural response factor given in terms of the fundamental elastic period of vibration of the structure in the direction under consideration, T, in seconds.

$$C = \frac{1}{15\sqrt{T}} \le 0.12 \tag{2.6}$$

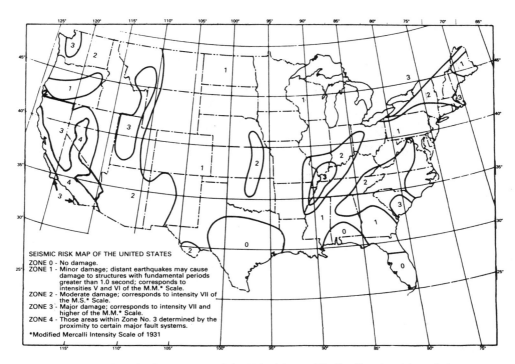

Figure 2.3 Seismic zone map of the United States (UBC). (Reprinted from Ref. 50 by permission.)

The fundamental period T can be estimated approximately from empirical equations. For buildings with bracings or shear walls, it is given in terms of the height of the structure h_n in feet, and the dimension of the structure in a direction parallel to the applied forces D in feet.

$$T = \frac{0.05 h_n}{\sqrt{D}} \qquad (2.7)$$

For buildings in which the lateral-force resisting system consists of ductile moment-resisting frames, the fundamental period can be calculated approximately from

$$T = 0.1 N \qquad (2.8)$$

where N is the total number of stories above the base.

Coefficient K depends on the type of the lateral-force resisting system. Its value ranges from 0.67 to 2.5, as given in Table 2.5. Coefficient I is the occupancy importance factor given in Table 2.6.

Coefficient S in Eq. (2.5) is the site-structure resonance factor. UBC prescribes two different methods for evaluating this coefficient.

TABLE 2.5 *K*-FACTOR (ABRIDGED FROM [50])

Type of structure	K
1. Buildings with a box system (shear walls or braced frames)	1.33
2. Buildings with ductile moment-resisting frames	0.67
3. Buildings with a dual system consisting of a ductile moment-resisting frame and shear walls or braced frames in which the moment-resisting frame is able to carry at least 25 percent of the required lateral force	0.8
4. One- to three-story buildings with stud wall framing and plywood horizontal diaphragms and vertical shear panels	1.0
5. All other building framing systems	1.0
6. Elevated tanks not supported by a building	2.5
7. Other structures	2.0

TABLE 2.6 OCCUPANCY IMPORTANCE FACTOR *I* [50]

Type of occupancy	I
Essential facilities such as hospitals, fire and police stations, and government disaster operation and communications centers	1.5
Any building where the primary occupancy is for assembly use with more than 300 persons in one room	1.25
Other structures	1.0

Method 1

According to this approach, the coefficient S is found from the following equation:

$$S = \begin{cases} 1.0 + \dfrac{T}{T_s} - 0.5\left(\dfrac{T}{T_s}\right)^2 & \text{for } \dfrac{T}{T_s} \leq 1.0 \\[2ex] 1.2 + 0.6\dfrac{T}{T_s} - 0.3\left(\dfrac{T}{T_s}\right)^2 & \text{for } \dfrac{T}{T_s} > 1.0 \end{cases} \tag{2.9}$$

where T_s is the characteristic site period. In using Eq. (2.9) T shall not be less than 0.3 second. The characteristic site period shall be within the following range:

$$0.5 \leq T_s \leq 2.5 \tag{2.10}$$

The value of coefficient S varies from 1.0 to 1.5. When T_s cannot be properly evaluated, S shall be taken equal to 1.5. Also, when T is greater than 2.5 seconds, the coefficient S may be found by assuming $T_s = 2.5$ seconds. The variation of S with T/T_s is shown in Fig. 2.4.

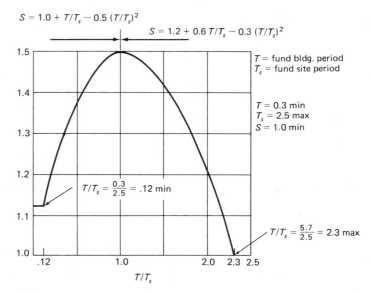

Figure 2.4 Variation of the site-structure resonance factor with T/T_s. (Reprinted from Ref. 39 by permission of the AISC.)

Method 2

This method appeared in the 1985 edition of UBC. In this approach, the site of the structure is classified into three categories according to the soil profile. For each soil profile, a fixed value of S is recommended as given in Table 2.7.

In Eq. (2.5), the product CS need not be greater than 0.14.

The total lateral force obtained from Eq. (2.5) is distributed over the height of the structure according to the following relations:

$$F_x = \frac{(V - F_t)w_x h_x}{\sum_{i=1}^{N} w_i h_i} \tag{2.11}$$

$$V = F_t + \sum_{i=1}^{N} F_i \tag{2.12}$$

$$F_t = \begin{cases} 0.07\,TV \leq 0.25\,V & \text{for } T > 0.7 \\ 0 & \text{for } T \leq 0.7 \text{ sec} \end{cases} \tag{2.13}$$

where

$$N = \text{total number of stories above the base}$$

$$F_i, F_x = \text{lateral force applied to level } i \text{ or } x, \text{ respectively}$$

$$F_t = \text{portion of } V \text{ placed at the top of the structure in addition to } F_N$$

$$h_i, h_x = \text{height of level } i \text{ or } x, \text{ above the base, respectively}$$

TABLE 2.7 SOIL PROFILE COEFFICIENT S [50]

Soil profile type	Description	S
1	Rock of any characteristic, either shale-like or crystalline in nature (such material may be characterized by a shear wave velocity greater than 2500 feet per second); or stiff soil conditions where the soil depth is less than 200 feet and the soil types overlying rock are stable deposits of sands, gravels, or stiff clays.	1.0
2	Deep cohesionless or stiff clay soil conditions, including sites where the soil depth exceeds 200 feet and the soil types overlying rock are stable deposits of sands, gravels, or stiff clays.	1.2
3	Soft to medium-stiff clays and sands, characterized by 30 feet or more of soft to medium-stiff clay with or without intervening layers of sand or other cohesionless soils. In locations where the soil properties are not known in sufficient detail to determine the soil profile or where the profile does not fit any of the three types, Soil Profile Type 3 shall be used.	1.5

2.7 LOAD COMBINATIONS

The last (1985) edition of UBC requires that any building be designed for the following load combinations [50]:

1. $D_n + L_n + L_r \text{ (or } S_n)$ (2.14)

2. $D_n + L_n + W_n \text{ (or } E_n)$ (2.15)

3. $D_n + L_n + W_n + S_n/2$ (2.16)

4. $D_n + L_n + S_n + W_n/2$ (2.17)

5. $D_n + L_n + S_n + E_n \text{ (only when } S_n > 30 \text{ psf)}$ (2.18)

where D_n is the dead load, L_n is the floor live load, L_r is the roof live load, S_n is the snow load, W_n is the wind load, and E_n is the earthquake load. In these load combinations, when the inclusion of the floor live load results in lower stresses, it should not be included.

In the Load and Resistance Factor Design approach, the following load factors and load combinations are recommended by the LRFDS [48]:

1. $1.4D_n$ (2.19)

2. $1.2D_n + 1.6L_n + 0.5L_r \text{ (or } S_n)$ (2.20)

3. $1.2D_n + 0.5L_n \text{ (or } 0.8W_n) + 1.6L_r \text{ (or } S_n)$ (2.21)

4. $1.2D_n + 0.5L_n + 0.5L_r \text{ (or } S_n) + 1.3W_n$ (2.22)

> **5.** $1.2D_n + 0.5L_n \text{ (or } 0.2S_n) + 1.5E_n$ (2.23)
>
> **6.** $0.9D_n - 1.3W_n \text{ (or } 1.5E_n)$ (2.24)

The last load combination is for wind or earthquake loads acting in the opposite direction of the dead load.

2.8 IMPACT

When a load is applied suddenly or dynamically, its effect on the structure may be considerably more severe than when the same load is applied gradually or statically. In buildings impact loading is considered in design of structures carrying cranes, machinery, and elevators and in design of hangers. To take into account the effect of impact, AISCS 1.3.3 recommends the following empirical rules for increasing the live load:

For supports of elevators	100 percent
For cab-operated traveling crane support girders and their connections	25 percent
For pendant-operated traveling crane support girders and their connections	10 percent
For supports of light machinery, shaft or motor driven, not less than	20 percent
For supports of reciprocating machinery or power-driven units, not less than	50 percent
For hangers supporting floors and balconies	33 percent

3

Microcomputers, BASIC Language, and the Approach of Interactive Design

3.1 OVERVIEW OF MICROCOMPUTERS

Microcomputers have gained substantial popularity among civil engineers. In addition to low cost, microcomputers appear to have a number of advantages over mainframe computers that make them appealing to civil engineering applications [4]:

1. They provide a user-friendly programming environment. Users need not worry about complicated operating systems and JCL commands.
2. They lend themselves to interactive programming very effectively.
3. The problem of turnaround time associated with a batch mode environment is eliminated.
4. They provide inexpensive interactive graphic capabilities.
5. The problems associated with computing through modem such as line interference are eliminated.
6. They can often be expanded due to their open architecture.

The programs presented in this book have been developed on an IBM Personal Computer (PC). A basic computer configuration for IBM PC includes the system unit with two floppy disk drives, the display monitor, the keyboard, and the 80-character dot matrix printer. The system unit includes an Intel 8088 Central Processing Unit (CPU), main memory, expansion slots, the power supply, and a speaker. The Intel 8088 is a 16-bit CPU chip. Main memory consists of random access memory (RAM) and read-only memory (ROM). RAM is used for temporary storage of programs or data files, while ROM is used for permanent storage of the system programs and data. In IBM PC, RAM chips can be added to the system board in units of 65,536 bytes on each chip or approximately 64 K, up to a total of 320 K. By plugging additional option boards into the expansion slots, up to $6 \times 64 = 384$ K more RAM can be added to the system.

Two different video display option boards can be used in expansion slots: the Monochrome Display and Printer Adapter board, and the Color Graphics Monitor Adapter board. The IBM Monochrome monitor may be used with the Monochrome board, which gives high-quality characters. A connector for connection to the IBM printer is also provided in the Monochrome board. The Color Graphics board can be used with a color monitor. Up to 16 colors can be used with this board.

The keyboard is the main input device to the IBM PC. It contains 83 typewriter-type keys, a 10-key numeric pad on the right, and 10 function keys on the left. The function keys are originally set for the most commonly used commands such as RUN, LIST, SAVE, and the like. Striking, say, the RUN key is equivalent to typing the word RUN. These function keys may be changed according to applications.

3.2 BASIC LANGUAGE

Programs in this book are all written in Advanced BASIC on an IBM PC. BASIC stands for *B*eginner's *A*ll-purpose *S*ymbolic *I*nstruction *C*ode. BASIC was originally intended to be a simple language for students in a beginning computer course. Today's Advanced BASIC, however, is not only an easy-to-learn language, but also it has emerged as a powerful and versatile high-level language. The advantages of BASIC and the reason for adopting this language in this book can be summarized as follows:

1. It is an easy-to-learn and easy-to-write high-level computer language. It uses plain English words such as INPUT, PRINT, and CHAIN.
2. Notwithstanding its simplicity, it can handle complex operations such as creating sequential and random access data files.
3. It has interactive design. This feature is particularly attractive in the design of structures where designer interaction is essential.
4. It can be modified and maintained easily.
5. It is the most popular language on microcomputers.
6. It has error routines which can be used to ensure data and file integrity and reduce the error margin.

3.3 PRINCIPLES OF STRUCTURED PROGRAMMING

Structured programming is a method of designing computer programs to minimize complexity [33]. Principles of structured programming are delineated in this section.

Generally speaking, microcomputer programs presented in this book have been developed with these principles in mind. Of course, these rules cannot always be followed rigidly, especially when efficiency or speed of computation is the major consideration.

1. The program should be "egoless," that is, easy to read and understand. Due to high manpower cost, clarity and simplicity should be valued more than complicated efficiency.
   a. Choose variable names close to standard and customary notations.
   b. Insert spaces within the program statements for better readability.
   c. Use comment statements generously throughout the program for internal documentation.
   d. Use blank statements to skip lines between blocks of the program.
   e. Indent statements within a loop.
   f. Define the variables within the program.
   g. Use one statement per line.
2. The program should be easy to modify and maintain.
   a. Use variables for constant values that may change in the future. Place the constant values at the beginning of the program.
   b. Begin with line number 100 and increment line numbers by 10 to improve readability and ease of modification.
   c. Initialize all the variables at the beginning of the program.
3. The program should be designed to maximize programmer efficiency [33].
   a. Use top-down programming; that is, design the program in stages from simple to complex.
   b. Use a flow chart in designing the logic of the program.
   c. Avoid GOTO statements as much as possible. This will prevent unnecessary complexity (the so-called "spaghetti" effect) and make the program logic easier to follow.
   d. Use modular programming. This approach will help to use fewer GOTO statements.
   e. The relationship between modules (coupling) should be weak. Each module should be as independent of the others as possible.
   f. A module should preferably have a single entry and single exit.
   g. A module should not be big in size. A good size is a single page of coded program.
4. The program should be reliable.
   a. State the limitations of the program.
   b. Incorporate as many error messages as possible.
   c. Use security to prevent modification of the program by unauthorized users.

3.4 HOW TO WRITE EFFICIENT PROGRAMS

Speed of microcomputer programs depends on [30]

1. Speed of the central processor
2. Efficiency of the compiler/interpreter
3. Central memory size
4. Disk storage size and access speed

Broadly speaking, microcomputers are an order of magnitude slower than typical mainframe computers. For analysis of large structures where a large number of equations must be solved, interpreted BASIC may not be adequate. For such cases it is recommended that the program be compiled. Also, use of a math coprocessor can substantially improve the processing power of a microcomputer. On an IBM PC, an 8087 coprocessor can be plugged into an empty socket on the PC's main board [29].

For large programs, the main memory storage may not be sufficient. In this case the program must be segmented. Segments are written on the peripheral storage and read into the main memory segment by segment. The program segmentation makes possible the execution of large programs within the 64 K memory limit of IBM's Advanced BASIC.

A list of guidelines for developing efficient software on microcomputers is presented in this section. Some of these guidelines are in contradiction with the principles of structured programming presented in Sec. 3.3. As the programs become larger, one may have to compromise some of the principles for the sake of increased efficiency.

1. On most available microcomputers, division and multiplication take 10 to 100 times more CPU time than addition and subtraction. Therefore, unnecessary divisions and multiplications should be avoided.
2. On microcomputers integer arithmetic is an order of magnitude faster than floating-point arithmetic. To increase the efficiency of the program, especially in graphics software, integer arithmetic should be preferred over the floating-point arithmetic.
3. In loops and iterations, the quantities that do not vary in the loop should be calculated outside the loop.
4. Microcomputers can perform a decrement operation more quickly than a comparison.
5. Although comment statements are necessary for understanding and maintaining the program, they use memory and make the program slow. Therefore, for large programs a good solution is having two versions of the same program: a documentation version with numerous comment statements and a run-time version with all the comment statements removed.

Speed of BASIC programs can also be improved by considering the following points (some of them may make the program difficult to read) [36]:

6. Combine several BASIC statements into a single logical line (with a maximum of 255 characters) by separating them with colons.

7. Delete all the unnecessary space (for example, use A+B instead of A + B). Each space in the program is treated as an alphanumeric symbol.

8. Declare all the variable types (integer, single precision, double precision, or string) with a DEF-type statement before they are used. Without this declaration, BASIC assumes that the variable is a single precision variable.

9. Place the frequently used subroutines near the top of the program. When the program encounters a GOSUB statement, the BASIC interpreter goes to the beginning of the program and checks each line until it finds the subroutine's line number.

3.5 INTERACTIVE COMPUTER-AIDED DESIGN

Computer programs have not been used extensively for design of structures. A number of reasons may be cited for the lack of interest in conventional programs for design of structures [21]:

1. In practical design cases there are a large number of alternatives whose selection needs the judgment of the experienced human designer.

2. Design specifications usually cover the general situations and leave the less frequent cases to the judgment of the human designer. In other words, they contain discontinuities and gaps to be filled by the human designer.

3. Human designers use their previously gained experience in design of new structures.

4. A human designer usually visualizes and sketches different structural forms and configurations before making the preliminary design, stress analysis, and the final design.

5. Design specifications change frequently—for example, every three or four years—even though not significantly.

6. Design specifications are based on years of experience gained by researchers and practicing engineers and contain rules of thumb and heuristics which may not be readily implemented in traditional computer languages.

7. Parts of design specifications and standards need interpretation by an experienced designer.

8. Design is an ill-defined and ill-structured problem, lacking a clearly defined goal, and not quite amenable to algorithmic procedures.

9. Design is a creative process.

The first five problems can be alleviated to some extent by using the approach of "interactive design." In this approach, the designer or user of the system is in charge, and the system works as an assistant to him. In this book, our approach is interactive design [3]. The interaction of the program with the user is of primary importance. An interactive program should ideally perform the following:

1. Carry all the numerical calculations.
2. Check for the consistency of design according to the specified design specification (AISCS or the new LRFDS).
3. Perform error checks.
4. Present possible alternatives to the user.
5. Prevent the user from entering the data in violation of the design specification.
6. Inform the user about the ranges of practical values.
7. Provide practical values for the final design (for example, in design of plate girders, dimensions are rounded to practical values).

The programs for design of beams, columns, and beam-columns also can come up with the minimum-weight design.

Some of the issues raised here touch the topics of intelligence and artificial intelligence. Artificial intelligence (AI) is a branch of computer science concerned with making computers act more like human beings. Computer programs using AI techniques to assist people in solving difficult problems involving knowledge, heuristics, and decision making are called *knowledge-based systems* or *expert systems*. In order to design a new structure, the human designer utilizes his previous experience. After the completion of each design he elicits useful information and stores it in his memory for use in future designs [21]. Thus, capturing the knowledge and expertise of an experienced designer in a computer program or expert system should have a significant impact on the structural design productivity. This very interesting subject is actively being researched by a number of researchers [5]. While the topic of expert systems for structural design is certainly beyond the scope of the present book, the author views an expert system for structural design as an extension of the interactive design approach presented in this book.

4

Design of Tension Members

4.1 TYPES OF TENSION MEMBERS

Tension members do not buckle. Therefore, steel can be used most efficiently as tension members. Different types of sections used as tension members are shown in Fig. 4.1. Steel cables are constructed of a number of wire ropes or strands. The wires are cold-drawn high-strength steel alloys. This combination of cold work and special steel alloys produces very high yield strength in the range of 200 to 250 ksi. Thus, cables are particularly suitable for covering large spans and are used in long-span suspension bridges, cable roofs, cable-stayed bridges, and ski lift cables. Cables, of course, are flexible. To provide stiffness, cable structures may be stiffened by adding stiffening members.

When the magnitude of tensile force is small in a tension member, solid round or rectanglar bars are used. For larger tensile forces or when more stiffness is required, round or rectangular tubes may be used. Round tubes or pipes might be preferred when the tension member is exposed to high-wind conditions. Connection details for round tubes, however, are cumbersome to construct

Single angles are commonly used as tension members, for example, as bracings for carrying lateral forces due to wind or earthquake. Angle end connection is simple but eccentric to its centroidal axis. The eccentric application of tensile force produces bending stresses in members which are often ignored in design practice. The eccentricity may be minimized by suitable detailing. For example, when only one leg of the angle is connected to the joint, it should be the larger one.

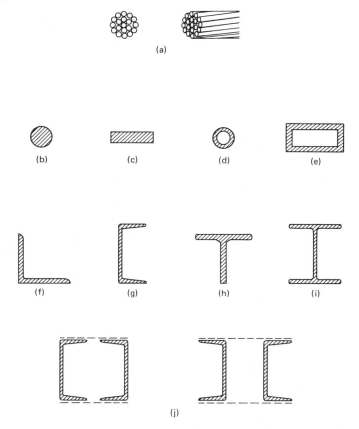

Figure 4.1 Types of tension members: (a) Cable with strands, (b) Round bar, (c) Rectangular bar, (d) Round tube, (e) Rectangular tube, (f) Angle (L), (g) Channel (C), (h) T section, (i) W section, (j) Channels with lacings.

Compared with an angle, a channel connected to the joint at its web often produces less eccentricity, since the centroid of most channels is close to their web. For carrying a large tensile force, W sections are used. For a very large tensile force, built-up sections—for example, channels with lacings—may be used. Dashed lines in Fig. 4.1(j) indicate the lacing bars. Inclined single or double lacing bars connect the two elements of the built-up section.

4.2 NET AND EFFECTIVE NET AREAS

When tension members are connected by welding, the total cross-sectional area is available for transferring the tension. When the connection is done by bolting (or riveting), holes must be made in the member. These holes evidently reduce the cross-sectional area available for transferring the tension. Thus, the net area of the section is the gross area minus deductions for the holes [Fig. 4.2(a)].

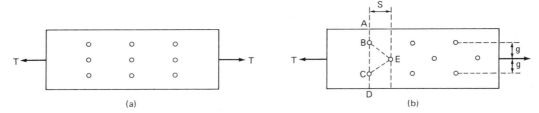

Figure 4.2   Holes in a tension member.

Holes are sometimes staggered, as shown in Fig. 4.2(b). Staggering of the holes increases the net area of the section. In Fig. 4.2(b), the plate may fail along section ABCD or section ABECD. How do we calculate the net area or the net width along a zigzag line of holes such as ABECD? According to the approximate procedure provided by AISCS 1.14.2.1, the net width (w_n) is obtained by deducting the sum of the diameters of all the holes located on the zigzag line from the gross width (w_g) and then adding for each inclined line such as BE the quantity $s^2/4g$.

$$w_n = w_g - \sum D + \sum \frac{s^2}{4g} \tag{4.1}$$

where

s = the center-to-center spacing of the two consecutive holes in the direction of stress (pitch)

g = the transverse center-to-center spacing of the same two holes (gage)

$\sum D$ = sum of the diameters of the holes in the line of holes

The critical net area A_n of the tension member is found from the line of holes with the minimum net width.

According to AISCS 1.14.4, in calculating the net area the width of a hole should be assumed $\frac{1}{16}$ in. greater than the nominal size of the hole.

When a tension member consists of more than one plate segment, the axial tension may not be transmitted through all the segments. For example, in the case of an angle, only one leg may be connected to the joint. In this case, when connection is through bolts or rivets, an effective net area should be calculated by using the following formula (AISCS 1.14.2.2):

$$A_e = C_t A_n \tag{4.2}$$

where C_t is a reduction factor given in Table 4.1 for different types of sections. This reduction is necessary to take into account the nonuniformly distributed transfer of stresses.

4.3 ALLOWABLE TENSILE STRESSES

1. Allowable tensile stress at the holes for pin-connected members and eye-bars (AISCS 1.5.1.1):

$$F_t = 0.45F_y \quad \text{on the actual net area, } A_n \tag{4.3}$$

TABLE 4.1   EFFECTIVE NET AREA COEFFICIENT C_t (AISCS 1.14.2.2)

Types of members	Minimum number of fasteners per line in the direction of tension	C_t
(a) All segments are connected to transmit the tension	1	1
(b) W, M, or S sections $\dfrac{b_f}{d} \geq \dfrac{2}{3}$ connection to flange(s)	3	0.90
(c) Tees $\dfrac{b_f}{d} \geq \dfrac{4}{3}$ connection to flange	3	0.90
(d) W, M or S sections not meeting the conditions of (b), tees not meeting the conditions of (c), and all other shapes, including built-up sections	3	0.85
(e) All sections	2	0.75

2. For threaded rods:

$$F_t = 0.33 F_u \quad \text{on the gross area, } A_g \qquad (4.4)$$

where F_u is the specified minimum ultimate tensile stress.

3. For other members, the allowable tensile stress is the smaller value obtained from the following two formulas:

$$F_t = 0.60 F_y \quad \text{on the gross area, } A_g \qquad (4.5)$$

$$F_t = 0.50 F_u \quad \text{on the effective net area, } A_e \qquad (4.6)$$

Equation (4.5) is based on the yield criterion, while Eq. (4.6) is based on the fracture criterion.

4.4 MAXIMUM SLENDERNESS RATIO

Since tension members do not buckle, there is no mandatory slenderness ratio limitations. In order to prevent the undesirable lateral movement and vibration, however, it is recommended that the slenderness ratio L/r of tension members except rods should not exceed 240 for main members and 300 for lateral bracing and other secondary members (AISCS 1.8.4).

4.5 EXAMPLES OF DESIGN OF TENSION MEMBERS

Example 1

Find the maximum allowable tensile load T for a channel section connected to a gusset plate as shown in Fig. 4.3, assuming that the gusset plate and the 1-in.-diameter bolts do not control the design. The yield stress and the ultimate strength of the steel are 50 ksi and 65 ksi, respectively.

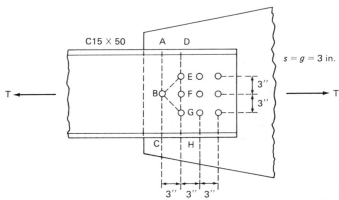

Figure 4.3

Solution   Properties of C15 × 50:

$$A_g = 14.7 \text{ in.}^2 \qquad t_w = 0.716 \text{ in.} \quad \text{(web thickness)}$$

For standard-size holes, the size of hole should be $\frac{1}{16}$ in. greater than the diameter of the fastener (AISCS 1.23.4.1).

Size of hole = $1 + \frac{1}{16} = \frac{17}{16}$ in.
Deduction for one hole = $\frac{17}{16} + \frac{1}{16} = \frac{9}{8}$ in.
Reduction factor: $C_t = 0.85$ (Table 4.1)
At section ABC: Tension = T

$$A_n = A_g - t_w\left(\tfrac{9}{8}\right) = 14.7 - 0.716\left(\tfrac{9}{8}\right) = 13.89 \text{ in.}^2$$

At section DEBGH: Tension = T

$$A_n = A_g - 3t_w\left(\frac{9}{8}\right) + 2\left(\frac{s^2}{4g}\right)t_w$$

$$= 14.7 - 3(0.716)\left(\tfrac{9}{8}\right) + 2\left(\tfrac{9}{12}\right)(0.716) = 13.36 \text{ in.}^2$$

At section DEFGH: Tension = $0.9T$

$$A_n = A_g - 3t_w\left(\tfrac{9}{8}\right) = 14.7 - 3(0.716)\left(\tfrac{9}{8}\right) = 12.28 \text{ in.}^2$$

$$\frac{A_n}{0.9} = \frac{12.28}{0.90} = 13.65 \text{ in.}^2$$

Therefore, the zigzag section DEBGH is the critical section and the governing net area is $A_n = 13.36$ in.^{2}

Maximum allowable tensile load based on the yield criterion:

$$T = 0.60 F_y A_g = 0.60(50)(14.7) = 441 \text{ K}$$

Maximum allowable tensile load based on the fracture criterion:

$$T = 0.50 F_u A_e = 0.50(65)(0.85)(13.36) = 369.1 \text{ K}$$

Therefore, the maximum allowable tensile load is

$$T = 369.1 \text{ K}$$

Example 2

Consider the *preliminary* design of a 12-story office building shown in Fig. 4.4. The building has a circular plan with a diameter of 50 ft and a constant story height of 10 ft. Floors 1 to 10 are supported at the center by a hollow circular reinforced concrete shear wall with a diameter of 16 ft and at the perimeter by 12 identical hangers suspended from 12 story-high truss systems located at the twelfth story.

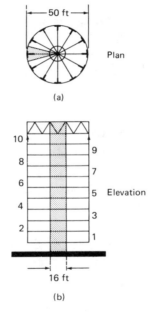

Figure 4.4 (a) Plan; (b) elevation.

Design the hangers for a dead load of 60 psf and a live load of 80 psf, choosing the lightest WT section and assuming uniform section throughout the height of each hanger. Use A36 steel with $Fy = 36$ ksi.

Solution   In order to take into account the effect of impact for hangers supporting floors, the assumed live load should be increased by 33 percent (AISCS 1.3.3).

Design load intensity:

$$w = 60 + 1.33(80) = 166.4 \text{ psf}$$

The load-carrying tributary area for each radial beam is shaded in the plan of Fig. 4.4.

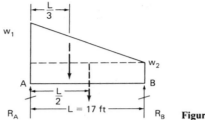

Figure 4.5

The load distribution on each radial beam is approximately trapezoidal, as shown in Fig. 4.5, with inner intensity w_2 and outer intensity w_1.

$$w_1 = \frac{\pi(50)}{12}(166.4) = 2178 \text{ lb/ft} = 2.178 \text{ K/ft}$$

$$w_2 = \frac{\pi(16)}{12}(166.4) = 697 \text{ lb/ft} = 0.697 \text{ K/ft}$$

The reaction at the outer end of any radial beam is (the supports condition of the radial beams is assumed simple):

$$R_A = \frac{w_2 L}{2} + \frac{2}{3}\left(\frac{1}{2}\right)(w_1 - w_2)L = \frac{(2w_1 + w_2)L}{6}$$

$$= \frac{(2 \times 2.178 + 0.697)(17)}{6} = 14.317 \text{ K}$$

Maximum tension in a hanger (neglecting its weight):

$$T_{max} = 10R_A = 143.17 \text{ K}$$

Allowable tensile stress:

$$F_t = 0.60\, F_y = 22 \text{ ksi}$$

Required area of the cross section:

$$\text{Area} = \frac{T_{max}}{F_t} = \frac{143.17}{22} = 6.51 \text{ in.}^2$$

Try WT8 × 22.5:

$$A = 6.63 \text{ in.}^2 \qquad r_x = 2.39 \text{ in.} \qquad r_y = 1.57 \text{ in.}$$

Check for the maximum tensile stress, including the weight of the hanger.

$$\text{Weight of the hanger} = 10(10)(0.0225) = 2.25 \text{ K}$$

$$T_{max} = 143.17 + 2.25 = 145.42 \text{ K}$$

Maximum tensile stress:

$$f_t = \frac{T_{max}}{A} = \frac{145.42}{6.63} = 21.93 \text{ ksi} < 22 \text{ K ksi} \qquad \underline{\underline{\text{O.K.}}}$$

Check the slenderness ratio (not mandatory).

$$\frac{L}{r_y} = \frac{(10)(12)}{1.57} = 76.4 < 240 \qquad \underline{\underline{\text{O.K.}}}$$

4.6 DESIGN OF CIRCULAR SUSPENSION CABLE ROOFS

Figure 4.6 shows an example of this type of structure. The roof carries its own weight and a small live load to allow for repairs. This type of structure is not economical for regions with heavy snowfalls, since the snow would accumulate in the dished roof surface and impose heavy loads [28]. The cables are supported by an outer reinforced concrete (R.C.) compression ring and a small steel tension ring at the center.

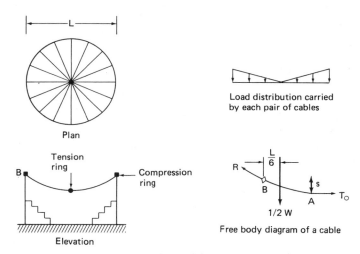

Figure 4.6

4.6.1 Design of Cables

Each cable carries a triangular slice of load. The maximum tension in the cable is equal to the resultant reaction at the supports. This reaction can easily be calculated by considering the equilibrium of one cable, such as cable AB in Fig. 4.6. Considering the moment equilibrium of AB about the point B (compression ring), we find the value of tension T_o at the lowest point A (tension ring).

$$T_o s = \tfrac{1}{2} W(\tfrac{1}{6}L)$$

$$T_o = \frac{WL}{12s} \tag{4.7}$$

where L is the span length, s is the sag of cables, and W is the total load carried by a pair of cables. Quantity T_o is the minimum tension on the cable. The horizontal component of tension at any point on the cable is equal to T_o. Noting that the vertical component of reaction at B is $W/2$, the magnitude of reaction at B, equal to the maximum tension in the cable, becomes

$$T_{\max} = R = \sqrt{T_o^2 + \left(\frac{W}{2}\right)^2} = \sqrt{\left(\frac{WL}{12s}\right)^2 + \left(\frac{W}{2}\right)^2} \tag{4.8}$$

It should be noted that in this approximate analysis the weight of cables has been neglected. Thus, the required cross-sectional area of cables is

$$A_{sc} = T_{\max}/F_t \qquad (4.9)$$

where F_t is the allowable tensile stress in steel cables.

Cables used in suspension roofs are made by twisting thin wires, and their cross-sectional area is approximately equal to $\frac{2}{3}(3.14/4)d^2$ [28], where d is the diameter of the cable. The load W carried by a pair of cables depends on the number of cables in the roof, while the number of cables in the roof is based on the spacing of cables around the circumference. Thus, for a given sag, the area of steel required depends on the spacing of cables around the circumference.

4.6.2 Design of R.C. Ring

If it is assumed that the compression ring is supported by vertical walls or columns, the vertical reaction will be carried by the vertical elements, and the compression ring needs to be designed for the horizontal reaction of T_o over a distance of s', the spacing of cables around the circumference. In other words, the compression ring is acted on by inward forces as in a pipe under suction. These inward forces produce roughly a uniform axial compressive force in the circular ring. We can find the magnitude of this compressive force C by considering the equilibrium of one-half of the compression ring as shown in Fig. 4.7.

$$2C = \frac{T_o}{s'}(L) \qquad (4.10)$$

Substituting for T_o from Eq. (4.7) in Eq. (4.10), we obtain

$$C = \frac{WL^2}{24ss'} \qquad (4.11)$$

The R.C. ring beam is designed as per American Concrete Institute (ACI) specification [45]. In the interactive BASIC program to be discussed in the following section, the user has the following two options for designing the reinforced concrete ring:

1. The cross-sectional area is found by neglecting the area of reinforcement and providing a nominal reinforcement of about 1 percent of the concrete cross-sectional area. Thus in this case, the area of compression ring is given by

$$A_c = C/f'_c \qquad (4.12)$$

where f'_c is the compressive strength of concrete.

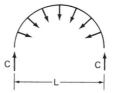

Figure 4.7

2. In this alternative, the program computes the cross-sectional area of the compression ring and the amount of reinforcement treating the ring as a reinforced concrete section. For a given reinforcement content, the area of concrete can be calculated by using the design strength equation for a tied column (ACI section 10.3.5.2) [45]:

$$\phi P_{n(\text{max})} = \phi\{0.80[0.85f'_c(A_g - A_{st}) + f_y A_{st}]\} \tag{4.13}$$

where

$$\phi = \text{strength reduction factor} = 0.7$$
$$P_{n(\text{max})} = \text{nominal maximum load}$$
$$A_g = \text{gross cross-sectional area of concrete}$$
$$A_{st} = \text{total area of longitudinal reinforcement}$$
$$f'_c = \text{compressive strength of concrete}$$
$$f_y = \text{yield stress of steel reinforcement}$$

The program checks for minimum and maximum amounts of reinforcement in the ring as per ACI specification Sec. 10.9.1. The lateral ties are designed as per ACI specification Sec. 7.10.5 [45].

4.7 INTERACTIVE MICROCOMPUTER-AIDED DESIGN OF CIRCULAR SUSPENSION ROOFS

4.7.1 Program Structure

The program consists of a main unit and twelve modules. The main program coordinates the whole process of design. The structure and the function of the modules used in the program are explained briefly in the following paragraphs.

1. Menu selection. In this module, the user will be provided with a menu as shown in Fig. 4.8. The user will be asked to select an option by keying in the

```
          ***  MAIN MENU  ***


   1.   START NEW PROBLEM

   2.   SUMMARY OF RESULTS

   3.   DISPLAY MENU

   4.   REDESIGN

   5.   QUIT



   ENTER THE NUMBER PLEASE  -->> ? 3█
```

Figure 4.8

option number of his choice. At the beginning of any new design problem it is obvious that the user has to pick the option "START NEW PROBLEM."

2. Basic input data. The basic input data for a design problem are those usually fixed for a particular design. In the design of a cable roof, the user has to input the following basic data interactively:

1. Span/diameter of the roof (L)
2. Height of the structure (H)
3. Compressive strength of concrete (f_c')
4. Allowable tensile stress of steel cable (F_t)
5. Yield stress of compression ring reinforcement (f_y)
6. Vertical dead load intensity (W_{dl})
7. Vertical live load intensity (W_{ll})

3. Design of cables. In this module the user will be asked to input the spacing of cables around the circumference (at the compression ring) as well as the sag-to-span-length ratio of the roof. This module computes the following:

1. Number of cables
2. Area of cross section and the diameter of cables
3. Length of a cable, assuming a parabolic profile for the cables
4. Total weight of all the cables
5. Volume of space covered by the roof

After design of cables, the above values will be displayed on the screen.

4. Design of R.C. ring. A rectangular cross section is chosen for the R.C. compression ring. The user will be asked to choose any one of the following design methods:

1. "Design concrete section by neglecting contribution of reinforcement and providing a nominal reinforcement."
2. "Design the ring as a reinforced concrete section as per ACI specification."

Further, the user will be asked to input the following parameters:

1. Width-to-depth ratio of R.C. ring cross section
2. Size (bar number) of the reinforcement bars
3. Amount of reinforcement (nominal for case 1) as a percentage of the gross cross-sectional area of the R.C. ring

After the design is completed, the cross-sectional dimensions, the total number of reinforcement bars of the size selected by the user, the spacing of lateral ties, and the total volume of concrete in the R.C. ring can be displayed on the screen.

5. Print final results.   This module displays the summary of results on the screen.

6. Continuation module.   The function of this module is to introduce breaking points in the program so that the user can observe the screen after each major step or option of the menu and to continue the program from that step by issuing the command "PRESS ANY KEY TO CONTINUE." Thus, the speed of the program execution is controlled, and the user can go through each step carefully.

7. Input error check.   This module helps to identify whether the input keyed by the user is legal or illegal. If the user either does not input value for a variable or inputs a negative value, then the warning message "PLEASE PROVIDE A POSITIVE NUMBER" will appear on the screen along with a beep sound from the terminal. If the user inputs an unrealistic value for a variable, the warning message "PLEASE PROVIDE A PRACTICAL VALUE" will appear on the screen. The following inputs given by the user are considered to be unrealistic:

1. Sag of the cable computed, using the sag-to-span ratio given by user, is greater than 0.8 times the height of the structure.
2. Spacing of cables at the circumference is >25 ft.
3. Bar size (number) of reinforcement is not a practical one.
4. Allowable/yield stress in steel is abnormally high or low with respect to its practical value.
5. Compressive strength of concrete is abnormally high or low with respect to its practical value.

8. Redesign menu.   This module helps the user change any one or all of the design parameters of the cable and R.C. ring, using the redesign menu of Fig.

```
            ***  REDESIGN MENU  ***
            ~~~~~~~~~~~~~~~~~~~~~~~~~

        1.  SAG-SPAN RATIO

        2.  SPACING OF CABLES

        3.  METHOD OF R.C. RING DESIGN

        4.  B/D RATIO OF R.C. RING

        5.  REINFORCEMENT CONTENT

        6.  SIZE (BAR #) OF REINFORCEMENT

        7.  NOTHING (START REDESIGN)

        8.  MAIN MENU



        ENTER THE NUMBER PLEASE -->>  ? ▮
CHANGES TO BE MADE                4
```

Figure 4.9

4.9. Based on the changes made the interactive program will redesign the structure accordingly.

9. Display menu.   This module provides the user with the display menu shown in Fig. 4.10.

```
    *** DISPLAY MENU ***

  1.  PLAN OF THE CABLE ROOF

  2.  ELEVATION OF THE STRUCTURE

  3.  CROSS-SECTION OF R.C. RING

  4.  MAIN MENU


  ENTER THE NUMBER PLEASE -->> ? ■
```

Figure 4.10

10. Display plan of the roof structure.   This module displays the plan of circular suspension cable roof showing the arrangement of cables, the tension ring, and the R.C. ring with all the dimensions, as shown in the example of Fig. 4.11.

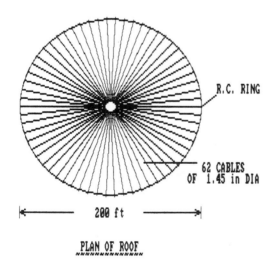

PRESS ANY KEY TO CONTINUE   Figure 4.11

11. Display elevation of the structure.   This module displays the sectional elevation of the cable roof structure with the vertical load distribution acting on a pair of cables as shown in the example of Fig. 4.12.

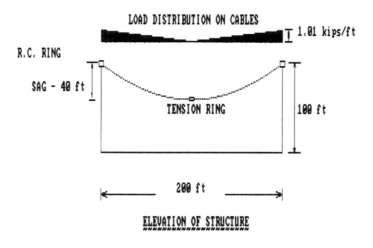

PRESS ANY KEY TO CONTINUE

Figure 4.12

12. Display the cross section of R.C. compression ring.   This module displays the cross section of R.C. ring, including the arrangement of reinforcement bars with all the necessary dimensions.

4.7.2 Sample Example and Interactive Session

4.7.2.1 Sample example.   A sample interactive session for design of a circular suspension cable roof is presented in Sec. 4.7.2.2. Figure 4.11 shows the plan of the roof structure. Figure 4.12 shows the cross-sectional elevation of the structure. Load distribution on a pair of cables is also shown in this figure. Figure 4.13 shows the first design for the cross section of the R.C. compression ring. By using the redesign menu (Fig. 4.9), the method of design for the R.C. ring beam was changed

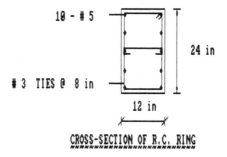

PRESS ANY KEY TO CONTINUE   Figure 4.13

from 1 to 2 (see Sec. 4.6.2). Also, the reinforcement content was changed from 1 percent to 2.8 percent. The revised design for the R.C. compression ring is shown in Fig. 4.14. In the first design, number 5 bars were chosen for the main reinforcement. In the second design, however, number 5 bars would not satisfy the minimum spacing requirement of the ACI specification. Therefore, the program automatically chose the next bar size available, that is, number 6.

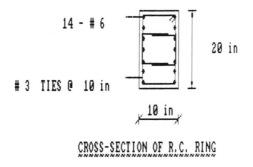

CROSS-SECTION OF R.C. RING

PRESS ANY KEY TO CONTINUE   **Figure 4.14**

Finally, by using the redesign menu once more, the bar size of the main reinforcement was changed from number 5 to number 7. The revised design for the R.C. compression ring is shown in Fig. 4.15. Note that in using the redesign menu (Fig. 4.9) one can make several changes simultaneously by typing the option numbers successively. However, for the program to start the process of redesigning, number 7 (START REDESIGN) must be typed at the end.

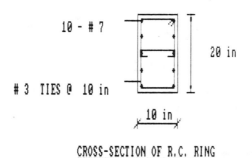

CROSS-SECTION OF R.C. RING

PRESS ANY KEY TO CONTINUE   **Figure 4.15**

4.7.2.2 Interactive session.

```
RUN
### MAIN MENU ###``````````````````````1.  START NEW PROBLEM
2.  SUMMARY OF RESULTS
3.  DISPLAY MENU
4.  REDESIGN
5.  QUIT
ENTER THE NUMBER PLEASE -->>
? 1

                ### CIRCULAR SUSPENSION CABLE ROOF DESIGN ###


                    ###  INPUT DATA  ###

DO YOU DESIRE TO USE THE INPUT DATA FROM A DISK ? n
ENTER PROBLEM HEADING ? example
SPAN SHOULD BE LESS THAN 400 ft.
SPAN OF ROOF (ft)? 200
HEIGHT SHOULD BE LESS THAN 150 ft.
HEIGHT OF ROOF (ft)? 100
COMPRESSIVE STRENGTH OF CONCRETE IS IN THE RANGE OF 2-10 Ksi
COMPRESSIVE STRENGTH OF CONCRETE (Ksi) ? 4.5
ALLOWABLE TENSILE STRESS OF STEEL CABLE IS IN THE RANGE OF 30-100 Ksi.
ALLOWABLE TENSILE STRESS OF STEEL CABLE  (Ksi) ? 60
YIELD STRESS OF REINFORCEMENTS IN IS IN THE RANGE OF 30-100 Ksi.
YIELD STRESS OF REINFORCEMENT (ksi) ? 60
VERTICAL DEAD LOAD INTENSITY IS IN THE RANGE OF 30-90 Psf.
VERTICAL DEAD LOAD INTENSITY (Psf) ? 85
VERTICAL LIVE LOAD INTENSITY IS IN THE RANGE OF 5-75 Psf.
VERTICAL LIVE LOAD INTENSITY (psf) ? 15
DO YOU WANT TO SAVE THE INPUT DATA IN THE DISK (Y/N)? y
NAME OF THE INPUT DATA FILE ? data1

                    PRESS ANY KEY TO CONTINUE

                 ###  CABLE DESIGN  ###

THE COMMON  SAG-TO-SPAN RATIO FOR CABLE ROOFS
IS IN THE RANGE : 1/6 - 1/4
SAG-TO-SPAN RATIO (in decimals)? .2
THE COMMON SPACING OF CABLES AT THE COMPRESSION RING
IS IN THE RANGE :  8 - 15 ft
SPACING OF CABLES (ft)? 10

SAG OF CABLES = 40.00 ft
NUMBER OF CABLES =  62
DIAMETER OF CABLES =  1.45 in
LENGTH OF A CABLE = 109.82 ft
TOTAL WEIGHT OF CABLES =  25.506 kips
VOLUME OF SPACE COVERED BY THE ROOF =  2.513E+06 cft

                    PRESS ANY KEY TO CONTINUE

                ### R.C. RING DESIGN ###

THE RING CAN BE DESIGNED BY FOLLOWING TWO METHODS

1.DESIGN AS A CONCRETE SECTION WITH NOMINAL REINFORCEMENT
2.DESIGN AS A REINFORCED CONCRETE SECTION
ENTER METHOD NUMBER (1/2) ? 1

THE COMMON WIDTH-TO-DEPTH RATIO FOR R.C. RING
CROSS-SECTION IS IN THE RANGE OF 1/3 TO 2/3
WIDTH-TO-DEPTH RATIO OF R.C. RING (in decimals)? .5
THE COMMON VALUE OF NOMINAL REINFORCEMENT IS AROUND:
1.0 % OF CONCRETE AREA
VALUE OF REINFORCEMENT AS A % OF CONCRETE AREA ? 1
AVAILABLE BAR SIZES (BAR #) ARE :
3  4  5  6  7  8  9  10  11  14
SIZE (BAR #) OF MAIN REINFORCEMENT ? 5
SIZE (BAR #) OF TIES ? 3
```

```
WIDTH OF R.C. COMPRESSION RING = 12.0 in
DEPTH OF R.C. COMPRESSION RING = 24.0 in
AREA OF REINFORCEMENT IN R.C. RING =  3.10 sqin
NUMBER OF # 5  BARS REQUIRED =  10
SPACING OF # 3  TIES =  8.00 in
VOLUME OF CONCRETE IN THE COMPRESSION RING = 1256.637 cft


                 ### DESIGN IS COMPLETE ###

                       PRESS ANY KEY TO CONTINUE
### MAIN MENU ###`````````````````````1.  START NEW PROBLEM
2.  SUMMARY OF RESULTS
3.  DISPLAY MENU
4.  REDESIGN
5.  QUIT
ENTER THE NUMBER PLEASE -->>
? 4

### REDESIGN MENU ###`````````````````````````1.  SAG-SPAN RATIO
2.  SPACING OF CABLES
3.  METHOD OF R.C. RING DESIGN
4.  B/D RATIO OF R.C. RING
5.  REINFORCEMENT CONTENT
6.  SIZE (BAR #) OF REINFORCEMENT
7.  NOTHING (START REDESIGN)
8.  MAIN MENU
ENTER THE NUMBER PLEASE -->>
? 3

CHANGES TO BE MADE 3

? 5

CHANGES TO BE MADE 5

? 7


                    PRESS ANY KEY TO CONTINUE


                 ### R.C. RING DESIGN ###

ENTER METHOD NUMBER (1/2) ? 2

THE ACCEPTABLE VALUE OF  REINFORCEMENT IS IN THE RANGE :
1.0 - 6.0 % OF CONCRETE AREA
VALUE OF REINFORCEMENT AS A % OF CONCRETE AREA ? 2.8
NEXT SIZE (BAR #) BAR IS CHOSEN

WIDTH OF R.C. COMPRESSION RING = 10.0 in
DEPTH OF R.C. COMPRESSION RING = 20.0 in
AREA OF REINFORCEMENT IN R.C. RING =  6.16 sqin
NUMBER OF # 6  BARS REQUIRED =  14
SPACING OF # 3  TIES = 10.00 in
VOLUME OF CONCRETE IN THE COMPRESSION RING =  872.665 cft


                 ### DESIGN IS COMPLETE ###

                       PRESS ANY KEY TO CONTINUE
### REDESIGN MENU ###`````````````````````````1.  SAG-SPAN RATIO
2.  SPACING OF CABLES
3.  METHOD OF R.C. RING DESIGN
4.  B/D RATIO OF R.C. RING
5.  REINFORCEMENT CONTENT
6.  SIZE (BAR #) OF REINFORCEMENT
7.  NOTHING (START REDESIGN)
8.  MAIN MENU
ENTER THE NUMBER PLEASE -->>
? 6

CHANGES TO BE MADE 6

? 7
```

```
                        PRESS ANY KEY TO CONTINUE

                    ‡‡‡  R.C. RING DESIGN  ‡‡‡

SIZE (BAR #) OF MAIN REINFORCEMENT ? 7
SIZE (BAR #) OF TIES ? 3

WIDTH OF R.C. COMPRESSION RING = 10.0 in
DEPTH OF R.C. COMPRESSION RING = 20.0 in
AREA OF REINFORCEMENT IN R.C. RING = 6.00 sqin
NUMBER OF # 7  BARS REQUIRED =  10
SPACING OF # 3  TIES = 10.00 in
VOLUME OF CONCRETE IN THE COMPRESSION RING =  872.665 cft


                    ‡‡‡  DESIGN IS COMPLETE  ‡‡‡

                        PRESS ANY KEY TO CONTINUE




‡‡‡  REDESIGN MENU  ‡‡‡`````````````````````````1.  SAG-SPAN RATIO
2.  SPACING OF CABLES
3.  METHOD OF R.C. RING DESIGN
4.  B/D RATIO OF R.C. RING
5.  REINFORCEMENT CONTENT
6.  SIZE (BAR #) OF REINFORCEMENT
7.  NOTHING (START REDESIGN)
8.  MAIN MENU
ENTER THE NUMBER PLEASE -->>
? 8

‡‡‡  MAIN MENU  ‡‡‡`````````````````````1.  START NEW PROBLEM
2.  SUMMARY OF RESULTS
3.  DISPLAY MENU
4.  REDESIGN
5.  QUIT
ENTER THE NUMBER PLEASE -->>
? 2

            ‡‡ SUMMARY OF FINAL DESIGN OF CABLE ROOF ‡‡

SAG OF CABLES = 40.00 ft
NUMBER OF CABLES =  62
DIAMETER OF CABLES =  1.45 in
LENGTH OF A CABLE = 109.82 ft
TOTAL WEIGHT OF CABLES =  25.506 kips
VOLUME OF SPACE COVERED BY THE ROOF =  2.513E+06 cft


                        PRESS ANY KEY TO CONTINUE
WIDTH OF R.C. COMPRESSION RING = 10.0 in
DEPTH OF R.C. COMPRESSION RING = 20.0 in
AREA OF REINFORCEMENT IN R.C. RING = 6.00 sqin
NUMBER OF # 7  BARS REQUIRED =  10
SPACING OF # 3  TIES = 10.00 in
VOLUME OF CONCRETE IN THE COMPRESSION RING =  872.665 cft


                        PRESS ANY KEY TO CONTINUE
‡‡‡  MAIN MENU  ‡‡‡`````````````````````1.  START NEW PROBLEM
2.  SUMMARY OF RESULTS
3.  DISPLAY MENU
4.  REDESIGN
5.  QUIT
ENTER THE NUMBER PLEASE -->>
? 5


                    ‡‡‡  DONE  ‡‡‡
Ok
```

4.8 PROBLEMS

4.1 An angle L9 × 4 × $\frac{1}{2}$ made of A441 steel ($F_y = 42$ ksi) is used as a tension member. Connection of the angle is through $\frac{7}{8}$-in.-diameter bolts. The pattern of holes in the angle is as shown in Fig. 4.16 (with two lines of holes in the long leg and a single line of holes in the short leg). Find the allowable tensile force for this member.

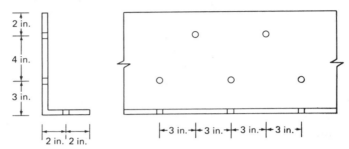

Figure 4.16

4.2 Two main plates 12 in. × 1 in. subjected to tension T have been connected to each other by two splice plates 12 in. × $\frac{7}{16}$ in. and one-inch-diameter bolts as shown in Fig. 4.17(a). The plates are made of A441 steel with yield stress of 40 ksi and ultimate stress of 60 ksi. Assuming that the bolts do not control the design of connections, determine the maximum tension capacity of the connection. Next, in order to increase the efficiency of the connection, holes are staggered as shown in Fig. 4.17(b). For what spacing s of the holes and bolts will the tension capacity of the connection be the largest? For this spacing, find the percentage increase in the capacity of the connection due to staggering of the bolts. (*Note.* You should refer to AISCS 1.14.2.3 and 1.16.4.1).

4.3 The fatigue-critical member of a bridge truss consists of 2L8 × 6 × 1 connected to a gusset plate by fillet welds, as shown in Fig. 4.18. Angles are made of A36 steel ($F_y = 36$ ksi). The member is subjected to a dead load tensile force of 50 K and a live load varying from 30 K in compression to 178 K in tension. Find the design lifetime of the bridge. The average daily number of standard vehicles passing over the bridge and producing the above-mentioned live load variation is 182. (*Note.* Refer to Appendix B of the AISCS.)

4.4 Design the members of the bottom chord of the truss of Fig. 4.19. Each member shall consist of two angles with a spacing of $\frac{1}{2}$ in. to be filled by gusset plates at joints. The two angles shall be connected to the gusset plate with two 1-in. bolts in the vertical leg of the angles. Use A36 steel ($F_y = 36$ ksi).

4.5 Find the size of the main cables and the cross-sectional dimensions of the rectangular concrete compression ring for a circular suspension roof covering an area of diameter 305.6 ft supporting a combined dead and live load of 60 psf. Allowable tensile stress of cable steel is 65 ksi. Compressive strength of concrete is 4 ksi. Use a sag-to-span ratio of 0.2 for the cables and a width-to-depth ratio of $\frac{1}{2}$ for the cross section of the compression ring. Solve the problem for three different cable spacings at the compression ring of about 8 ft, 12 ft, and 15 ft.

4.6 Solve Problem 4.5 with a sag-to-span ratio of 0.25 for the cables.

4.7 Solve Example 2 of this chapter, assuming that the diameter of the structure is 60 ft, the dead load is 70 psf, and the live load is 75 psf. Select the lightest channel (C section) available in the AISCM for the hangers.

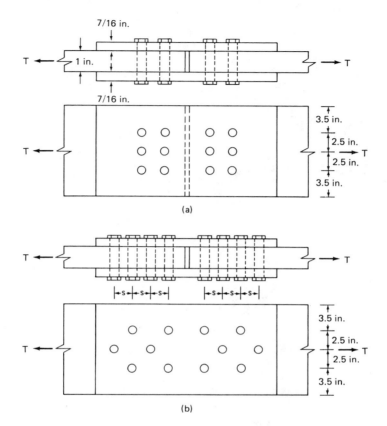

(a)

(b)

Figure 4.17

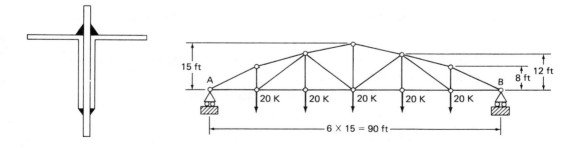

Figure 4.18

5
Design
of Beams

5.1 TYPES OF BEAMS

5.1.1 Standard Hot-Rolled Sections [Fig. 5.1(a)]

Standard hot-rolled sections in general and wide-flange (or W) sections in particular are the most popular types of beam sections. For a given weight, W sections have larger section moduli than S sections and consequently in general are more economical. S sections may be used in the following situations:

1. For crane rails where larger flange thickness near the web may be advantageous for lateral bending.
2. For clearance or other reasons, narrow flanges may be required.
3. Where shear forces are considerably high.

A W shape bent about its minor axis may be as little as 5 percent as strong as the same shape bent about its major axis.
Channels are used in the following situations:

1. Where the load on the beam is light.
2. Where narrow flanges are desirable for clearance or other reasons.

Channels should be avoided where lateral loading is present.

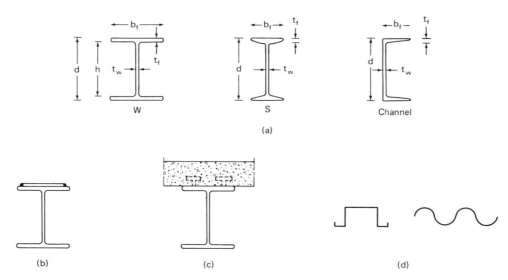

Figure 5.1 Different types of beams. a) Standard hot-rolled sections, b) plate covered section, c) composite beam, d) sheet beams.

Note that in Fig. 5.1(a), d is the depth of the cross section, h and t_w are depth and thickness of the web, respectively, and b_f and t_f are the width and the thickness of the flange, respectively.

5.1.2 Plate-Covered Sections [Fig. 5.1(b)]

The bending capacity of available rolled sections may be increased by adding cover plate(s) to the section.

5.1.3 Plate and Box Girders (Figs. 9.2 and 9.5)

For girders covering large spans and/or carrying heavy loads, the available standard rolled sections may be inadequate. In this case, the designer may use a plate or box girder made of steel plates. These girders may be homogeneous, made of a single grade of steel, or hybrid, made of high-strength flanges. Design of plate girders is covered in detail in Chapter 9.

5.1.4 Composite Beams [Fig. 5.1(c)]

If the steel beam is covered with a concrete deck, the designer may take advantage of the additional strength of the concrete, provided that the concrete deck is properly fastened to the steel beam (for example, by shear studs).

5.1.5 Sheet Beams [Fig. 5.1(d)]

These beams are made of thin cold-formed steel sheets in a variety of shapes. They are useful for very light loads (for example, light roofs).

5.2 REVIEW OF BEAM THEORY

5.2.1 Introduction

In this section, and in fact throughout this volume, we assume that the cross section of the beam has at least one axis of symmetry. The plane containing the axis of symmetry and the longitudinal axis of the beam is called the *plane of symmetry.* Beams are subjected to loads with resultants acting in their plane of symmetry and transverse to their longitudinal axis. Thus, beams carry the loads basically through bending.

In the design of beams, the following considerations are necessary:

1. Bending stresses
2. Shearing stresses
3. Local buckling
4. Lateral torsional buckling
5. Web crippling
6. Deflections

Initial selection of the beam is usually made on the basis of the maximum bending stresses. The other design requirements are checked subsequently.

5.2.2 Elastic Bending of Beams

In elastic theory of bending, the following assumptions are usually made:

1. Deflections are small and change of geometry is negligible.
2. Plane sections remain plane after bending.
3. Shear deformations are small.
4. Interaction of axial forces and bending is negligible.
5. Buckling and stability of the beam is not a problem.
6. The material is linearly elastic; that is, the stress is proportional to the strain.

Assumption 2 implies that the variation of the strain ε over the cross section is linear [Fig. 5.2(b)], and we can write:

$$\varepsilon = Ky = \frac{y}{\rho} \tag{5.1}$$

where ρ is the radius of curvature and K is the curvature given in terms of the bending moment at the section M, the modulus of elasticity E, and the moment of inertia of the cross section about the axis of bending I_x.

$$K = \frac{1}{\rho} = \frac{M}{EI_x} \tag{5.2}$$

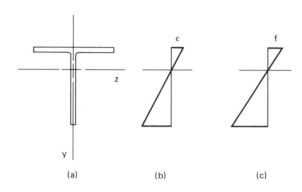

(a)   (b)   (c)

Figure 5.2 Strain ε and elastic stress f distribution over a cross section of a beam.

Based on assumption 6 and using Eqs. (5.1) and (5.2), the normal stress due to bending is

$$f_b = E\varepsilon = EKy = \frac{My}{I_x} \tag{5.3}$$

Denoting the maximum value of the bending moment over the beam length by M_{max} and the maximum y distance from the elastic neutral axis (E.N.A.) or centroidal axis by c, the maximum bending stress will be

$$(f_b)_{max} = \frac{M_{max}c}{I_x} = \frac{M_{max}}{S_x} \tag{5.4}$$

where S_x is called the minimum section modulus of the section in the case of beams with one axis of symmetry and the section modulus for the case of doubly symmetric beams. If the allowable stress as specified by the AISCS is F_b, the required section modulus will be

$$S_x = \frac{M_{max}}{F_b} \tag{5.5}$$

5.2.3 Shear Stresses

Shear stresses are usually not a controlling factor in the design of beams, except for the following cases:

1. The beam is very short.
2. There are holes in the web of the beam. These holes may be for passing electrical and mechanical ducts or for increasing the bending strength in case of castellated beams (Fig. 5.3).
3. The beam is subjected to a very heavy concentrated load near one of the supports.
4. The beam is coped, as shown in Fig. 5.4.

Generally speaking, the flanges of the beam carry the normal stresses due to bending moments, and the web carries the shear stresses due to shear forces. The

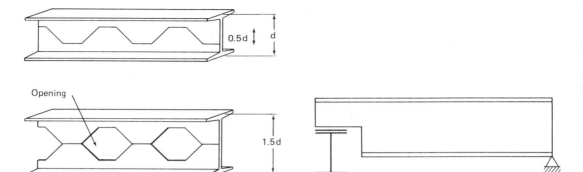

Figure 5.3   Castellated beam.   **Figure 5.4**   Beam with a coped end.

distribution of shear stresses over the depth of an I-section is shown in Fig. 5.5. The shear stress at any point in the web is found from

$$f_v = \frac{VQ}{I_x t} \tag{5.6}$$

where

V = transverse shear on the cross section

Q = first moment of the area of the cross section above (or below) the point at which the shear stress is computed, about the neutral axis

t = thickness of the web where the shear stress is calculated

Considering that the variation of the shear stress over the depth of the web is small, it is a common practice to use the following simple equation for computing the shear stress in the web:

$$f_v = \frac{V}{ht} \tag{5.7}$$

where h is the depth of the web, given in terms of the total depth of the beam, d, and the thickness of the flange, t_f.

$$h = d - 2t_f \tag{5.8}$$

The maximum shear stress over the span of the beam should be less than the allowable shear stress F_v.

Figure 5.5   Shear stress distribution over the cross section of an I-beam.

5.3 LOCAL BUCKLING

The hot-rolled steel sections are thin-walled sections consisting of a number of thin plates. When normal stresses due to bending and/or direct axial forces are large, each plate (for example, flange plate or web plate) may buckle locally in a plane perpendicular to its plane.

In order to prevent this undesirable phenomenon, the width-to-thickness ratios of the thin flange and the web plates are limited by AISCS. These limitations, expressed in terms of the yield stress of steel, F_y, are summarized in Table 5.1. We divide the hot-rolled thin-walled sections into three categories:

1. Compact sections
2. Noncompact sections
3. Sections with slender compression elements

TABLE 5.1   LOCAL BUCKLING WIDTH-THICKNESS LIMITATIONS

(1) Type of element	(2) Width-thickness ratio	(3) Compact	(4) Noncompact
Flanges of I-sections, tees, and channels	$\dfrac{b_f}{t_f}$	$\dfrac{130}{\sqrt{F_y}}$	$\dfrac{190}{\sqrt{F_y}}$
Webs of I-sections, channels, and hollow rectangular sections	$\dfrac{h}{t_w}$	$\dfrac{640^{a}}{\sqrt{F_y}}$	$\dfrac{14{,}000^{b}}{\sqrt{F_y(F_y + 16.5)}}$
Stems of tees	$\dfrac{d}{t_w}$	N.A.	$\dfrac{127}{\sqrt{F_y}}$
Flanges of hollow rectangular sections	$\dfrac{b_f}{t_f}$	$\dfrac{190}{\sqrt{F_y}}$	$\dfrac{238}{\sqrt{F_y}}$
Single-angle struts, double-angle struts with separators	$\dfrac{b}{t}$	N.A.	$\dfrac{76}{\sqrt{F_y}}$
Hollow circular sections	$\dfrac{D}{t}$	$\dfrac{3300}{F_y}$	$\dfrac{3300}{F_y}$

[a]When, in addition to bending, the section is also acted on by an axial compressive stress f_a, the limiting width-thickness ratio becomes:

$$\frac{h}{t_w} = \begin{cases} \dfrac{640}{\sqrt{F_y}}\left(1 - 3.74\dfrac{f_a}{F_y}\right) & \text{when } \dfrac{f_a}{F_y} \le 0.16 \\[2ex] \dfrac{257}{\sqrt{F_y}} & \text{when } \dfrac{f_a}{F_y} > 0.16 \end{cases}$$

[b]For uniformly compressed webs of columns: $\dfrac{h}{t_w} = \dfrac{253}{\sqrt{F_y}}$

Compact sections are expected to develop their full plastic moment capacity, M_p. The width-thickness ratios for these sections are in the third column of Table 5.1. When the width-thickness ratios for a section exceed these limits, the section is called *noncompact*. The limiting ratios for the noncompact sections are given in column 4 of Table 5.1. When the width-thickness ratios exceed these values, the section is referred to as a section with slender compression elements.

5.4 LATERAL TORSIONAL BUCKLING

The compression flange of a beam behaves like an axially loaded column. Thus, in beams covering long spans the compression flange may tend to buckle. This tendency, however, is resisted by the tension flange to a certain extent. The overall effect is a phenomenon known as lateral torsional buckling, in which the beam tends to twist and displace laterally. A theoretical treatment of the subject may be found in Salmon and Johnson [37]. Lateral torsional buckling may be prevented through the following provisions:

1. Lateral supports at intermediate points in addition to lateral supports at the vertical supports
2. Using torsionally strong sections (for example, box sections)
3. I-sections with relatively wide flanges

There are a number of methods for providing lateral supports. If a concrete slab is used, the steel beam may be embedded in the concrete slab as shown in Fig. 5.6. Resting the concrete slab on the beam does not provide sufficient lateral support. In this case, the beam should be anchored to the concrete slab by studs as shown in Fig. 5.7. When light floor decks are used, lateral bracing may be provided by cross-bracing in the plane of the floor, as shown in Fig. 5.8.

When sufficient lateral support is not provided, the phenomenon of lateral torsional buckling is prevented in actual design of steel structures by reducing the

Figure 5.6 Steel beam embedded in the concrete slab.

Figure 5.7 Steel beam connected to the concrete slab by studs.

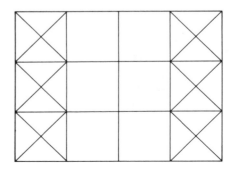

Figure 5.8 Cross-bracing in the plane of the floor.

allowable bending stresses. In other words, the allowable bending stress is specified as a function of the unbraced length of the beam L_u, as discussed in the following section. The larger the unbraced length, the lower the allowable bending stress, and vice versa.

5.5 ALLOWABLE BENDING STRESSES

5.5.1 Bending About the Major Axis

5.5.1.1 Compact sections.
For a member to qualify as a compact section, it must satisfy the following requirements (AISCS Sec. 1.5.1.4.1):

1. The beam shall not be hybrid.
2. The beam shall not be made of A514 steel.
3. The flanges shall be continuously connected to the web(s).
4. The width-thickness ratio of the compression flange and the web shall not exceed the limiting values given in the third column of Table 5.1.
5. The laterally unsupported length L_u of the compression flange shall satisfy the following requirements:
   a. For I and T sections and channels

$$L_u \leq \frac{76b_f}{\sqrt{F_y}} \tag{5.9}$$

$$L_u \leq \frac{20,000}{(d/A_f)F_y} \tag{5.10}$$

   b. For rectangular box members

$$\frac{d}{b} \leq 6 \tag{5.11}$$

$$\frac{t_f}{t_w} \leq 2 \tag{5.12}$$

$$L_u \leq \left(1950 + 1200\frac{M_1}{M_2}\right)\frac{b}{F_y} \tag{5.13}$$

where d is the depth of the cross section, b is its width, M_1 is the smaller and M_2 is the larger bending moment at the ends of the unbraced length with respect to the strong axis of bending. The ratio M_1/M_2 is positive when M_1 and M_2 have the same sign (double curvature bending) and negative otherwise (single curvature bending). The unbraced length L_u need not be less than $1200b/F_y$.

The allowable bending stress for a compact section is given by:

$$F_b = 0.66F_y \tag{5.14}$$

Note that when local flange and web buckling is not a problem and sufficient lateral support is provided to prevent lateral torsional buckling, the 'basic' allowable bending stress is $0.60F_y$. For compact sections, however, this value is increased by 10 percent, because compact sections can develop their full plastic moment capacity.

5.5.1.2 Noncompact sections.

For sections with width-thickness ratio not exceeding the limiting values given in the fourth column of Table 5.1, the allowable bending stress is the larger value computed from Eqs. (5.15) and (5.16), but not exceeding $0.60F_y$ (AISCS Sec. 1.5.1.4.5).

$$F_b = \begin{cases} 0.60F_y & \dfrac{L_u}{r_T} < \sqrt{\dfrac{102,000\,C_b}{F_y}} \\[2ex] \left[\dfrac{2}{3} - \dfrac{F_y(L_u/r_T)^2}{1,530,000\,C_b}\right]F_y & \sqrt{\dfrac{102,000\,C_b}{F_y}} \leq \dfrac{L_u}{r_T} \leq \sqrt{\dfrac{510,000\,C_b}{F_y}} \\[2ex] \dfrac{170,000\,C_b}{(L_u/r_T)^2} & \dfrac{L_u}{r_T} \geq \sqrt{\dfrac{510,000\,C_b}{F_y}} \end{cases} \quad (5.15)$$

$$F_b = \frac{12,000\,C_b}{L_u d / A_f} \quad (5.16)$$

where

r_T = radius of gyration of the portion of the cross section consisting of the compression flange plus one-third of the compression web area (shaded area in Fig. 5.9) about the axis in the web plane (y-axis in Fig. 5.9)

A_f = area of the compression flange

$$C_b = 1.75 + 1.05(M_1/M_2) + 0.3(M_1/M_2)^2 \leq 2.3 \quad (5.17)$$

The ratio M_1/M_2 has been defined in the previous section. When the bending moment at a section within an unbraced length is greater than the two end moments, C_b shall be taken as unity.

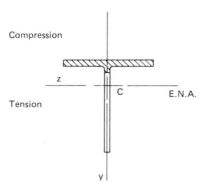

Figure 5.9 Portion of the cross section (shaded) for calculating r_T.

Note 1. For channels, only Eq. (5.16) applies.

Note 2. When the area of the compression flange is less than that of the tension flange, only Eq. (5.15) applies.

Note 3. C_b can be conservatively assumed equal to one.

Note 4. For cantilever beams, C_b is conservatively assumed equal to one.

Note 5. For sections satisfying all the requirements of the compact sections, except the width-thickness ratio for the flange, but satisfying the width-thickness limitation for noncompact sections given in the fourth column of Table 5.1, the allowable bending stress may be calculated from:

$$F_b = [0.79 - 0.002(b_f/2t_f)\sqrt{F_y}]F_y \qquad (5.18)$$

Note 6. Lateral torsional buckling for box sections with d/b less than or equal to 6 is not a problem. The allowable bending stress for these sections is $F_b = 0.60F_y$, provided that the limiting width-thickness ratios given in column 4 of Table 5.1 are satisfied.

5.5.1.3 Sections with slender compression elements.   In this section, we cover only the requirements for I-sections and tees. For angles, channels, and hollow rectangular sections, the reader may refer to the AISCS Appendix C.

When the width-thickness ratio of the compression elements of a beam exceeds the values given in the last column of Table 5.1, a reduction factor Q_s is used in order to prevent local buckling (AISCS Appendix C). In this case, the maximum allowable bending stress shall not exceed the applicable value given in Sec. 5.5.1.2 nor the following value:

$$F_b = 0.6F_yQ_s \qquad (5.19)$$

where Q_s is given by the following expressions:

For compression flanges of I-beams:

$$Q_s = \begin{cases} 1.415 - 0.00437(b_f/2t_f)\sqrt{F_y} & \dfrac{190}{\sqrt{F_y}} < \dfrac{b_f}{t_f} < \dfrac{352}{\sqrt{F_y}} \\[3mm] \dfrac{20,000}{F_y(b_f/2t_f)^2} & \dfrac{b_f}{t_f} \geq \dfrac{352}{\sqrt{F_y}} \end{cases} \qquad (5.20)$$

For stems of tees:

$$Q_s = \begin{cases} 1.908 - 0.00715(d/t_w)\sqrt{F_y} & \dfrac{127}{\sqrt{F_y}} < \dfrac{d}{t_w} < \dfrac{176}{\sqrt{F_y}} \\[3mm] \dfrac{20,000}{F_y(d/t_w)^2} & \dfrac{d}{t_w} \geq \dfrac{176}{\sqrt{F_y}} \end{cases} \qquad (5.21)$$

Channels and tees whose flange width-to-thickness ratio exceeds the value given in the fourth column of Table 5.1 should conform to the additional limits summarized in Table 5.2.

TABLE 5.2   LIMITING PROPORTIONS FOR CHANNELS
AND TEES WITH SLENDER ELEMENTS

Section	$\dfrac{b_f}{d}$	$\dfrac{t_f}{t_w}$
Rolled channel	≤0.50	≤2.0
Built-up channel	≤0.25	≤3.0
Rolled tee	≥0.50	≥1.10
Built-up tee	≥0.50	≥1.25

5.5.2 Bending About the Minor Axis

5.5.2.1 Compact sections (AISCS Sec. 1.5.1.4.3).   For I sections:

$$F_b = 0.75 F_y \tag{5.22}$$

For rectangular tubular sections:

$$F_b = 0.66 F_y \tag{5.23}$$

5.5.2.2 Noncompact sections.   For I sections and rectangular tubular sections, in general,

$$F_b = 0.60 F_y \tag{5.24}$$

For I sections satisfying all the requirements of the compact sections except the width-thickness ratio for the flange, but satisfying the width-thickness limitation for noncompact sections given in the fourth column of Table 5.1, the allowable bending stress may be calculated from:

$$F_b = [1.075 - 0.005(b_f/2t_f)\sqrt{F_y}]F_y \tag{5.25}$$

5.5.2.3 Sections with slender compression elements.   For I sections and tees, Eq. (5.19) shall be used for determining the allowable bending stress.

5.6 ALLOWABLE SHEAR STRESS

The allowable shear stress for hot-rolled sections is given by (AISC Sec. 1.5.1.2.1)

$$F_v = 0.4 F_y \tag{5.26}$$

For very thin webs transverse stiffeners may be required. In this case design of beam will be similar to design of a plate girder. Design of plate girders is covered in Chapter 9. The allowable shear stress for stiffened and unstiffened plate girders is given by Eq. (9.6). This equation at the limit of $F_v = 0.4F_y$ yields $C_v = 1.156$.

When the spacing of the stiffeners is very large, the coefficient k in Eq. (9.8) becomes $k = 5.34$. From Eq. (9.7) the limiting h/t_w ratio becomes $380/\sqrt{F_y}$. Therefore, in order to use the full allowable shear stress value $F_v = 0.40F_y$, we must have

$$\frac{h}{t_w} \leq \frac{380}{\sqrt{F_y}}$$

If this condition is not satisfied, the allowable shear stress will be less than $0.40F_y$ and shall be found from Eq. (9.6).

5.7 WEB CRIPPLING

At the supports and the points of concentrated loads, the web behaves like a short column, and the beam must transfer compression from the wide flange to the narrow web. When the magnitude of the reaction or concentrated load is excessive, high compressive stresses may cause yielding at the junction of the web and the flange. As a result, the flange and web tend to fold over each other at the toe of the fillet. This phenomenon is called web crippling.

According to AISCS (Sec. 1.10.10.1), it is conservatively assumed that the load is distributed at 45-degree angles into the critical section at the toe of the fillet at a distance k from the face of the beam (Fig. 5.10). (Experiments indicate that the load is in fact distributed over a length of $N + 5k$ to $N + 7k$.)

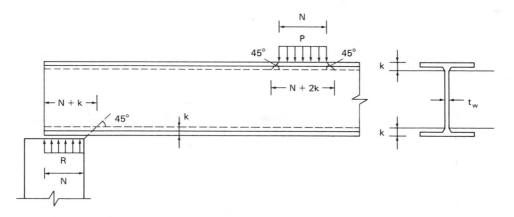

Figure 5.10 Load distribution for web crippling.

The compressive stress f_c at the web toe of the fillets due to concentrated loads shall satisfy the following equations (Fig. 5.10):

For interior loads:

$$f_c = \frac{P}{t_w(N + 2k)} \leq 0.75F_y \tag{5.27}$$

For end reactions:

$$f_c = \frac{R}{t_w(N + k)} \le 0.75 F_y \qquad (5.28)$$

where, as indicated in Fig. 5.10,

P = concentrated load

R = reaction

N = bearing length (at least equal to k for end reactions)

k = distance from outer face of the flange to the web toe of fillet

When the compressive stress exceeds the allowable value of $0.75 F_y$, the following solutions may be adopted:

1. Increase the bearing length
2. Use web stiffeners

5.8 DEFLECTIONS

Deflections of beams are sometimes limited to satisfy the aesthetic or comfort requirements or to prevent damage to nonstructural elements. For example, in beams supporting plastered ceilings in order to prevent cracking of plaster, the maximum deflection due to live load is limited to $\frac{1}{360}$ of the span (AISCS 1.13.1).

Deflection limitations are usually a matter of the designer's judgment, depending on the type of the structure, nonstructural materials, and loadings. As a general guideline, AISCS commentary recommends the following rules (Sec. 1.13.1):

1. Depth of fully stressed beams should possibly be at least equal to $F_y/800$ times the span, where F_y is in ksi. If the depth of the member is less than this value, the allowable bending stress should be decreased in the same proportion as the depth is decreased from the recommended value.
2. Depth of fully stressed purlins in sloping roofs should be at least equal to $F_y/1000$ times the span, where F_y is in ksi.

5.9 EXAMPLES OF DESIGN OF BEAMS

Example 1

Portion AB of a continuous beam is made of a W30 × 108 section and a steel with yield stress of $F_y = 65$ ksi. It has the bending moment diagram shown in Fig. 5.11.

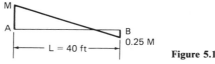

Figure 5.11

Assuming lateral supports at A and B only, determine the maximum bending moment, M_{max}, this portion of the beam can support. Neglect the weight of the beam.

Solution   Properties of the W30 × 108 from the AISCM:

$$d = 29.83 \text{ in.} \qquad S_x = 299 \text{ in.}^3 \qquad t_w = 0.545 \text{ in.}$$

$$b_f = 10.475 \text{ in.} \qquad t_f = 0.760 \text{ in.} \qquad r_T = 2.61 \text{ in.}$$

$$\frac{d}{A_f} = 3.75 \text{ 1/in.}$$

We should first determine whether the section is compact or not. We had better start with the requirements for unbraced length [Eqs. (5.9) and (5.10)] and then check the other requirements.

$$\text{Unbraced length} = L_u = 40(12) = 480 \text{ in.}$$

Check Eq. (5.9):

$$\frac{76b_f}{\sqrt{F_y}} = \frac{76(10.475)}{\sqrt{65}} = 98.7 \text{ in.} < 480 \text{ in.} \quad \therefore \text{ noncompact}$$

The reader can verify that the section is not a section with slender compression elements. Now, we can calculate the allowable bending stress, which is the larger value computed from Eqs. (5.15) and (5.16).

$$\frac{L_u}{r_T} = \frac{480}{2.61} = 183.9; \qquad \frac{M_1}{M_2} = 0.25 \text{ (reverse curvature)}$$

$$C_b = 1.75 + 1.05(M_1/M_2) + 0.3(M_1/M_2)^2$$

$$= 1.75 + 1.05(0.25) + 0.3(0.25)^2 = 2.03 < 2.3$$

$$F_b' = \frac{12,000C_b}{L_u d/A_f} = \frac{(12,000)(2.03)}{(480)(3.75)} = 13.53 \text{ ksi} < 0.6F_y = 39 \text{ ksi}$$

$$\sqrt{\frac{510,000C_b}{F_y}} = \sqrt{\frac{(510,000)(2.03)}{65}} = 126.2 < \frac{L_u}{r_T}$$

$$F_b'' = \frac{170,000C_b}{(L_u/r_T)^2} = \frac{(170,000)(2.03)}{(183.9)^2} = 10.20 \text{ ksi} < F_b'$$

$$\text{Allowable bending stress} = F_b = 13.53 \text{ ksi}$$

$$M_{max} = S_x F_b = (299)(13.53)/12 = 337.12 \text{ K-ft}$$

To speed up the design process, as a general guideline, it is expedient to find the allowable bending stress first from Eq. (5.16). If this value happens to be greater than $0.60F_y$, the allowable bending stress is in fact $0.60F_y$, and there is no need to check the more involved Eq. (5.15).

Example 2

Select the lightest W section available in the AISCM for the beam ABCDE shown in Fig. 5.12 using A36 steel. Lateral supports are provided at supports and at points B and D specified on the figure. The weight of the beam is included in the distributed

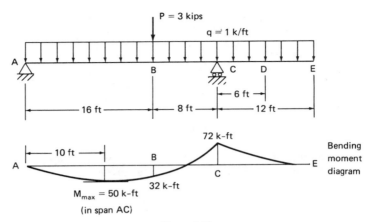

Figure 5.12

load of 1 K/ft. Consider only bending and do not check for shear stresses and deflections.

Solution   To start the iterative design process, initially assume the allowable bending stress to be $F_b = 0.60F_y$. (An experienced designer will probably assume a somewhat lesser value because of the relatively long unbraced lengths.)

Maximum bending moment along the beam:

$$M_{max} = (72)(12) = 864 \text{ K-in.}$$

$$S_x = \frac{M_{max}}{F_b} = \frac{864}{0.6(36)} = 40 \text{ in.}^3$$

Try W16 × 31.

$$S_x = 47.2 \text{ in.}^3 \qquad \frac{d}{A_f} = 6.53 \text{ 1/in.} \qquad r_T = 1.39 \text{ in.} \qquad b_f = 5.525 \text{ in.}$$

a. Check for lateral support from *A* to *B*.

$$L_u = 192 \text{ in.}$$

$$C_b = 1$$

(The bending moment at a point within the unbraced length is larger than at both ends.)

Check compactness, Eq. (5.9):

$$\frac{76b_f}{\sqrt{F_y}} = \frac{76(5.525)}{\sqrt{36}} = 70 \text{ in.} < L_u \quad \therefore \text{ noncompact}$$

$$\sqrt{\frac{102,000C_b}{F_y}} = 53; \qquad \sqrt{\frac{510,000C_b}{F_y}} = 119$$

$$\frac{L_u}{r_T} = \frac{192}{1.39} = 138.1 > \sqrt{\frac{510,000C_b}{F_y}}$$

$$F_b' = \frac{12,000C_b}{L_u d/A_f} = \frac{12,000}{(192)(6.53)} = 9.57 \text{ ksi}$$

$$F_b'' = \frac{170,000 C_b}{(L_u/r_\gamma)^2} = \frac{170,000}{(138.1)^2} = 8.91 \text{ ksi}$$

$$\therefore F_b = 9.57 \text{ ksi}$$

$$f_b = \frac{M_{max}}{S_x} = \frac{600}{47.2} = 12.71 \text{ ksi} > F_b \qquad \underline{\underline{\text{N.G.}}}$$

Try W14 × 34.

$$S_x = 48.6 \text{ in.}^3 \qquad \frac{d}{A_f} = 4.56 \text{ 1/in.}$$

$$r_T = 1.76 \text{ in.} \qquad b_f = 6.745 \text{ in.}$$

$$\frac{L_u}{r_T} = \frac{192}{1.76} = 109.1$$

$$F_b' = \frac{12,000 C_b}{L_u d / A_f} = \frac{12,000}{(192)(4.56)} = 13.71 \text{ ksi}$$

$$F_b'' = \left[\frac{2}{3} - \frac{F_y (L_u/r_T)^2}{1,530,000 C_b}\right] F_y = \left[\frac{2}{3} - \frac{36(109.1)^2}{1,530,000}\right](36) = 13.92 \text{ ksi}$$

$$F_b = 13.92 \text{ ksi}$$

$$f_b = \frac{M_{max}}{S_x} = \frac{600}{48.6} = 12.35 \text{ ksi} < F_b = 13.92 \text{ ksi} \qquad \underline{\underline{\text{O.K.}}}$$

b. Check for lateral support from B to C.

$$L_u = 8 \text{ ft} = 96 \text{ in.}$$

$$\frac{76 b_f}{\sqrt{F_y}} = \frac{76(6.745)}{\sqrt{36}} = 85.4 \text{ in.} < 96 \text{ in.} \quad \therefore \text{ noncompact}$$

$$C_b = 1.75 + 1.05\left(\frac{M_1}{M_2}\right) + 0.3\left(\frac{M_1}{M_2}\right)^2$$

$$= 1.75 + 1.05(\tfrac{32}{72}) + 0.3(\tfrac{32}{72})^2 = 2.27 < 2.3$$

$$F_b' = \frac{12,000 C_b}{L_u d / A_f} = \frac{(12,000)(2.27)}{(96)(4.56)} = 62.2 > 0.60 F_y = 22 \text{ ksi}$$

$$\therefore F_b = 22 \text{ ksi}$$

$$f_b = \frac{864}{48.6} = 17.78 \text{ ksi} < 22 \text{ ksi} \qquad \underline{\underline{\text{O.K.}}}$$

c. Check for lateral support from C to D.

$$L_u = 6 \text{ ft} = 72 \text{ in.}$$

$$\frac{76 b_f}{\sqrt{F_y}} = \frac{76(6.745)}{\sqrt{36}} = 85.4 \text{ in.} > L_u$$

$$\frac{20,000}{(d/A_f) F_y} = \frac{20,000}{(4.56)(36)} = 121.8 > L_u$$

∴ Compact Section

$$F_b = 0.66\,F_y = 24\ \text{ksi} > f_b = 17.78\ \text{ksi}$$ <u>O.K.</u>

$$\boxed{\text{USE W14} \times 34}$$

Note that for all W sections given in the AISCM made of A36 steel except W6 × 15 the local flange and web buckling requirements for compactness are always satisfied.

Example 3

Select the lightest W16 for the cantilever beam shown in Fig. 5.13 using A36 steel. Load P_1 is applied in the vertical plane of symmetry. Load P_2 passes through the centroid of the section and makes an angle of 30 deg with the vertical plane of symmetry. (It is located in a plane perpendicular to the vertical plane of symmetry.) Assume total lateral support and neglect the weight of the beam. Specify the point where the normal stress has the largest absolute value.

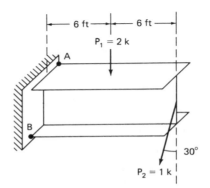

Figure 5.13

Solution This beam is subjected to biaxial bending, that is, simultaneous bending about major and minor axes. If we denote these two axes by x and y, the resultant normal stress due to bending M_x about the x-axis and M_y about the y-axis for a doubly symmetric section is found from

$$f_b = \pm \frac{M_x}{S_x} \pm \frac{M_y}{S_y}$$

where S_x and S_y are section moduli with respect to x (major axis) and y (minor axis), respectively.

Denoting the maximum bending stress due to bending about the x-axis by $f_{bx} = M_x/S_x$ and the maximum bending stress due to bending about the y-axis by $f_{by} = M_y/S_y$ and noting that the allowable bending stresses about the major and minor axes are different, we must satisfy the following interaction equation for design of beams subjected to biaxial bending:

$$\frac{f_{bx}}{F_{bx}} + \frac{f_{by}}{F_{by}} \leq 1.0 \tag{5.29}$$

where F_{bx} and F_{by} are the allowable bending stresses with respect to major and minor axes, respectively. To start the iterative design process it may initially be assumed that

$F_{bx} = 0.60F_y$ and $F_{by} = 0.75F_y$. Thus, the following approximate equation for the section modulus S_x may be used for preliminary selection of the section:

$$S_x = \frac{M_x}{0.60F_y} + \frac{R_s M_y}{0.75F_y} \tag{5.30}$$

where $R_s = S_x/S_y$. This ratio can be estimated approximately for different ranges of sections. The most frequent values for R_s are 7 for W27 to W36 sections, 5 to 6 for W16 to W24 sections, and 2 to 3 for W10 to W14 sections.

For the example of Fig. 5.13, the maximum bending moment occurs at the fixed support.

$$M_x = 2(6) + (1)\cos 30°(12) = 22.39 \text{ K-ft} = 268.71 \text{ K-in.}$$

$$M_y = (1)\cos 60°(12) = 6 \text{ K-ft} = 72 \text{ K-in.}$$

The section is compact. Assume $R_s = 5$.

$$S_x = \frac{M_x}{24} + \frac{R_s M_y}{27} = \frac{268.71}{24} + \frac{5(72)}{27} = 24.53 \text{ in.}^3$$

Try W16 × 26.

$$S_x = 38.4 \text{ in.}^3 \quad \text{and} \quad S_y = 3.49 \text{ in.}^3$$

$$\frac{f_{bx}}{F_{bx}} + \frac{f_{by}}{F_{by}} = \frac{M_x}{24S_x} + \frac{M_y}{27S_y} = \frac{268.71}{24(38.4)} + \frac{72}{27(3.49)}$$

$$= 0.291 + 0.764 = 1.06 > 1.0 \qquad \underline{\text{N.G.}}$$

Try W16 × 31.

$$S_x = 47.2 \text{ in.}^3 \quad \text{and} \quad S_y = 4.49 \text{ in.}^3$$

$$\frac{f_{bx}}{F_{bx}} + \frac{f_{by}}{F_{by}} = \frac{268.71}{(24)(47.2)} + \frac{72}{27(4.49)}$$

$$= 0.237 + 0.594 = 0.831 < 1 \qquad \underline{\text{O.K.}}$$

$$\boxed{\text{USE W16} \times 31}$$

Points A and B indentified on Fig. 5.13 are the points of maximum stress.

5.10 INTERACTIVE MICROCOMPUTER-AIDED DESIGN OF BEAMS ACCORDING TO AISCS

5.10.1 Program Capabilities and Limitations

The program presented in this section is an interactive BASIC program for design of simply supported beams according to the AISCS. It can easily be extended for design of beams with other end conditions by simply modifying the analysis portion of the program. The loading on the beam may consist of a uniformly distributed load plus any number of concentrated loads. Concentrated loads are numbered from left to right. A portion of the beam between two adjacent concentrated loads or between a reaction and a concentrated load is called a *segment*. A portion of the beam between two adjacent lateral supports is called a *region*. Segments and regions are numbered from left to right. The beam may have full-lateral support or lateral supports at any given number of locations.

The input data may be read from an existing file or may be inputted by the user interactively. For a given loading condition on the beam, the interactive program can either check the adequacy of a given design selected by the user or design and select a section from the AISC database file, which includes all the W, M, S, HP, WT, C, and MC shapes available in the AISCM.

For design, the user may select the nominal depth of the section. For example, if the user selects the W12 series from among the W shapes, the program will try to come up with the lightest W12 section. If no W12 section in the AISC database satisfies the design requirements, the program automatically moves to the next groups of W shapes—that is, W14—and gives the message NEXT LARGER SEC- TION DEPTH IS CHOSEN. The program then tries to select the lightest W14 section. If no W14 section satisfies the design requirements, the program starts to search among W16 shapes. This process is continued until the program finds a W shape satisfying all the design requirements. If no W shape is adequate for the given loading condition, the program will give the message THE STRONGEST SECTION AVAILABLE W36 × 300 CANNOT SUSTAIN THE LOADS. YOU NEED TO USE ANOTHER SHAPE.

If the user does not select the nominal depth, the program will come up with the lightest section within the specified shape, say W. The program is designed to be highly interactive. If the user wants to change any portion of the input data, he can simply enter this change by selecting the appropriate option in the redesign menu (Fig. 5.16) without having to reenter all the input data from scratch. The redesign menu in this program as well as other programs is an attractive feature of these programs.

The program can handle the design of seven types of hot-rolled sections (Fig. 5.17). The user can choose the axis of bending, that is, major or minor axis. He will be asked whether the beam is used in a roof or floor. This information is used for checking the deflection (see Sec. 5.8).

The analysis portion of the program calculates the support reactions, the maximum shearing force, the maximum bending moment, bending moments at the locations of concentrated loads, and the points of lateral supports.

If all the design requirements are satisfied, the program will display the properties of the selected section, together with the total weight of the beam and the maximum ratio of actual bending stress to the allowable bending stress, and whether the section is compact, noncompact, or with slender elements.

The interactive program can display the loading diagram, shearing force diagram, bending moment diagram, and the cross section of the section selected (W, M, S, HP, WT, C, and MC). The cross-sectional display includes all the necessary dimensions.

5.10.2 Program Structure and Menus

The program is modular. Different modules perform various tasks. An error check module gives error messages for unrealistic values inputted by the user. If the user chooses a number which does not exist on the menu, the program will provide an error message. If the user selects a section not available in the AISCM, the message ENTER AN ACTUAL SECTION PLEASE! will be given. If the distance of a lateral support or a concentrated load from the left support is less than zero or greater than the length of the span, a message reminding the user about the actual length of the beam will appear.

The program is menu-driven. This makes the program very easy to use without a manual. Menus have a hierarchical order. There are five menus in the program. Figure 5.14 shows the main menu. This menu allows the user to move to various parts of the program. At the beginning of a design, the user must choose the first option. If he starts with other options, the message YOU HAVE TO PROVIDE DATA FIRST! will appear.

```
     * BEAM DESIGN-AISC *
     ***  MAIN  MENU  ***

     1. NEW PROBLEM

     2. SUMMARY OF RESULTS

     3. MAIN DISPLAY MENU

     4. REDESIGN

     5. SAVE DATA ON A DISK

     6. QUIT


     ENTER A NUMBER ==>>? ■
```

Figure 5.14 Main menu.

From the main menu, the user can move to the display menu shown in Fig. 5.15 and the redesign menu shown in Fig. 5.16. The remaining menus are the section menu shown in Fig. 5.17 and the steel-type menu shown in Fig. 5.18.

```
     *** DISPLAY  MENU ***


     1. DISPLAY LOADING

     2. SHEAR DIAGRAM

     3. MOMENT DIAGRAM

     4. CROSS-SECTION

     5. MAIN MENU



     ENTER A NUMBER ==>>? ■
```

Figure 5.15 Display menu.

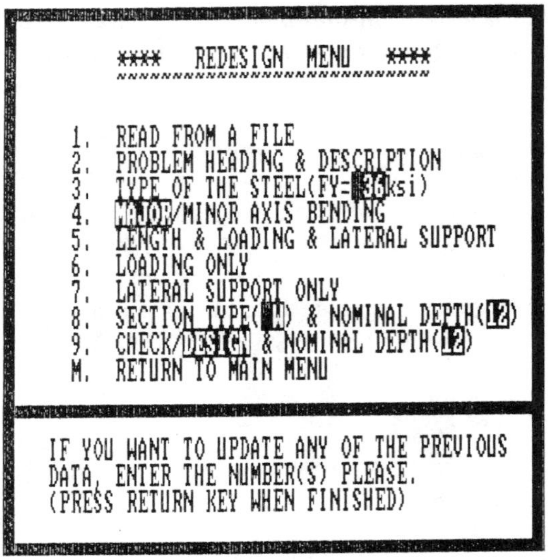

```
**** REDESIGN MENU ****
~~~~~~~~~~~~~~~~~~~~~~~~~~

1.  READ FROM A FILE
2.  PROBLEM HEADING & DESCRIPTION
3.  TYPE OF THE STEEL(FY=▮36ksi)
4.  MAJOR/MINOR AXIS BENDING
5.  LENGTH & LOADING & LATERAL SUPPORT
6.  LOADING ONLY
7.  LATERAL SUPPORT ONLY
8.  SECTION TYPE(▮W) & NOMINAL DEPTH(▮12)
9.  CHECK/DESIGN & NOMINAL DEPTH(▮12)
M.  RETURN TO MAIN MENU


IF YOU WANT TO UPDATE ANY OF THE PREVIOUS
DATA, ENTER THE NUMBER(S) PLEASE.
(PRESS RETURN KEY WHEN FINISHED)
```

Figure 5.16 Redesign menu.

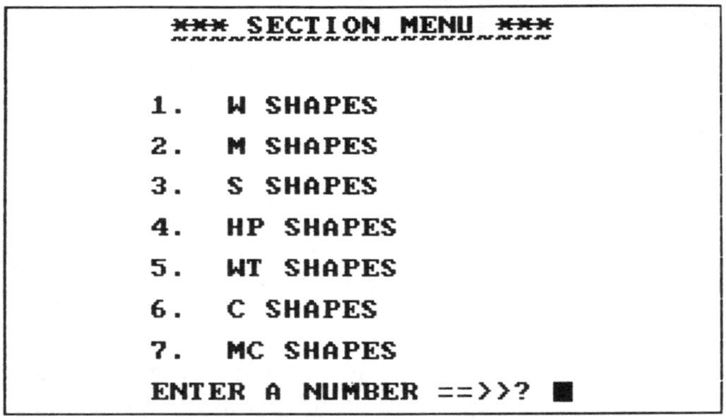

```
*** SECTION MENU ***
~~~~~~~~~~~~~~~~~~~~~~

1.  W SHAPES

2.  M SHAPES

3.  S SHAPES

4.  HP SHAPES

5.  WT SHAPES

6.  C SHAPES

7.  MC SHAPES

ENTER A NUMBER ==>>? ▮
```

Figure 5.17 Section menu.

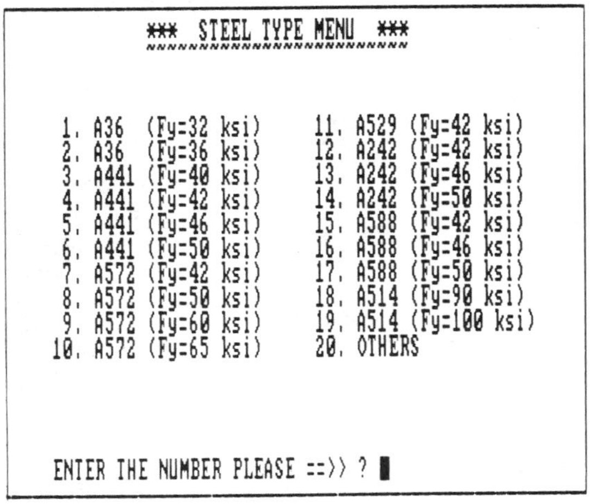

```
*** STEEL TYPE MENU ***
~~~~~~~~~~~~~~~~~~~~~~~~~

 1. A36  (Fy=32 ksi)    11. A529 (Fy=42 ksi)
 2. A36  (Fy=36 ksi)    12. A242 (Fy=42 ksi)
 3. A441 (Fy=40 ksi)    13. A242 (Fy=46 ksi)
 4. A441 (Fy=42 ksi)    14. A242 (Fy=50 ksi)
 5. A441 (Fy=46 ksi)    15. A588 (Fy=42 ksi)
 6. A441 (Fy=50 ksi)    16. A588 (Fy=46 ksi)
 7. A572 (Fy=42 ksi)    17. A588 (Fy=50 ksi)
 8. A572 (Fy=50 ksi)    18. A514 (Fy=90 ksi)
 9. A572 (Fy=60 ksi)    19. A514 (Fy=100 ksi)
10. A572 (Fy=65 ksi)    20. OTHERS



ENTER THE NUMBER PLEASE ==>> ? ▮
```

Figure 5.18 Steel type menu.

The program and the AISC sections database are saved on one floppy disk. The name of the program primary file is BEAMAISC. In order to obtain hard copies of the graphic output and have access to the AISC sections database, the following commands should be used in the system mode before loading the program:

>GRAPHICS
>BASICA /S:186

5.10.3 Sample Example and Interactive Session

5.10.3.1 Sample example.   As an example, the simple beam shown in Fig. 5.19 has been solved by the interactive microcomputer program. The beam is first subjected to the loading shown in this figure. The interactive session is given in Sec. 5.10.3.2. The shear force and bending moment diagrams as plotted by the microcomputer program are shown in Figs. 5.20 and 5.21, respectively. The steel type and lateral support conditions are given in the interactive session. The beam is bent about its major axis.

Initially, the program was asked to design a W10 section. But no W10 section can carry the specified load safely. Therefore, the program automatically moves to the next groups of W shapes, that is, W12. The selected section, W12 × 136, is shown in Fig. 5.22. Next, using the redesign menu (Fig. 5.16), the type of the section is changed to an S shape. For this case, the program was asked to find the lightest S shape. The answer is S24 × 90, which is plotted in Fig. 5.23.

Using the redesign menu (Fig. 5.16) again, we change some of the input data. The span length is reduced to 10 ft, the loading is changed to uniformly distributed loading of 2 K/ft only, with no concentrated load, and full lateral support is assumed for the beam. The adequacy of a C15 × 50, shown in Fig. 5.24 bent about its minor axis, is checked.

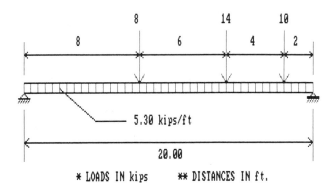

* LOADS IN kips   ** DISTANCES IN ft.

PRESS ANY KEY TO GO BACK TO DISPLAY MENU

Figure 5.19 Loading diagram.

SHEAR FORCE DIAGRAM

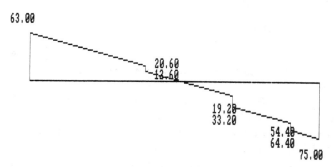

* VALUES OF SHEAR FORCES ARE IN Kips.

PRESS ANY KEY TO CONTINUE

Figure 5.20   Shear force diagram.

BENDING MOMENT DIAGRAM

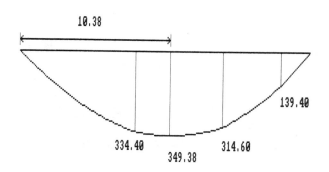

* VALUES OF BENDING MOMENTS ARE IN Kips.ft

PRESS ANY KEY TO CONTINUE

Figure 5.21   Bending moment diagram.

BEAM SECTION

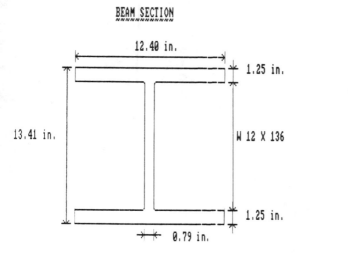

12.40 in.

1.25 in.

13.41 in.

W 12 X 136

1.25 in.

0.79 in.

PRESS ANY KEY TO CONTINUE

Figure 5.22

BEAM SECTION

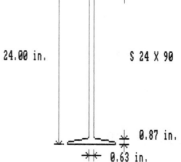

7.13 in.

0.87 in.

24.00 in.

S 24 X 90

0.87 in.

0.63 in.

PRESS ANY KEY TO CONTINUE

Figure 5.23

BEAM SECTION

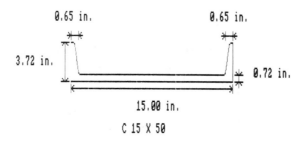

0.65 in.   0.65 in.

3.72 in.

0.72 in.

15.00 in.

C 15 X 50

PRESS ANY KEY TO CONTINUE

Figure 5.24

5.10.3.2 Interactive session.

```
run
# BEAM DESIGN-AISC #
### MAIN MENU ###
^^^^^^^^^^^^^^^^^^
1. NEW PROBLEM
2. SUMMARY OF RESULTS
3. MAIN DISPLAY MENU
4. REDESIGN
5. SAVE DATA ON A DISK
6. QUIT
ENTER A NUMBER ==>>
? 1


       ##### DESIGN OF SIMPLY SUPPORTED BEAMS  #####
       ^^^^^^^^^^^^^^^^^^^^^^^^^^^^^^^^^^^^^^^^^^^^^

       ((( Based on AISC SPECIFICATION 1980 )))
           ^^^^^^^^^^^^^^^^^^^^^^^^^^^^^^^^^^


              ### BASIC INPUT DATA ###
              ^^^^^^^^^^^^^^^^^^^^^^^^

DO YOU DESIRE TO USE THE DATA FROM A DISK (Y/N) ? n

PROBLEM HEADING ? sample problem

PROBLEM DESCRIPTION ?

LENGTH OF THE SPAN (ft) ? 20

### STEEL TYPE MENU ###
^^^^^^^^^^^^^^^^^^^^^^^
  1. A36  (Fy=32 ksi)     11. A529 (Fy=42 ksi)
  2. A36  (Fy=36 ksi)     12. A242 (Fy=42 ksi)
  3. A441 (Fy=40 ksi)     13. A242 (Fy=46 ksi)
  4. A441 (Fy=42 ksi)     14. A242 (Fy=50 ksi)
  5. A441 (Fy=46 ksi)     15. A588 (Fy=42 ksi)
  6. A441 (Fy=50 ksi)     16. A588 (Fy=46 ksi)
  7. A572 (Fy=42 ksi)     17. A588 (Fy=50 ksi)
  8. A572 (Fy=50 ksi)     18. A514 (Fy=90 ksi)
  9. A572 (Fy=60 ksi)     19. A514 (Fy=100 ksi)
 10. A572 (Fy=65 ksi)     20. OTHERS
ENTER THE NUMBER PLEASE ==>>
? 2


IS THE BEAM USED FOR FLOOR OR ROOF (F/R) ? f

IS THE BEAM BENT ABOUT MAJOR AXIS OR MINOR AXIS (MAJ/MIN)? maj



              ### LOADS ###
              ^^^^^^^^^^^^^

INTENSITY OF DISTRIBUTED DEAD LOAD (Kips/ft) ? 3.4

IS WEIGHT OF BEAM INCLUDED IN THE ABOVE DEAD LOAD  (Y/N) ? y

INTENSITY OF DISTRIBUTED LIVE LOAD (Kips/ft) ? 1.9


NUMBER OF CONCENTRATED LOADS ? 3

DEAD LOAD PORTION OF CONCENTRATED LOAD No. 1 FROM LEFT SUPPORT (KIPS) ? 6
LIVE LOAD PORTION OF CONCENTRATED LOAD No. 1 FROM LEFT SUPPORT (KIPS) ? 2
DISTANCE OF LOAD NO. 1 FROM LEFT SUPPORT (ft) ? 8
```

DEAD LOAD PORTION OF CONCENTRATED LOAD No. 2 FROM LEFT SUPPORT (KIPS) ? 14
LIVE LOAD PORTION OF CONCENTRATED LOAD No. 2 FROM LEFT SUPPORT (KIPS) ? 0
DISTANCE OF LOAD NO. 2 FROM LEFT SUPPORT (ft) ? 14

DEAD LOAD PORTION OF CONCENTRATED LOAD No. 3 FROM LEFT SUPPORT (KIPS) ? 8
LIVE LOAD PORTION OF CONCENTRATED LOAD No. 3 FROM LEFT SUPPORT (KIPS) ? 2
DISTANCE OF LOAD NO. 3 FROM LEFT SUPPORT (ft) ? 18

‡‡‡ LATERAL SUPPORT ‡‡‡

DOES THE BEAM HAVE TOTAL LATERAL SUPPORT (Y/N) ? n

NUMBER OF POINTS OF LATERAL SUPPORT ? 3

ENTER DISTANCE OF LATERAL SUPPORT NO. 1
(FROM LEFT SUPPORT IN ft) ? 5

ENTER DISTANCE OF LATERAL SUPPORT NO. 2
(FROM LEFT SUPPORT IN ft) ? 10

ENTER DISTANCE OF LATERAL SUPPORT NO. 3
(FROM LEFT SUPPORT IN ft) ? 15

‡‡‡‡ STRUCTURAL ANALYSIS ‡‡‡‡

‡‡   REACTIONS AND MAXIMUM SHEAR FORCE   ‡‡

REACTION AT LEFT SUPPORT 63.00 Kips.

REACTION AT RIGHT SUPPORT 75.00 Kips.

MAXIMUM SHEAR FORCE 75.00 Kips.
PRESS ANY KEY TO CONTINUE

‡‡   VALUES OF BENDING MOMENTS   ‡‡

BENDING MOMENT AT LOAD NO. 1   334.40 Kips.ft

BENDING MOMENT AT LOAD NO. 2   314.60 Kips.ft

BENDING MOMENT AT LOAD NO. 3   139.40 Kips.ft

MAXIMUM BENDING MOMENT 349.38 Kips.ft

LOCATION OF MAXIMUM BENDING MOMENT 10.38 ft. FROM LEFT SUPPORT

PRESS ANY KEY TO CONTINUE

DO YOU DESIRE TO SAVE THE DATA ON A DISK (Y/N) ? n

‡‡‡ SECTION MENU ‡‡‡
1.   W SHAPES
2.   M SHAPES
3.   S SHAPES
4.   HP SHAPES
5.   WT SHAPES
6.   C SHAPES
7.   MC SHAPES
ENTER A NUMBER ==>>
? 1

DO YOU WANT TO CHECK A SECTION OR DESIGN (CHK/DSN) ? dsn

              ‡‡‡   DESIGN   OF BEAM   ‡‡‡

DO YOU WANT TO CHOOSE THE NOMINAL DEPTH?
(IF YOU DON'T CHOOSE, THE PROGRAM WILL FIND THE
LIGHTEST SECTION POSSIBLE FOR THE GIVEN SHAPE) (Y/N)? y

AVAILABLE NOMINAL DEPTHS FOR 'W' SECTIONS ARE :
  4   5   6   8   10   12   14   16   18   21   24   27   30   33   36
NOMINAL DEPTH OF THE BEAM ? 10

                  searching ...

NEXT LARGER SECTION DEPTH IS CHOSEN

SELECTED BEAM :
               ‡‡‡‡‡   W 12 X 136   ‡‡‡‡‡

A	= 39.900 in.^2		bf	= 12.400 in.
d	= 13.410 in.		tf	= 1.250 in.
tw	= 0.790 in.		rT	= 3.410 in.
Ix	= 1240.00 in.^4		Iy	= 398.00 in.^4
Sx	= 186.00 in.^3		Sy	= 64.20 in.^3
rx	= 5.58 in.		ry	= 3.16 in.

THE MAXIMUM RATIO OF THE ACTUAL STRESS TO THE ALLOWABLE STRESS IS 0.949

THE SECTION IS COMPACT
TOTAL WEIGHT OF THE BEAM : 2.72 Kips.
PRESS ANY KEY TO CONTINUE

‡ BEAM DESIGN-AISC ‡
‡‡‡ MAIN MENU ‡‡‡

1. NEW PROBLEM
2. SUMMARY OF RESULTS
3. MAIN DISPLAY MENU
4. REDESIGN
5. SAVE DATA ON A DISK
6. QUIT
ENTER A NUMBER ==>>
? 4

         ‡‡‡‡   REDESIGN MENU   ‡‡‡‡

    1.   READ FROM A FILE
    2.   PROBLEM HEADING & DESCRIPTION
    3.   TYPE OF THE STEEL(FY= 36ksi)
    4.   MAJOR/MINOR AXIS BENDING
    5.   LENGTH & LOADING & LATERAL SUPPORT
    6.   LOADING ONLY
    7.   LATERAL SUPPORT ONLY
—→ 8.   SECTION TYPE(W) & NOMINAL DEPTH(12)
    9.   CHECK/DESIGN & NOMINAL DEPTH(12)
    M.   RETURN TO MAIN MENU

   IF YOU WANT TO UPDATE ANY OF THE PREVIOUS
   DATA, ENTER THE NUMBER(S) PLEASE.
   (PRESS RETURN KEY WHEN FINISHED)

```
‡‡‡ SECTION MENU ‡‡‡
^^^^^^^^^^^^^^^^^^^^^^
1.  W SHAPES
2.  M SHAPES
3.  S SHAPES
4.  HP SHAPES
5.  WT SHAPES
6.  C SHAPES
7.  MC SHAPES
ENTER A NUMBER ==>>
? 3


                  ‡‡‡  DESIGN  OF BEAM  ‡‡‡
                  ^^^^^^^^^^^^^^^^^^^^^^^^^

DO YOU WANT TO CHOOSE THE NOMINAL DEPTH?
(IF YOU DON'T CHOOSE, THE PROGRAM WILL FIND THE
LIGHTEST SECTION POSSIBLE FOR THE GIVEN SHAPE) (Y/N)? n


                        searching ...


SELECTED BEAM :

                  ‡‡‡‡‡    S 24 X 90    ‡‡‡‡‡

            A   =  26.500 in.^2        bf  =   7.125 in.
            d   =  24.000 in.          tf  =   0.870 in.
            tw  =   0.625 in.          rT  =   1.600 in.
            Ix  = 2250.00 in.^4        Iy  =    44.90 in.^4
            Sx  =  187.00 in.^3        Sy  =   12.60 in.^3
            rx  =    9.21 in.          ry  =    1.30 in.

THE MAXIMUM RATIO OF THE ACTUAL STRESS TO THE ALLOWABLE  STRESS IS  0.944

THE SECTION IS COMPACT
TOTAL WEIGHT OF THE BEAM :  1.80 Kips.
PRESS ANY KEY TO CONTINUE



            ‡‡‡‡   REDESIGN  MENU   ‡‡‡‡
            ^^^^^^^^^^^^^^^^^^^^^^^^^^^^^^
      1.  READ FROM A FILE
      2.  PROBLEM HEADING & DESCRIPTION
      3.  TYPE OF THE STEEL(FY= 36ksi)
-->> 4.  MAJOR/MINOR AXIS BENDING
-->> 5.  LENGTH & LOADING & LATERAL SUPPORT
      6.  LOADING ONLY
      7.  LATERAL SUPPORT ONLY
-->> 8.  SECTION TYPE( S) & NOMINAL DEPTH(12)
-->> 9.  CHECK/DESIGN & NOMINAL DEPTH(12)
      M.  RETURN TO MAIN MENU




    IF YOU WANT TO UPDATE ANY OF THE PREVIOUS
    DATA, ENTER THE NUMBER(S) PLEASE.
    (PRESS RETURN KEY WHEN FINISHED)




IS THE BEAM BENT ABOUT MAJOR AXIS OR MINOR AXIS (MAJ/MIN)? ain

LENGTH OF THE SPAN (ft) ? 10


IS THE BEAM USED FOR FLOOR OR ROOF (F/R) ? f
```

```
                        ### LOADS ###

INTENSITY OF DISTRIBUTED DEAD LOAD (Kips/ft) ? 1
IS WEIGHT OF BEAM INCLUDED IN THE ABOVE DEAD LOAD  (Y/N) ? y
INTENSITY OF DISTRIBUTED LIVE LOAD (Kips/ft) ? 1

NUMBER OF CONCENTRATED LOADS ? 0


                    ### LATERAL SUPPORT ###

DOES THE BEAM HAVE TOTAL LATERAL SUPPORT (Y/N) ? y


                  #### STRUCTURAL ANALYSIS ####


            ## REACTIONS AND MAXIMUM SHEAR FORCE ##

REACTION AT LEFT SUPPORT  10.00 Kips.

REACTION AT RIGHT SUPPORT  10.00 Kips.

MAXIMUM SHEAR FORCE  10.00 Kips.
PRESS ANY KEY TO CONTINUE



              ## VALUES OF BENDING MOMENTS ##

MAXIMUM BENDING MOMENT   25.00 Kips.ft

LOCATION OF MAXIMUM BENDING MOMENT  5.00 ft. FROM LEFT SUPPORT


PRESS ANY KEY TO CONTINUE

### SECTION MENU ###

1.  W SHAPES
2.  M SHAPES
3.  S SHAPES
4.  HP SHAPES
5.  WT SHAPES
6.  C SHAPES
7.  MC SHAPES
ENTER A NUMBER ==>>
? 6




DO YOU WANT TO CHECK A SECTION OR DESIGN (CHK/DSN) ? chk


SECTION DESIGNATION : DEPTH, WEIGHT
(e.g. W,36,300)        C,15,50



                    searching ...


CHECK DEFLECTION CRITERIA IS NOT FULLFILLEDFOR SECTION : C  15  X  50
```

```
SELECTED BEAM :
                  $$$$$    C 15 X 50    $$$$$
           A   =  14.700 in.^2        bf  =  3.716 in.
           d   =  15.000 in.          tf  =  0.650 in.
           tw  =   0.716 in.          rT  =  0.000 in.
           Ix  = 404.00 in.^4         Iy  =   11.00 in.^4
           Sx  =  53.80 in.^3         Sy  =   3.78 in.^3
           rx  =   5.24 in.           ry  =   0.87 in.

THE MAXIMUM RATIO OF THE ACTUAL STRESS TO THE ALLOWABLE STRESS IS  2.939

THE SECTION IS UNSAFE
TOTAL WEIGHT OF THE BEAM :  0.50 Kips.
PRESS ANY KEY TO CONTINUE
```

5.11 LOAD AND RESISTANCE FACTOR DESIGN OF BEAMS

5.11.1 Design for Flexure

In design of beams according to LRFDS, the following limit states are considered:

1. Lateral torsional buckling (LTB)
2. Flange local buckling (FLB)
3. Web local buckling (WLB)

Similar to Sec. 5.3, the beam sections are divided into three categories: compact sections, noncompact sections, and sections with slender compression elements.

For a section to be considered as compact, the width-thickness ratios of its compression elements must not be greater than the limiting width-thickness ratios λ_p given in Table 5.3. When the width-thickness ratio of any one of the compression elements is larger than λ_p it is considered noncompact. When the width-thickness ratio of a compression element in a noncompact section exceeds λ_r given in Table 5.3, the element is called a slender compression element and the corresponding section is classified as a section with slender compression elements. Note that in Table 5.3, b is equal to $b_f/2$ for I-sections but equal to b_f for channels and hollow rectangular sections. The limiting ratios in Table 5.3 are similar to those in Table 5.1; however, there are differences.

According to LRFDS, the flexural design strength is $\phi_b M_n$, where

$$\phi_b = \text{resistance factor for flexure} = 0.9$$

$$M_n = \text{nominal flexural strength}$$

The nominal flexural strength is the smallest value computed based on the following limit states of LTB, FLB, and WLB.

1. For $\lambda \leq \lambda_p$

$$M_n = M_p \tag{5.31}$$

TABLE 5.3 LIMITING WIDTH-THICKNESS RATIOS FOR COMPRESSION ELEMENTS (LIMIT STATES OF FLB AND WLB)

Type of element	Width-thickness ratio (λ)	Limiting width-thickness ratios	
		λ_p	λ_r
Flanges of I sections and channels in flexure	$\dfrac{b}{t}$	$\dfrac{65}{\sqrt{F_y}}$	$\dfrac{141}{\sqrt{F_y - F_r}}$
Flanges of I sections and channels in pure compression	$\dfrac{b}{t}$	N.A.	$\dfrac{95}{\sqrt{F_y}}$
Webs of I sections, channels, and hollow rectangular sections in flexural compression	$\dfrac{h_c}{t_w}$	$\dfrac{640^*}{\sqrt{F_y}}$	$\dfrac{970}{\sqrt{F_y}}$
Uniformly compressed stiffened elements	$\dfrac{b}{t}$ or $\dfrac{h_c}{t_w}$	N.A.	$\dfrac{253}{\sqrt{F_y}}$
Stems of tees	$\dfrac{b}{t}$	N.A.	$\dfrac{127}{\sqrt{F_y}}$
Flanges of hollow rectangular sections	$\dfrac{b}{t}$	$\dfrac{190}{\sqrt{F_y}}$	$\dfrac{238}{\sqrt{F_y - F_r}}$
Hollow circular sections in axial compression	$\dfrac{D}{t}$	$\dfrac{2070}{F_y}$	$\dfrac{3300}{F_y}$
Hollow circular sections in flexure	$\dfrac{D}{t}$	$\dfrac{2070}{F_y}$	$\dfrac{8970}{F_y}$

*When, in addition to bending, the section is also acted on by an axial compressive force, the limiting width-thickness ratio becomes

$$\frac{h_c}{t_w} = \begin{cases} \dfrac{640}{\sqrt{F_y}}\left[1 - \dfrac{2.75P_u}{\phi_b P_y}\right] & \text{when } \dfrac{P_u}{\phi_b P_y} \leq 0.125 \\[3ex] \dfrac{191}{\sqrt{F_y}}\left[2.33 - \dfrac{P_u}{\phi_b P_y}\right] \geq \dfrac{253}{\sqrt{F_y}} & \text{when } \dfrac{P_u}{\phi_b P_y} > 0.125 \end{cases}$$

where P_u is the required axial strength and $P_y = A_y F_y$ is the yield strength.

2. For $\lambda_p < \lambda \leq \lambda_r$

For the limit state of LTB:

$$M_n = C_b\left[M_p - (M_p - M_r)\left(\frac{\lambda - \lambda_p}{\lambda_r - \lambda_p}\right)\right] \leq M_p \qquad (5.32)$$

For the limit state of FLB and WLB:

$$M_n = M_p - (M_p - M_r)\left(\frac{\lambda - \lambda_p}{\lambda_r - \lambda_p}\right) \qquad (5.33)$$

3. For $\lambda > \lambda_r$

For LTB and FLB:

$$M_n = M_{cr} = S_x F_{cr} \leq C_b M_r \qquad (5.34)$$

For WLB:

For λ of the web $> \lambda_r$ design as for a plate girder (see Chapter 9), where

M_p = plastic moment = ZF_y

M_{cr} = buckling moment

M_r = limiting buckling moment, M_{cr}, when $\lambda = \lambda_r$

λ = controlling slenderness parameter

  = minor axis slenderness ratio L_u/r_y for lateral-torsional buckling

  = flange width-thickness ratio b/t for flange local buckling

  = web depth-thickness ratio h/t_w for web local buckling

λ_p = limiting slenderness parameter for compact sections
(largest value of λ for which $M_n = M_p$)

λ_r = limiting slenderness parameter for sections with slender elements
(largest value of λ for which buckling is inelastic)

F_{cr} = critical stress

S_x = section modulus

Z = plastic section modulus

L_u = unbraced length

r_y = radius of gyration about minor axis

The quantity C_b has been defined by Eq. (5.17).

The λ_p and λ_r values corresponding to FLB and WLB have been summarized in Table 5.3. The slenderness parameter λ_p corresponding to LTB for doubly symmetric I shapes bent about the major axis is

$$\lambda_p = 300/\sqrt{F_{yf}} \qquad (5.35)$$

The slenderness parameter λ_r corresponding to LTB for I shapes bent about the major axis is

$$\lambda_r = \frac{X_1}{F_{yw} - F_r} \sqrt{1 + \sqrt{1 + X_2(F_{yw} - F_r)^2}} \qquad (5.36)$$

where

$$X_1 = \frac{\pi}{S_x} \sqrt{\frac{EGJA}{2}} \qquad (5.37)$$

$$X_2 = \frac{4C_w}{I_y}\left(\frac{S_x}{GJ}\right)^2 \tag{5.38}$$

S_x = section modulus about major axis (in.^{3})

E = modulus of elasticity of steel (29,000 ksi)

G = shear modulus of steel (11,200 ksi)

A = cross-sectional area (in.^{2})

F_r = compressive residual stress in flange (10 ksi for rolled shapes, 16.5 ksi for welded shapes)

J = torsional constant (in.^{4})

C_w = warping constant (in.^{6}), $C_w \simeq \frac{1}{4}I_y(d - t_f)^2$

I_y = moment of inertia about minor axis (in.^{4})

It should be noted that there are no LTB λ_p and λ_r limits for any I shape bent about its minor axis (LTB is not a problem in this case).

The critical stress F_{cr} for LTB for bending about the major axis is given by

$$F_{cr} = \frac{C_b X_1 \sqrt{2}}{\lambda}\sqrt{1 + \frac{X_1^2 X_2}{2\lambda^2}} \tag{5.39}$$

and for FLB of the rolled shapes is given by

$$F_{cr} = \frac{20,000}{\lambda^2} \tag{5.40}$$

When the bending is about the major axis, the limiting buckling moment M_r for LTB and FLB is given by

$$M_r = (F_{yw} - F_r)S_x \tag{5.41}$$

and for WLB is given by

$$M_r = F_{yf}S_x \tag{5.42}$$

For bending about the minor axis, M_r needs to be evaluated for FLB only, and its value is specified by

$$M_r = F_y S_y \tag{5.43}$$

5.11.2 Design for Shear

The design shear strength of the web is $\phi_v V_n$, where

$$\phi_v = \text{resistance factor for shear} = 0.9$$

$$V_n = \text{nominal shear strength}$$

The nominal shear strength is determined as follows:

$$V_n = \begin{cases} 0.60 F_{yw} A_w & \dfrac{h_c}{t_w} \le 187 \sqrt{\dfrac{k}{F_{yw}}} \\[3ex] 0.60 F_{yw} A_w \dfrac{187 \sqrt{\dfrac{k}{F_{yw}}}}{h_c / t_w} & 187 \sqrt{\dfrac{k}{F_{yw}}} < \dfrac{h_c}{t_w} \le 234 \sqrt{\dfrac{k}{F_{yw}}} \\[3ex] \dfrac{26{,}400 k A_w}{(h_c / t_w)^2} & \dfrac{h_c}{t_w} > 234 \sqrt{\dfrac{k}{F_{yw}}} \end{cases} \qquad (5.44)$$

where

k = web plate buckling coefficient = 5

F_{yw} = yield stress of the web (ksi)

A_w = web area = dt_w (in.^{2})

h_c = twice the distance from the neutral axis to the inside face of the compression flange less the fillet or corner radius (in.)

t_w = thickness of web

d = overall depth of the member

5.11.3 Design Considerations for Concentrated Loads

5.11.3.1 Local web yielding.   The design compressive strength of the web at the toe of the fillet under concentrated loads is ϕR_n, where $\phi = 1.0$ and R_n is computed as follows:

a. When the concentrated load is applied to the interior of a span (that is, at a distance of at least the beam's depth from its end):

$$R_n = (5k + N) F_{yw} t_w \qquad (5.45)$$

b. When the concentrated load is applied at or near the end of the member (that is, at a distance not more than the beam's depth from its end):

$$R_n = (2.5k + N) F_{yw} t_w \qquad (5.46)$$

where

N = length of bearing

k = distance from the outer face of the flange to the web toe of fillet

Equation (5.45) or (5.46) can be used to find the minimum required bearing length.

5.11.3.2 Web crippling. The design compressive strength for unstiffened portions of webs subjected to concentrated load is ϕR_n, where $\phi = 0.75$ and R_n is computed as follows:

a. When the concentrated load is at a distance not less than $d/2$ from the end of the member (d is the overall depth of the cross section):

$$R_n = 135t_w^2\left[1 + 3\left(\frac{N}{d}\right)\left(\frac{t_w}{t_f}\right)^{1.5}\right]\sqrt{F_{yw}t_f/t_w} \qquad (5.47)$$

b. When the concentrated load is at a distance less than $d/2$ from the end of the member:

$$R_n = 68t_w^2\left[1 + 3\left(\frac{N}{d}\right)\left(\frac{t_w}{t_f}\right)^{1.5}\right]\sqrt{F_{yw}t_f/t_w} \qquad (5.48)$$

When stiffeners are provided and they extend at least one-half of the web depth, Eqs. (5.47) and (5.48) need not be checked.

5.11.3.3 Sidesway web buckling. When flanges of a beam subjected to concentrated loads are not restrained against relative movement by stiffeners or lateral bracings, the design compressive strength is ϕR_n, where $\phi = 0.85$ and R_n is computed as follows:

a. When the loaded flange is restrained against rotation and $d_c b_f/L_u t_w < 2.3$:

$$R_n = \frac{12{,}000t_w^3}{h}\left[1 + 0.4\left(\frac{d_c b_f}{L_u t_w}\right)^3\right] \qquad (5.49)$$

where

L_u = largest laterally unbraced length at the point of application of concentrated load

$d_c = d - 2k$ = web depth of section clear of fillets

h = web depth

When $d_c b_f/L_u t_w > 2.3$, Eq. (5.49) does not have to be checked.

b. When the loaded flange is not restrained against rotation and $d_c b_f/L_u t_w < 1.7$:

$$R_n = \frac{12{,}000t_w^3}{h}\left[0.4\left(\frac{d_c b_f}{L_u t_w}\right)^3\right] \qquad (5.50)$$

When $d_c b_f/L_u t_w > 1.7$, Eq. (5.50) does not have to be checked.

5.11.3.4 Shear strength of the web in beam-columns subjected to high shear. When the beam-column is subjected to concentrated loads, the design shear strength of the web shall be ϕR_v, where $\phi = 0.90$ and

$$R_v = \begin{cases} 0.7F_y d_c t_w & \text{for } P_u \leq 0.75P_n \\ 0.7F_y d_c t_w(1.9 - 1.2P_u/P_n) & \text{for } P_u > 0.75P_n \end{cases} \qquad (5.51)$$

where

$$P_u = \text{required axial strength}$$

$$P_n = \text{nominal axial strength}$$

Note that in the case of beams $P_n = 0$, but for beam-columns $P_n \neq 0$.

5.11.3.5 Compression buckling of the web.   When concentrated loads are applied to both flanges of the beam, the design compressive strength for unstiffened portions of webs is ϕR_n, where $\phi = 0.90$ and

$$R_n = \frac{4,100 t_w^3 \sqrt{F_{yw}}}{d_c} \tag{5.52}$$

5.12 INTERACTIVE MICROCOMPUTER-AIDED LOAD AND RESISTANCE FACTOR DESIGN OF BEAMS

5.12.1 Program Capabilities and Limitations

In this section, an interactive BASIC program for design of simply supported beams according to LRFDS is presented. This program is rather similar to the one presented in Sec. 5.10. The following load combinations and the corresponding load factors are considered (see Sec. 2.7):

$$1.4 D_n$$

$$1.2 D_n + 1.6 L_n + 0.5 (L_r \text{ or } S_n)$$

where

$$D_n = \text{dead load acting on the beam}$$

$$L_n = \text{live load acting on the beam}$$

$$L_r = \text{roof live load}$$

$$S_n = \text{snow load}$$

The type of loads acting on the beam, the definition of segment and region, the capability of either to design or to check a section, and displays of cross section of beam, shear force diagram, and bending moment diagram are the same as those described in Sec. 5.10.1. Further, the program has the capability of displaying the beam loading for dead load, live load, and the aforementioned load combinations. It is assumed that the concentrated loads are applied to the top flange only.

If the weight of the beam is not included in the dead load value inputted by the user, the program will estimate the beam weight as five percent of the intensity of the dead load and add to the dead load to obtain the total intensity of the dead load. After the design of a section, the program will check to see whether the estimated weight of beam is within 10 percent of the actual beam weight. If this condition is not satisfied, the program will perform a reanalysis with the actual beam weight and check the adequacy of the section selected.

The program can select a W, M, S or HP shape from the AISC database. After selecting the cross-sectional shape of the beam from the section menu (Fig. 5.28), two options are available to the user. He can select the nominal depth of the beam section (for example, 14 for W14). In this case, the program will find the lightest section for the beam within the specified shape and nominal depth. If the user does not choose the nominal depth, the program will find the lightest available section within the given shape. To find the lightest section, the program uses four data files for the four types of structural shapes (W, M, S, and HP). In these data files, sections are stored in order of increasing weight. This is different from the main AISC sections database, in which sections are saved in the same order presented in the AISCM.

To start the design process, an approximate value for the required plastic modulus is determined by using the equation

$$Z_{req} = M_{max}/F_y$$

where M_{max} is the maximum bending moment due to the critical load combination.

Using the approximate value of Z, we choose a tentative section from the AISC sections database for a given shape and nominal depth (for example, W12) or from one of the four data files mentioned previously if only the cross-sectional shape is specified by the user. Then for the case of major axis bending, the section is checked for the following by computing the minimum flexural strength:

1. LTB (for both load cases)
2. FLB (for maximum design bending moment)
3. WLB (for maximum design bending moment)
4. Shear strength (for maximum design shear force)

For minor axis bending, the section is checked only for flange local buckling.

When the beam is subjected to concentrated loads, the following design considerations are also checked:

1. Local web yielding
2. Web crippling
3. Sidesway web buckling

Based on the local web yielding and web crippling requirements, the minimum bearing length is calculated for each location of concentrated loads. If any one of the above checks is not satisfied, the next larger section within the given shape and nominal depth will be chosen. The design will be complete when all the above checks are satisfied.

After the completion of a design, the user has the flexibility of changing either the loads acting on the beam or the lateral support system or the type of shape or the nominal depth or all of them by using the "REDESIGN MENU." The interactive program will come up with a new design accommodating all the changes.

5.12.2 Program Structure and Menus

The program consists of several subprograms chained together through the CHAIN statement. Each subprogram is divided into smaller modules. An error check module helps to identify whether the input keyed by the user is acceptable or not. For example, an error message will be displayed when the nominal depth inputted by the user does not exist in the AISC sections database, or when the yield stress of steel is abnormally high or low with respect to its range of practical values.

There are six menus in the program. The main menu is shown in Fig. 5.25. From this menu, the user can move to the main display menu shown in Fig. 5.26

```
  * BEAM DESIGN-LRFD *
  ***   MAIN  MENU  ***
~~~~~~~~~~~~~~~~~~~~~~~~

1. NEW PROBLEM

2. SUMMARY OF RESULTS

3. MAIN DISPLAY MENU

4. REDESIGN

5. SAVE DATA ON A DISK

6. QUIT


ENTER A NUMBER ==>>? ■
```

Figure 5.25 Main menu.

```
 *** MAIN DISPLAY MENU ***
~~~~~~~~~~~~~~~~~~~~~~~~~~~~

1. LOADING DISPLAY MENU

2. SHEAR DIAGRAM

3. MOMENT DIAGRAM

4. CROSS-SECTION

5. MAIN MENU



ENTER A NUMBER ==>>? ■
```

Figure 5.26 Main display menu.

and redesign menu shown in Fig. 5.16. The first option in the main display menu leads the user to the loading display menu shown in Fig. 5.27. Dead load, live load, load combination case 1, or load combination case 2 may be displayed individually. The remaining menus are the section menu and steel type menu shown in Figs. 5.28 and 5.18, respectively. Note that the redesign and steel-type menus in this program

```
 ***   LOADING DISPLAY MENU   ***
~~~~~~~~~~~~~~~~~~~~~~~~~~~~~~~~~~~

 1. DEAD LOAD

 2. LIVE LOAD

 3. LOAD CASE 1 (1.4 D.L.)

 4. LOAD CASE 2 (1.2 D.L.+1.6
    L.L.+0.5 (S.L. OR R.L.))

 5. MAIN DISPLAY MENU


ENTER A NUMBER ==>>? ■
```

Figure 5.27 Loading display menu.

```
┌────────────────────────────────────────┐
│    *** SECTION MENU ***                 │
│    ~~~~~~~~~~~~~~~~~~~~~~~                │
│                                         │
│                                         │
│    1.   W SHAPES                        │
│                                         │
│    2.   M SHAPES                        │
│                                         │
│    3.   S SHAPES                        │
│                                         │
│    4.   HP SHAPES                       │
│                                         │
│    ENTER A NUMBER ==>>? ■               │
└────────────────────────────────────────┘
```

Figure 5.28 Section menu.

have the same format as those in the program for design of beams according to the AISCS described in Sec. 5.10.

The program and the AISC sections database are saved on one floppy disk. The name of the program primary file is BEAMLRFD. In order to obtain hard copies of the graphic output and have access to the AISC sections database, the following commands should be used in the system mode before loading the program:

>GRAPHICS
>BASICA /S:186

5.12.3 Sample Example and Interactive Session

5.12.3.1 Sample example.   As an example, the simple beam shown in Fig. 5.29 has been designed by the interactive microcomputer program. The beam is

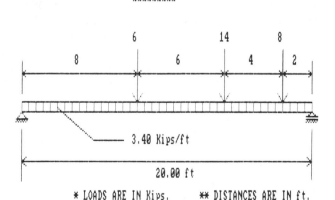

PRESS ANY KEY TO CONTINUE   **Figure 5.29**

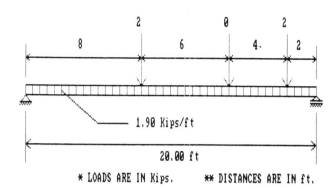

PRESS ANY KEY TO CONTINUE   **Figure 5.30**

subjected to the dead load shown in Fig. 5.29 and the live load shown in Fig. 5.30. The beam is analyzed for load case 1 (Fig. 5.31) and load case 2 (Fig. 5.32). The span of the beam, the lateral support system, and the type of loading are the same as those for the beam designed in Sec. 5.10.3 using the AISCS. The shear force diagrams and bending moment diagrams plotted by the program are shown in Figs. 5.33 through 5.36. Initially, a W12 section is designed, which is shown in Fig. 5.37.

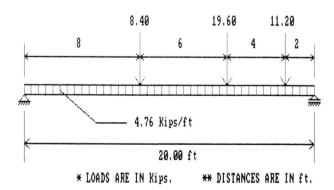

PRESS ANY KEY TO CONTINUE   **Figure 5.31**

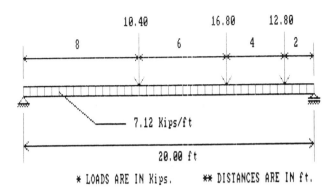

BEAM LOADING

LOAD CASE 2 (1.2 D.L.+1.6 L.L.+ 0.5 (S.L. OR R.L.))

10.40   16.80   12.80

8   6   4   2

7.12 Kips/ft

20.00 ft

* LOADS ARE IN Kips.   ** DISTANCES ARE IN ft.

PRESS ANY KEY TO CONTINUE   **Figure 5.32**

Then an S section is designed for the beam as shown in Fig. 5.38. In both cases, the beam is subjected to major axis bending. It should be noted that designs obtained by using LRFDS often give a lighter section compared to AISCS. In the example presented in this section and Sec. 5.10.3, LRFDS yielded W12 × 120 and S20 × 86, while AISCS yielded W12 × 136 and S24 × 90.

SHEAR FORCE DIAGRAM

LOAD CASE 1 (1.4 D.L.)

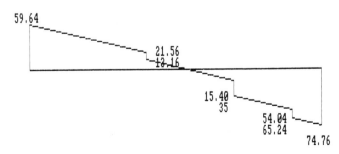

59.64

21.56
18.16

15.40
35

54.04
65.24

74.76

* VALUES OF SHEAR FORCES ARE IN Kips.

PRESS ANY KEY FOR SFD OF LOAD CASE 2   **Figure 5.33**

SHEAR FORCE DIAGRAM

LOAD CASE 2 (1.2 D.L.+1.6 L.L.+0.5 (S.L. OR R.L.))

83.76

26.80
16.40

26.32
43.12

71.60
84.40

98.64

* VALUES OF SHEAR FORCES ARE IN Kips.

PRESS ANY KEY TO CONTINUE   Figure 5.34

Finally, the interactive program is used to find the lightest available section using a W shape. The answer is W24 × 68, which is shown in Fig. 5.39. It should be noted that this section is much lighter than the W12 shape selected previously for the same beam. Since a W12 shape has a nominal depth of only 12 inches, it is not the best choice for a beam. The reader can observe the difference by looking at Figs. 5.37 and 5.39.

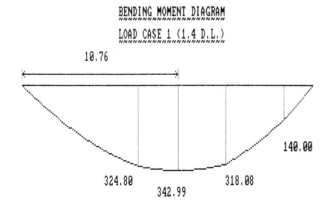

BENDING MOMENT DIAGRAM

LOAD CASE 1 (1.4 D.L.)

10.76

140.00

324.80   318.08
          342.99

* VALUES OF BENDING MOMENTS ARE IN Kips.ft

PRESS ANY KEY FOR BMD OF LOAD CASE 2   Figure 5.35

BENDING MOMENT DIAGRAM

LOAD CASE 2 (1.2 D.L.+1.6 L.L.+0.5 (S.L. OR R.L.))

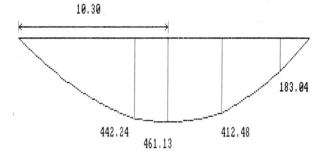

10.30

183.04

442.24   412.48

461.13

* VALUES OF BENDING MOMENTS ARE IN Kips.ft

PRESS ANY KEY TO CONTINUE

Figure 5.36

BEAM SECTION

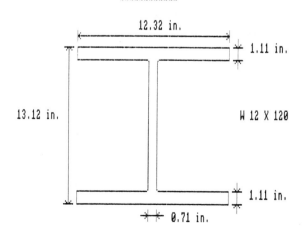

12.32 in.

1.11 in.

13.12 in.   W 12 X 120

1.11 in.

0.71 in.

PRESS ANY KEY TO CONTINUE

Figure 5.37

BEAM SECTION

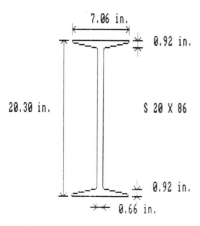

PRESS ANY KEY TO CONTINUE

Figure 5.38

BEAM SECTION

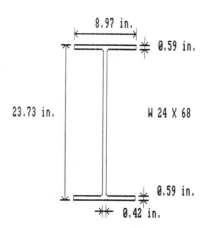

PRESS ANY KEY TO CONTINUE

Figure 5.39

5.12.3.2 Interactive session.

```
RUN
# BEAM DESIGN-LRFD #
### MAIN MENU ###
~~~~~~~~~~~~~~~~~~~
1. NEW PROBLEM
2. SUMMARY OF RESULTS
3. MAIN DISPLAY MENU
4. REDESIGN
5. SAVE DATA ON A DISK
6. QUIT
ENTER A NUMBER ==>>
? 1




        ##### DESIGN OF SIMPLY SUPPORTED BEAMS  #####
        ~~~~~~~~~~~~~~~~~~~~~~~~~~~~~~~~~~~~~~~~~~~~~~

     ((( Based on AISC Proposed LRFD SPECIFICATION 1986  )))
         ~~~~~~~~~~~~~~~~~~~~~~~~~~~~~~~~~~~~~~~~~~~~~


            ### BASIC INPUT DATA ###
            ~~~~~~~~~~~~~~~~~~~~~~~~~

DO YOU DESIRE TO USE THE DATA FROM A DISK (Y/N) ? N

PROBLEM HEADING ? SAMPLE PROBLEM

PROBLEM DESCRIPTION ? BEAM WEIGHT IS INCLUDED IN DEAD LOAD

LENGTH OF THE SPAN (ft) ? 20
```

```
‡‡‡  STEEL TYPE MENU  ‡‡‡
˄˄˄˄˄˄˄˄˄˄˄˄˄˄˄˄˄˄˄˄˄˄˄˄˄˄
  1. A36   (Fy=32 ksi)      11. A529  (Fy=42 ksi)
  2. A36   (Fy=36 ksi)      12. A242  (Fy=42 ksi)
  3. A441  (Fy=40 ksi)      13. A242  (Fy=46 ksi)
  4. A441  (Fy=42 ksi)      14. A242  (Fy=50 ksi)
  5. A441  (Fy=46 ksi)      15. A588  (Fy=42 ksi)
  6. A441  (Fy=50 ksi)      16. A588  (Fy=46 ksi)
  7. A572  (Fy=42 ksi)      17. A588  (Fy=50 ksi)
  8. A572  (Fy=50 ksi)      18. A514  (Fy=90 ksi)
  9. A572  (Fy=60 ksi)      19. A514  (Fy=100 ksi)
 10. A572  (Fy=65 ksi)      20. OTHERS
ENTER THE NUMBER PLEASE ==))
? 2


IS THE BEAM USED FOR FLOOR OR ROOF (F/R) ? F

IS THE BEAM BENT ABOUT MAJOR AXIS OR MINOR AXIS (MAJ/MIN)? MAJ


                      ‡‡‡  LOADS  ‡‡‡
                      ˄˄˄˄˄˄˄˄˄˄˄˄˄

INTENSITY OF DISTRIBUTED DEAD LOAD (Kips/ft) ? 3.4

IS WEIGHT OF BEAM INCLUDED IN THE ABOVE
DEAD LOAD   (Y/N) ? Y

INTENSITY OF DISTRIBUTED LIVE LOAD (Kips/ft) ? 1.9


NUMBER OF CONCENTRATED LOADS ? 3

DEAD LOAD PORTION OF CONCENTRATED LOAD No.  1  FROM LEFT SUPPORT (Kips) ? 6
LIVE LOAD PORTION OF CONCENTRATED LOAD No.  1  FROM LEFT SUPPORT (Kips) ? 2
DISTANCE OF LOAD NO. 1 FROM LEFT SUPPORT (ft) ? 8

DEAD LOAD PORTION OF CONCENTRATED LOAD No.  2  FROM LEFT SUPPORT (Kips) ? 14
LIVE LOAD PORTION OF CONCENTRATED LOAD No.  2  FROM LEFT SUPPORT (Kips) ? 0
DISTANCE OF LOAD NO. 2 FROM LEFT SUPPORT (ft) ? 14

DEAD LOAD PORTION OF CONCENTRATED LOAD No.  3  FROM LEFT SUPPORT (Kips) ? 8
LIVE LOAD PORTION OF CONCENTRATED LOAD No.  3  FROM LEFT SUPPORT (Kips) ? 2
DISTANCE OF LOAD NO. 3 FROM LEFT SUPPORT (ft) ? 18


                    ‡‡‡  LATERAL SUPPORT  ‡‡‡
                    ˄˄˄˄˄˄˄˄˄˄˄˄˄˄˄˄˄˄˄˄˄˄˄˄˄

DOES THE BEAM HAVE TOTAL LATERAL SUPPORT (Y/N) ? N

NUMBER OF POINTS OF LATERAL SUPPORT ? 3

ENTER DISTANCE OF LATERAL SUPPORT NO. 1
(FROM LEFT SUPPORT IN ft) ? 5

ENTER DISTANCE OF LATERAL SUPPORT NO. 2
(FROM LEFT SUPPORT IN ft) ? 10

ENTER DISTANCE OF LATERAL SUPPORT NO. 3
(FROM LEFT SUPPORT IN ft) ? 15


                  ‡‡‡‡  STRUCTURAL ANALYSIS  ‡‡‡‡
                  ˄˄˄˄˄˄˄˄˄˄˄˄˄˄˄˄˄˄˄˄˄˄˄˄˄˄˄˄˄˄

NOTE: ONLY TWO LOAD COMBINATIONS ARE CONSIDERED:
      1.   1.4 D.L.
      2.   1.2 D.L. + 1.6 L.L. + 0.5 (S.L. OR R.L.)
    WHERE
        D.L. = DEAD LOAD
        L.L. = LIVE LOAD
        S.L. = SNOW LOAD
        R.L. = ROOF LIVE LOAD
PRESS ANY KEY TO CONTINUE
```

⁑⁑ LOAD CASE 1 : 1.4 D.L. ⁑⁑

⁑ REACTIONS AND MAXIMUM SHEAR FORCE ⁑

REACTION AT LEFT SUPPORT 59.64 Kips.

REACTION AT RIGHT SUPPORT 74.76 Kips.

MAXIMUM SHEAR FORCE 74.76 Kips.
PRESS ANY KEY TO CONTINUE

⁑ VALUES OF BENDING MOMENTS ⁑

BENDING MOMENT AT LOAD NO. 1   324.80 Kips.ft

BENDING MOMENT AT LOAD NO. 2   318.08 Kips.ft

BENDING MOMENT AT LOAD NO. 3   140.00 Kips.ft

MAXIMUM BENDING MOMENT 342.99 Kips.ft

LOCATION OF MAXIMUM BENDING MOMENT 10.76 ft. FROM LEFT SUPPORT

PRESS ANY KEY TO CONTINUE

⁑⁑ LOAD CASE 2 : 1.2 D.L.+1.6 L.L. ⁑⁑

⁑ REACTIONS AND MAXIMUM SHEAR FORCE ⁑

REACTION AT LEFT SUPPORT 83.76 Kips.

REACTION AT RIGHT SUPPORT 98.64 Kips.

MAXIMUM SHEAR FORCE 98.64 Kips.
PRESS ANY KEY TO CONTINUE

⁑ VALUES OF BENDING MOMENTS ⁑

BENDING MOMENT AT LOAD NO. 1   442.24 Kips.ft

BENDING MOMENT AT LOAD NO. 2   412.48 Kips.ft

BENDING MOMENT AT LOAD NO. 3   183.04 Kips.ft

MAXIMUM BENDING MOMENT 461.13 Kips.ft

LOCATION OF MAXIMUM BENDING MOMENT 10.30 ft. FROM LEFT SUPPORT

PRESS ANY KEY TO CONTINUE

⁑⁑ DESIGN VALUES FOR SHEAR FORCE AND BENDING MOMENT ⁑⁑

MAXIMUM SHEAR FORCE 98.64 Kips.

MAXIMUM BENDING MOMENT 461.13 Kips.ft

LOCATION OF MAXIMUM BENDING MOMENT 10.30 ft. FROM LEFT SUPPORT

PRESS ANY KEY TO CONTINUE

DO YOU DESIRE TO SAVE THE DATA ON A DISK (Y/N) ? Y

NAME OF THE DATA FILE ==) ? DATA1

‡‡‡ SECTION MENU ‡‡‡
1. W SHAPES
2. M SHAPES
3. S SHAPES
4. HP SHAPES
ENTER A NUMBER ==))
? 1

DO YOU WANT TO CHECK A SECTION OR DESIGN (CHK/DSN) ? DSN

                    ‡‡‡ DESIGN OF BEAM ‡‡‡

DO YOU WANT TO CHOOSE THE NOMINAL DEPTH?
(IF YOU DON'T CHOOSE, THE PROGRAM WILL FIND THE
LIGHTEST SECTION POSSIBLE FOR THE GIVEN SHAPE) (Y/N)? Y

AVAILABLE NOMINAL DEPTHS FOR 'W' SECTIONS ARE :
 4 5 6 8 10 12 14 16 18 21 24 27 30 33 36
NOMINAL DEPTH OF THE BEAM ? 10

                        searching ...

NEXT LARGER SECTION DEPTH IS CHOSEN

SELECTED BEAM :
                    ‡‡‡‡‡   W 12 X 120   ‡‡‡‡‡

           A   = 35.300 in.^2   bf = 12.320 in.
           d   = 13.120 in.   tf = 1.105 in.
           tw = 0.710 in.   rT = 3.380 in.
           Ix = 1070.00 in.^4   Iy = 345.00 in.^4
           Zx = 186.00 in.^3   Zy =   85.40 in.^3
           rx =   5.51 in.   ry =   3.13 in.

THE RATIO OF THE MAXIMUM FACTORED FLEXURAL
MOMENT TO THE NOMINAL FLEXURAL STRENGTH IS 0.918
THE SECTION IS A COMPACT SECTION
TOTAL WEIGHT OF THE BEAM : 2.40 Kips.

PRESS ANY KEY TO CONTINUE

‡ BEAM DESIGN-LRFD ‡
‡‡‡ MAIN MENU ‡‡‡

1. NEW PROBLEM
2. SUMMARY OF RESULTS
3. MAIN DISPLAY MENU
4. REDESIGN
5. SAVE DATA ON A DISK
6. QUIT
ENTER A NUMBER ==))
? 4

            ‡‡‡‡ REDESIGN MENU ‡‡‡‡

       1. READ FROM A FILE
       2. PROBLEM HEADING & DESCRIPTION
       3. TYPE OF THE STEEL(FY= 36ksi)
       4. MAJOR/MINOR AXIS BENDING
       5. LENGTH & LOADING & LATERAL SUPPORT
       6. LOADING ONLY
       7. LATERAL SUPPORT ONLY

→→8.   SECTION TYPE(W) & NOMINAL DEPTH(12)
    9.   CHECK/DESIGN & NOMINAL DEPTH(12)
    M.   RETURN TO MAIN MENU

    IF YOU WANT TO UPDATE ANY OF THE PREVIOUS
    DATA, ENTER THE NUMBER(S) PLEASE.
    (PRESS RETURN KEY WHEN FINISHED)

!!! SECTION MENU !!!
~~~~~~~~~~~~~~~~~~~~~

1.   W SHAPES
2.   M SHAPES
3.   S SHAPES
4.   HP SHAPES
ENTER A NUMBER ==>>
? 3

                    !!! DESIGN OF BEAM !!!
                    ~~~~~~~~~~~~~~~~~~~~~~~

DO YOU WANT TO CHOOSE THE NOMINAL DEPTH?
(IF YOU DON'T CHOOSE, THE PROGRAM WILL FIND THE
LIGHTEST SECTION POSSIBLE FOR THE GIVEN SHAPE) (Y/N)? Y

AVAILABLE NOMINAL DEPTHS FOR 'S' SECTIONS ARE :
  3 4 5 6 7 8 10 12 15 18 20 24
NOMINAL DEPTH OF THE BEAM ? 10

                    searching ...

NEXT LARGER SECTION DEPTH IS CHOSEN
NEXT LARGER SECTION DEPTH IS CHOSEN
NEXT LARGER SECTION DEPTH IS CHOSEN
NEXT LARGER SECTION DEPTH IS CHOSEN

SELECTED BEAM :
               !!!!!   S 20 X 86   !!!!!

        A   = 25.300 in.^2   bf =   7.060 in.
        d   = 20.300 in.   tf =   0.920 in.
        tw =   0.660 in.   rT =   1.630 in.
        Ix = 1580.00 in.^4   Iy = 46.80 in.^4
        Zx = 183.00 in.^3   Zy = 23.00 in.^3
        rx =   7.89 in.   ry =   1.36 in.

THE RATIO OF THE MAXIMUM FACTORED FLEXURAL
MOMENT TO THE NOMINAL FLEXURAL STRENGTH IS 0.933
THE SECTION IS A COMPACT SECTION
TOTAL WEIGHT OF THE BEAM : 1.72 Kips.

PRESS ANY KEY TO CONTINUE

            !!!!   REDESIGN MENU   !!!!
            ~~~~~~~~~~~~~~~~~~~~~~~~~~~~

    1.   READ FROM A FILE
    2.   PROBLEM HEADING & DESCRIPTION
    3.   TYPE OF THE STEEL(FY= 36ksi)
    4.   MAJOR/MINOR AXIS BENDING
    5.   LENGTH & LOADING & LATERAL SUPPORT
    6.   LOADING ONLY
    7.   LATERAL SUPPORT ONLY
→→8.   SECTION TYPE(S) & NOMINAL DEPTH(20)
    9.   CHECK/DESIGN & NOMINAL DEPTH(20)
    M.   RETURN TO MAIN MENU

```
      IF YOU WANT TO UPDATE ANY OF THE PREVIOUS
      DATA, ENTER THE NUMBER(S) PLEASE.
      (PRESS RETURN KEY WHEN FINISHED)




!!! SECTION MENU !!!
~~~~~~~~~~~~~~~~~~~~
1.  W SHAPES
2.  M SHAPES
3.  S SHAPES
4.  HP SHAPES
ENTER A NUMBER ==>>
? 1




DO YOU WANT TO CHECK A SECTION OR DESIGN (CHK/DSN) ? DSN


            !!!  DESIGN  OF  BEAM  !!!
            ~~~~~~~~~~~~~~~~~~~~~~~~~

DO YOU WANT TO CHOOSE THE NOMINAL DEPTH?
(IF YOU DON'T CHOOSE, THE PROGRAM WILL FIND THE
LIGHTEST SECTION POSSIBLE FOR THE GIVEN SHAPE) (Y/N)? N


                    searching ...


SELECTED BEAM :
                     !!!!!   W 24 X 68   !!!!!

            A   =  20.100 in.^2        bf  =  8.965 in.
            d   =  23.730 in.          tf  =  0.585 in.
            tw  =   0.415 in.          rT  =  2.260 in.
            Ix  = 1830.00 in.^4        Iy  =   70.40 in.^4
            Zx  =  177.00 in.^3        Zy  =   24.50 in.^3
            rx  =    9.55 in.          ry  =    1.87 in.

THE RATIO OF THE MAXIMUM FACTORED FLEXURAL
MOMENT TO THE NOMINAL FLEXURAL STRENGTH IS  0.965
THE SECTION IS A COMPACT SECTION
TOTAL WEIGHT OF THE BEAM :  1.36 Kips.




PRESS ANY KEY TO CONTINUE

! BEAM DESIGN-LRFD !
!!! MAIN MENU !!!
~~~~~~~~~~~~~~~~~~~
1. NEW PROBLEM
2. SUMMARY OF RESULTS
3. MAIN DISPLAY MENU
4. REDESIGN
5. SAVE DATA ON A DISK
6. QUIT
ENTER A NUMBER ==>>
? 6

            !!!!! D O N E !!!!!
            ~~~~~~~~~~~~~~~~~~~~

Ok
```

5.13 PROBLEMS

5.1 A W21 × 83 made of A36 steel with yield stress of 36 ksi is used as a simply supported beam with a span of 15 ft carrying a uniformly distributed load of intensity 5 Kips/ft including its own weight. The end supports make an inclination of 10 degrees with the horizontal, as shown in Fig. 5.40. Check the adequacy of the beam according to AISCS. Assume that the load acts through the centroid of the cross section. Lateral support is provided at the two ends of the beam only.

10°   **Figure 5.40**

5.2 A structural tube is used as a beam to span a highway bridge of span length 32 ft. The beam is simply supported and subjected to a vertical (dead + live) load of 1 K/ft and a horizontal wind load of 60 psf acting on the face MN of the tube (Fig. 5.41). In addition to supports A and C, the beam has lateral support at the midpoint B. Using A441 steel with yield stress of 50 ksi, select the lightest tube with a depth of 12 in.

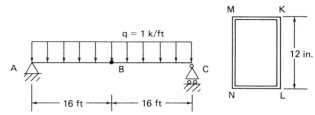

Cross section of the beam   **Figure 5.41**

5.3 A castellated beam has been built out of a W33 × 118 section as shown in Fig. 5.42. If A36 steel is used (F_y = 36 ksi), what uniformly distributed load can this simply supported

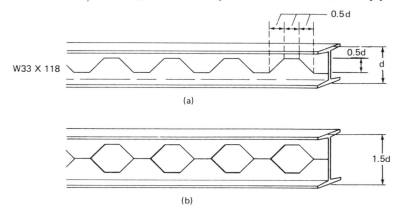

(a)

(b)

Note : d is the total depth

Figure 5.42

castellated beam carry for a span of 20 ft when the compression flange is braced laterally? You may not consider this beam as a compact section. How can we increase the loading capacity of this beam by only minor modifications?

5.4 The floor of an indoor balcony consists of W beams of length L placed at spacing of 10 ft. Each beam is connected to the column at one end through a shear connection and is supported by a hanger at a distance x from the other end (Fig. 5.43). What

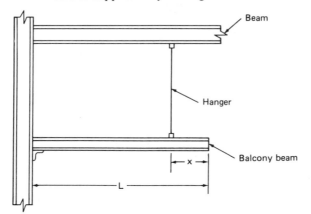

Figure 5.43

should the distance x be in order to obtain the minimum weight beam? Using this distance, find the lightest W shape for the beam. Use A36 ($F_y = 36$ ksi) and assume total lateral support. Dead and live loads of the floor are 60 psf and 100 psf, respectively. $L = 30$ ft.

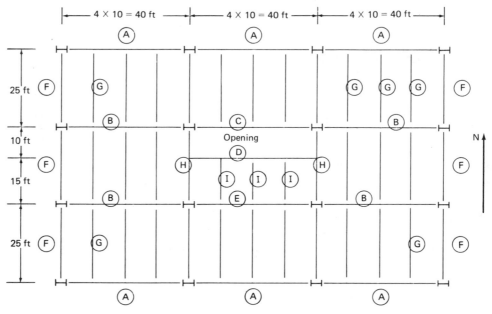

Figure 5.44

5.5 The floor arrangement shown in Fig. 5.44 should be designed for a live load of 100 psf and the dead load of a 6-in. reinforced concrete slab and the beam weights. Design the typical beams in the N–S direction (identified by F, G, H, and I) and the girders in the E–W direction (identified by A, B, C, D, and E), assuming all of them are simply supported. Use 150 pcf for the weight of concrete, $F_y = 36$ ksi, and assume full lateral support.

5.6 Solve Problem 5.4, using A572 steel with yield stress of 60 ksi.

5.7 The three-span continuous beam shown in Fig. 5.45 has a total length of $L = 60$ ft and is subjected to a distributed load of intensity $q = 2$ K/ft. Where should the internal supports be located (that is, find the distance x shown in the figure) in order to obtain the minimum weight beam? Assume total lateral support. Using A36 steel with yield stress of 36 ksi, find the lightest W shape for the beam.

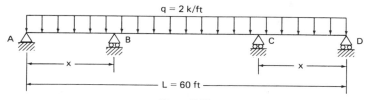

Figure 5.45

5.8 Solve Problem 5.7, using A572 steel with yield stress of 65 ksi.

5.9 Find the lightest W shape for the beam shown in Fig. 5.46, using A36 steel ($F_y = 36$ ksi) and assuming
a. full lateral support
b. lateral supports at points A and B only

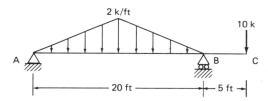

Figure 5.46

5.10 Solve Problem 5.9, using A441 steel ($F_y = 50$ ksi).

5.11 Find the lightest S shape for the beam shown in Fig. 5.46, using A36 steel ($F_y = 36$ ksi) assuming
a. full lateral support
b. lateral supports at points A and B only

5.12 Solve Problem 5.11, using A441 steel ($F_y = 50$ ksi).

5.13 Find the lightest double channels for the beam shown in Fig. 5.46, using A36 steel ($F_y = 36$ ksi) and assuming full lateral support.

5.14 Find the lightest rectangular tube for the beam shown in Fig. 5.46, using A529 steel and assuming
a. full lateral support
b. lateral supports at points A and B only

5.15 A built-up simply supported beam consists of a W24 × 68 and a C15 × 33.9 as shown in Fig. 5.47. The beam has a span of 30 ft and full lateral support. Using A36 steel (F_y = 36 ksi), find the maximum uniformly distributed load this beam can carry.

Figure 5.47

5.16 The cross section of the cantilever beam of Fig. 5.48 is shown in Fig. 5.49. Find the maximum load-carrying capacity of this beam (q_{max}), using A36 steel (F_y = 36 ksi) and assuming
a. full lateral support
b. lateral supports at points A and B only

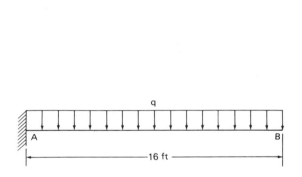

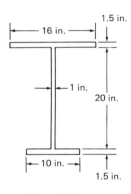

Figure 5.48   **Figure 5.49**

5.17 A simple beam with a span of 15 ft is subjected to two concentrated vertical loads, as shown in Fig. 5.50, and a uniformly distributed horizontal load of 0.5 K/ft. Find the lightest W12 section for the beam, using A36 steel (F_y = 36 ksi) and assuming
a. full lateral support
b. lateral supports at A, B, C, and D only

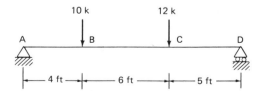

Figure 5.50

5.18 Solve Problem 5.17, substituting the uniformly distributed horizontal load by a concentrated horizontal load of magnitude 7.5 K applied at the middle point of the span. Assume that all the loads pass through the centroid of the cross section and thus produce no torsional effect.

5.19 In Example 2 of this chapter, suppose a cover-plated W14 × 30 is used for the beam. What should be the size of the cover plates in order that the beam carry the specified load? Assume that the cover plates are welded properly to the beam flanges.

5.20 Find the lightest W18 for Example 3 of this chapter.

Design
of Compression Members

6.1 INTRODUCTION

When a member is subjected to compressive axial forces, it is referred to as a *compression member*. Compression members are found as:

1. Columns in buildings
2. Piers in bridges
3. Top chords of trusses
4. Bracing members

Compression and tension members differ in the following ways:

1. Slender compression members can buckle.
2. In tension members, bolt holes reduce the effective cross-sectional area for carrying the loads. In compression members, however, the bolts tend to fill the holes and the entire gross area of the cross section is normally assumed to resist the loads.

In this chapter we cover the design of compression members subjected to concentric loads. These members are also referred to as *columns*. In practice, the dead or live load is usually not centered over an exterior building column and the line of action of the resultant gravity load usually is within the inner side of the column. Even in the case of interior building columns, the dead and live loads may not be centered. In such cases, the member is subjected to both axial compressive forces and bending moments due to the eccentric loading. Design of these members, referred to as *beam-columns*, is covered in the following chapter.

6.2 THE EULER FORMULA FOR BUCKLING LOAD OF COLUMNS

The Euler formula is derived for an ideal or perfect column. That is, it is assumed that the column is long, slender, straight, homogeneous, and elastic and is subjected to concentric axial compressive loads. It is also assumed that the two ends of the columns are hinged. Suppose that this ideal column in equilibrium is laterally

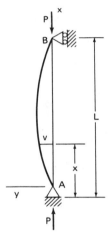

Figure 6.1 Column with hinged ends.

disturbed as shown in Fig. 6.1. We can write the following differential equation for the lateral displacement v:

$$EI\frac{d^2v}{dx^2} = M = -Pv \tag{6.1}$$

where E is the modulus of elasticity, I is the moment of inertia about the axis of bending in the cross section, P is the axial compressive force, and M is the bending moment at a distance x from support A. Denoting

$$\frac{P}{EI} = k^2 \tag{6.2}$$

Equation (6.1) can be written as

$$\frac{d^2v}{dx^2} + k^2v = 0 \tag{6.3}$$

This is a second-order homogeneous linear differential equation with constant coefficients. The solution of this differential equation is

$$v = C_1 \cos kx + C_2 \sin kx \tag{6.4}$$

The integration constants C_1 and C_2 may be found by applying the following geometric boundary conditions:

$$\text{At } x = 0: \quad v = 0 \rightarrow C_1 = 0; \quad v = C_2 \sin kx \tag{6.5}$$

$$\text{At } x = L: \quad v = 0 \rightarrow C_2 \sin kL = 0 \tag{6.6}$$

Equation (6.6) indicates that either $C_2 = 0$, which means no lateral displacement and instability problem, or $\sin kL = 0$ with solution

$$kL = n\pi \qquad n = 1, 2, \ldots \tag{6.7}$$

Substituting for k from Eq. (6.7) into Eq. (6.2) and solving for P, we obtain

$$P = \frac{n^2 \pi^2 EI}{L^2} \qquad n = 1, 2, \ldots \tag{6.8}$$

The smallest value obtained from Eq. (6.8) is known as critical load, buckling load, or Euler formula:

$$P_{cr} = \frac{\pi^2 EI}{L^2} \tag{6.9}$$

This equation can be interpreted as follows. If we increase the magnitude of the load P from zero, for small values of P, there is no lateral displacement and $v = 0$. But as P approaches the critical value P_{cr}, the lateral displacement becomes

$$v = C_2 \sin \frac{\pi x}{L} \tag{6.10}$$

which is indefinite. In other words, the column shows unstable behavior.

It should be noted that the critical buckling load given by Eq. (6.9) is independent of the strength of the material (say, F_y, the yield stress). The results obtained from Eq. (6.9) compare very well with experimental results for long slender columns.

Equation (6.9) was found for a column with hinged ends. This equation can be used for columns with other end conditions, provided that it is modified as follows:

$$P_{cr} = \frac{\pi^2 EI}{(KL)^2} \tag{6.11}$$

where KL is the distance between the points of zero moment, or inflection points. The length KL is known as the effective length of the column. The dimensionless coefficient K is called the *effective length factor*. The values of this factor for different types of columns are given in Fig. 6.2. These values are ideal values obtained from theoretical analyses similar to the one presented in this section. In practice, these values are modified somewhat. The modified values as recommended by AISCS (commentary Sec. 1.8) are given inside parentheses in Fig. 6.2.

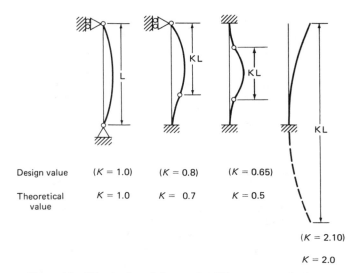

Figure 6.2   Effective length factors for different types of columns.

Dividing the critical load P_{cr} by the cross-sectional area of the column A we can find the critical stress f_{cr}.

$$f_{cr} = \frac{P_{cr}}{A} = \frac{\pi^2 EI}{(KL)^2 A} = \frac{\pi^2 E}{(KL/r)^2} \qquad (6.12)$$

where r is the radius of gyration of the cross section about the axis of bending and KL/r is called the slenderness ratio of the column. Note that if the column is not restricted to bend in a particular plane, it will buckle in a plane perpendicular to the minor axis of the cross section. Hence, the moment of inertia in Eq. (6.11) and the radius of gyration in Eq. (6.12) are with respect to the minor axis of the cross section and therefore the minimum values for the cross section. The corresponding critical load and stress are the minimum values causing buckling of the column.

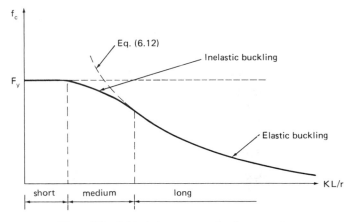

Figure 6.3   Critical stress versus slenderness ratio.

The variation of the critical stress f_c with the slenderness ratio KL/r is shown qualitatively in Fig. 6.3. The range of slenderness ratios is divided into three regions, broadly called *short*, *medium*, and *long* columns. By short columns, it is simply meant columns with small *slenderness ratios*, and so on. Short columns do not buckle and simply fail by excessive yielding. Long columns usually fail by elastic buckling, for which Euler buckling stress, Eq. (6.12), is valid. Between short and long regions, there is an intermediate region in which the failure of the column occurs through inelastic buckling.

6.3 RESIDUAL STRESSES

Residual stresses due to uneven cooling of standard sections after hot-rolling, and also welding, can adversely affect the resistance of columns against buckling. As an example, in an I-section, the outer tips of the flanges and the middle portion of the web cool more quickly than the relatively thick portions at the intersection of the flanges and the web. The result of this uneven cooling is that the areas cooled more quickly develop residual compressive stresses, while the areas cooled more slowly develop residual tensile stresses. The magnitude of the residual stresses can be as large as 10–15 ksi.

6.4 ALLOWABLE COMPRESSIVE STRESSES

In AISCS equations for allowable compressive stresses, various imperfections such as the effect of residual stresses, the actual end restraint conditions, crookedness, and small unavoidable eccentricities are empirically taken into account. The AISCS assumes that the elastic buckling holds valid when the stress in the column is not greater than one-half of the yield stress ($F_y/2$) (Fig. 6.4). Defining

C_c = the slenderness ratio dividing elastic from inelastic buckling

we can write

$$\frac{1}{2} F_y = \frac{\pi^2 E}{C_c^2} \rightarrow C_c = \sqrt{\frac{2\pi^2 E}{F_y}} \tag{6.13}$$

For columns with slenderness ratio less than C_c, AISCS assumes a parabolic variation and the allowable axial compressive stress is given by (Fig. 6.4).

$$F_a = \frac{1 - \dfrac{(KL/r)^2}{2C_c^2}}{\text{F.S.}} F_y \qquad \frac{KL}{r} \leq C_c \tag{6.14}$$

$$\text{F.S.} = \text{Factor of Safety} = \frac{5}{3} + \frac{3(KL/r)}{8C_c} - \frac{(KL/r)^3}{8C_c^3} \tag{6.15}$$

Note that at very small slenderness ratios, $KL/r \approx 0$ and the F.S. in compression is 1.67, which is the same as in tension ($F_t = 0.60F_y$). As KL/r increases, the F.S.

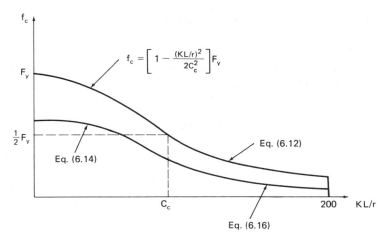

Figure 6.4   Variation of critical stress and allowable stress as specified by AISCS.

also increases until at $KL/r = C_c$ it becomes $\frac{23}{12} = 1.92$. This variation of F.S. is rational because the longer the column, the more sensitive it becomes to the imperfections in the evaluation of the effective length and the critical stress.

For columns with slenderness ratio greater than C_c, the Euler formula [Eq. (6.12)] is employed with a F.S. of $\frac{23}{12} = 1.92$ (Fig. 6.4).

$$F_a = \frac{12\pi^2 E}{23(KL/r)^2} \qquad C_c < KL/r \leq 200 \qquad (6.16)$$

According to AISCS Sec. 1.8.4, the slenderness ratio of compression members shall not be greater than 200.

For axially loaded bracing and secondary members, when L/r exceeds 120, another, less conservative equation is given by AISCS because of the relative unimportance of these members.

$$F_{as} = \frac{F_a}{1.6 - \dfrac{L}{200r}} \qquad L/r > 120 \qquad (6.17)$$

where F_a is obtained from Eq. (6.14) or Eq. (6.16) as applicable. Note that for these members K is taken as unity. Also, note that $F_{as} = F_a$ for $L/r < 120$.

Note that in the design of compression members, local buckling of plate elements should be prevented. In other words, the allowable stress values given in this section may not be used unless the width-thickness ratios of the elements are limited to the values given in the fourth column of Table 5.1. When the width-thickness ratios of the plate elements of the section exceed these limiting values, the allowable axial compressive stress should be modified according to Appendix C of the AISCS. Since these cases are not encountered frequently in practice, they are not covered in this book.

6.5 EXAMPLES OF DESIGN OF COLUMNS

Example 1

The diagonally braced frame shown in Fig. 6.5 should be designed for a lateral earthquake load of 36 Kips. Three different openings are required as shown in the figure. The diagonal must fit within the thickness of a 4-in.-thick wall. Also, the width of the diagonal in the plane of the frame must not exceed 4 in. Design a rolled section for the diagonal from the AISCM, using A36 steel.

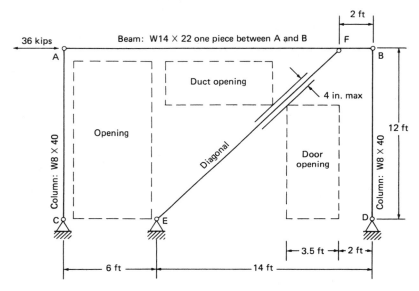

Figure 6.5

Solution   The structure is statically determinate. The diagonal must be designed as a compression member.

$$P = \text{axial force in the diagonal} = \frac{36}{\cos 45°} = 50.91 \text{ K}$$

The requirements of the problem will be satisfied best by selecting a tube. Try TUBE 4×4 with wall thickness $t = 0.5$ in., $A = 6.36$ in.^{2}, and $r_{\min} = 1.39$ in.

$$\frac{KL}{r} = \frac{(1.)(12)(\sqrt{2})(12)}{1.39} = 146.5$$

$$C_c = \sqrt{\frac{2\pi^2 E}{F_y}} = \sqrt{\frac{2\pi^2(29,000)}{36}} = 126.1 < \frac{KL}{r}$$

$$F_a = \frac{12\pi^2 E}{23(KL/r)^2} = 6.96 \text{ ksi}$$

$$F_{as} = \frac{F_a}{1.6 - L/(200r)} = 8.02 \text{ ksi}$$

Compression capacity of the section =

$$AF_{as} = (6.36)(8.02) = 51.0 > P = 50.91 \text{ K} \qquad \underline{\underline{\text{O.K.}}}$$

Check local flange buckling (column four of Table 5.1):

$$\frac{b}{t} = \frac{4}{0.5} = 8 < \frac{238}{\sqrt{F_y}} = 39.6 \qquad \underline{\underline{\text{O.K.}}}$$

$$\boxed{\text{USE TUBE } 4 \times 4 \text{ with } t = 0.5 \text{ in.}} \qquad \text{weight} = 21.63 \text{ lb/ft}$$

Note: In this example, bracing is considered a secondary member as recommended by the AISCS Sec. 1.5.1.3.3. However, some earthquake engineers consider the bracing a primary member for resisting the earthquake forces.

Example 2

The cross section of a built-up column is shown in Fig. 6.6. It consists of two L8 × 8 × 1 angles and is made of A36 steel. Using AISCS, find the maximum allowable axial load for this column for a length of 15 ft. Assume that there exist ideal hinged conditions at the column ends and that the angles are properly connected so that they may not buckle individually.

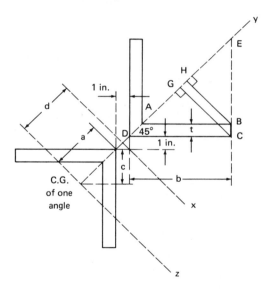

Figure 6.6

Solution   Properties of L8 × 8 × 1: A_1 = area of cross section = 15 in.^{2},

$$r_z = 1.56 \text{ in.}, \qquad c = 2.37 \text{ in.}, \qquad w = 51 \text{ lb/ft}$$

We must first calculate the moments of inertia of the built-up section with respect to x and y (axes of symmetry of the built-up section). From the geometry we can write (Fig. 6.6)

$$a = \sqrt{2}\, c = 3.35 \text{ in.}$$

$$d = a + \frac{\sqrt{2}}{2} = 4.06 \text{ in.}$$

The moment of inertia of one angle about the z-axis shown in Fig. 6.6 is

$$I_{z1} = A_1 r_z^2 = (15)(1.56)^2 = 36.50 \text{ in.}^4$$

Thus,

$$I_x = 2[I_{z1} + A_1 d^2] = 2[36.50 + 15(4.06)^2] = 567.51 \text{ in.}^4$$

$$I_y = 4(I_{ABCD})_y = 4[(I_{CDE})_y - (I_{ABE})_y]$$

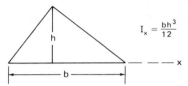

$$I_x = \frac{bh^3}{12}$$

Figure 6.7

Using the relation for moment of inertia of a triangle about its base (Fig. 6.7), we have

$$I_y = 4[\tfrac{1}{12}\overline{DE}(\overline{CG})^3 - \tfrac{1}{12}\overline{AE}(\overline{BH})^3]$$

$$\overline{DE} = \sqrt{2}b \qquad \overline{CG} = \frac{\sqrt{2}}{2}b \qquad \overline{AE} = \sqrt{2}(b-t) \qquad \overline{BH} = \frac{\sqrt{2}}{2}(b-t)$$

$$I_y = \tfrac{1}{6}[b^4 - (b-t)^4] = \tfrac{1}{6}[8^4 - (8-1)^4] = 282.5 \text{ in.}^4$$

$$I_{\min} = I_y = 282.5 \text{ in.}^4$$

$$r_{\min} = \sqrt{I_y/A} = \sqrt{(282.5)/(2 \times 15)} = 3.07 \text{ in.}$$

$$\frac{KL}{r} = \frac{(1.)(15)(12)}{3.07} = 58.7 < C_c = 126.1$$

From Eq. (6.14): $F_a = 17.55 \text{ ksi}$

$$P = AF_a = (30)(17.55) = 526.5 \text{ K}$$

$$P_{\text{all}} = P - \text{weight} = 526.5 - \frac{2(15)(51)}{1000} = 525 \text{ K}$$

Note: Moments of inertia of the built-up section could also be computed by using rotation of axes and the parallel-axis theorem.

Example 3

A column is built of two tubes 16×8 and one angle L8 $\times$ 8 $\times$ 1 available in a shop. In order to utilize the material efficiently a designer has proposed the arrangement shown in Fig. 6.8. The sections are made of A36 steel. Using AISCS, determine the maximum allowable axial load for this column for a length of 14 ft. Assume ideal hinged conditions at the column ends. The sections are properly connected so that they may not buckle individually. Thickness of the tubes is $t = 0.5$ in.

Solution   For L8 $\times$ 8 $\times$ 1: $I_{x1} = I_{y1} = 89 \text{ in.}^4$,

$$A = 15.0 \text{ in.}^2, \qquad r_z = 1.56 \text{ in.}, \qquad I_z = 15(1.56)^2 = 36.50 \text{ in.}^4$$

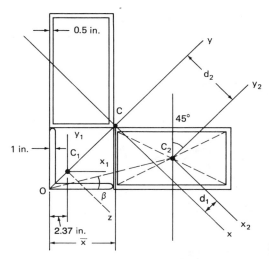

Figure 6.8

For a tube 16×8 ($t = 0.5$ in.): $A = 22.4$ in.^{2}

C_1: centroid of the L8 $\times$ 8 $\times$ 1

C: centroid of the built-up section

$$\bar{x} = \frac{(15)(2.37) + (22.4)(4 + 16)}{15 + 2(22.4)} = 8.086 \text{ in.}$$

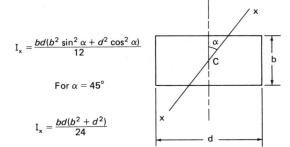

$$I_x = \frac{bd(b^2 \sin^2 \alpha + d^2 \cos^2 \alpha)}{12}$$

For $\alpha = 45°$

$$I_x = \frac{bd(b^2 + d^2)}{24}$$

Figure 6.9

For one tube (Fig. 6.9):

$$I_{x2} = I_{y2} = \frac{8(16)(8^2 + 16^2)}{24} - \frac{7(15)(7^2 + 15^2)}{24} = 507.92 \text{ in.}^4$$

$$\overline{OC} = \bar{x}\sqrt{2} = 11.44 \text{ in.}$$

$$\overline{OC_2} = \sqrt{16^2 + 4^2} = 16.49 \text{ in.}$$

$$\beta = \arctan \tfrac{4}{16} = 14.04 \text{ deg}$$

$$d_1 = \overline{OC_2} \cos (45° - \beta) - \overline{OC} = 2.70 \text{ in.}$$

$$d_2 = \overline{OC_2} \sin (45° - \beta) = 8.48 \text{ in.}$$

Moment of inertia of the built-up section about the weak axis (x-axis):

$$I_{min} = I_x = 36.50 + 15(11.44 - 2.37\sqrt{2})^2 + 2[507.92 + 22.4(2.70)^2]$$

$$= 2360.24 \text{ in.}^4$$

Cross-sectional area of the built-up section:

$$A = 15 + (2)(22.4) = 59.8 \text{ in.}^2$$

$$r_x = \sqrt{I_x/A} = \sqrt{2360.24/59.8} = 6.28 \text{ in.}$$

$$\frac{KL}{r} = \frac{(1.0)(14 \times 12)}{6.28} = 26.75 < C_c = 126.1$$

From Eq. (6.14): $F_a = 20.17$ ksi

$$P = AF_a = (59.8)(20.17) = 1206.2 \text{ K}$$

6.6 INTERACTIVE MICROCOMPUTER-AIDED DESIGN OF COLUMNS ACCORDING TO AISCS

6.6.1 Program Description

The program presented in this section is an interactive BASIC program for design of axially loaded columns with various end conditions according to the AISCS. The various column types considered in this program are shown in the column-type menu of Fig. 6.11. The input data may be read from an existing file or may be inputted by the user interactively. For a given axial compressive force, the program can either check the adequacy of a given section selected by the user or design a section selected from the section menu of Fig. 6.14. This menu includes W shapes, HP shapes, double C sections, double W sections, and double HP sections.

For design, the user may select the nominal depth of the cross section. For example, if the user selects W12 series among the W sections, the program will try to come up with the lightest W12 section. If no W12 section available in the AISC sections database satisfies the design requirements, the program will automatically

```
*COLUMN DESIGN-AISC*
*** MAIN MENU ***
~~~~~~~~~~~~~~~~~~~~~~~

1. NEW PROBLEM

2. SUMMARY OF RESULTS

3. DISPLAY CROSS-SECTION

4. REDESIGN

5. SAVE DATA ON A DISK

6. QUIT


ENTER A NUMBER ==>>? ■
```

Figure 6.10   Main menu.

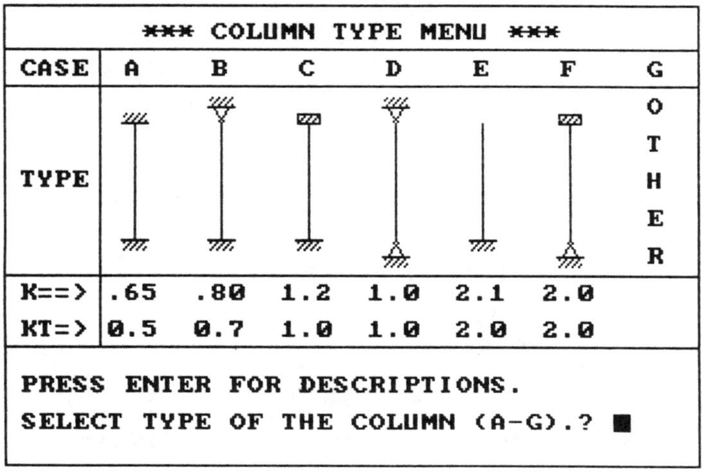

Figure 6.11

move to W14 series and will try to select the lightest W14 section. This process is
continued until the program finds a W section satisfying all the design requirements.
A message will be displayed when no W section is adequate for the given loading
condition. When the user does not select any nominal depth, the program will come
up with the lightest section within the type of the cross section selected by the user.

There are five menus in the program. They are the main menu (Fig. 6.10),
column type menu (Fig. 6.11), section menu (Fig. 6.14), steel type menu (Fig. 5.18),
and the redesign menu (Fig. 6.15). From the column type menu, the user can see
the K-value description (Fig. 6.12) and the end condition code (Fig. 6.13). Note

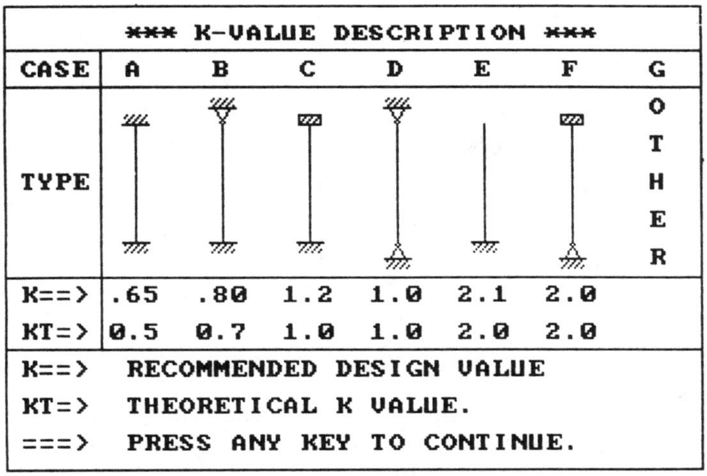

Figure 6.12

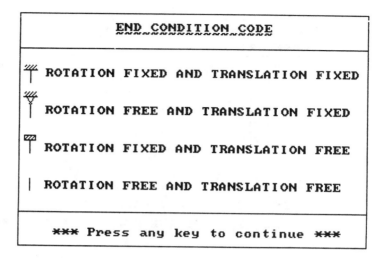

Figure 6.13

that in the column type menu, the user can also choose any *K*-value other than those specified in the menu. The menus make the program highly interactive. Using the redesign menu (Fig. 6.15), the user can change any portion of the input data without having to reenter all the input data from scratch.

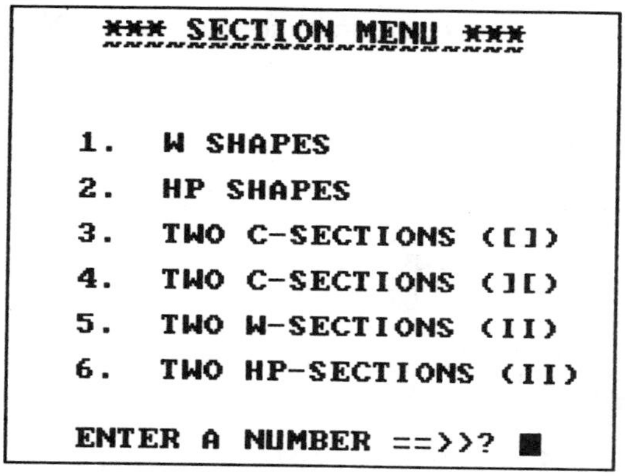

Figure 6.14

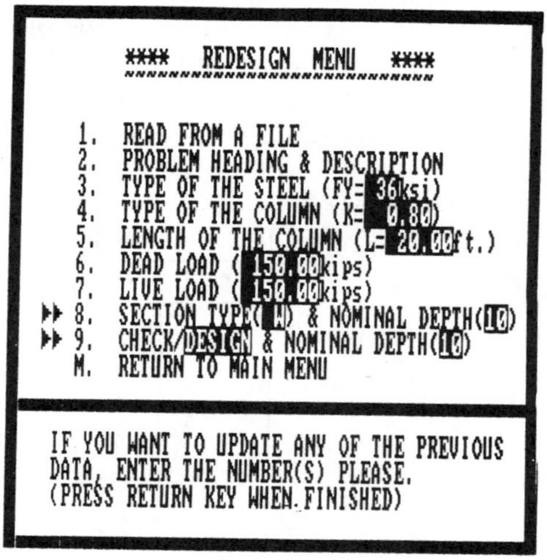

Figure 6.15

The interactive program can display any one of the cross sections shown on the section menu (Fig. 6.14). The cross-sectional display includes the necessary dimensions. The program provides error messages for unrealistic input given by the user.

The program and the AISC sections database are saved on one floppy disk. The name of the program primary file is AISCCOL. In order to obtain hard copies of the graphic output and have access to the AISC sections database, the following commands should be used in the system mode before loading the program:

>GRAPHICS
>BASICA/S:186

6.6.2 Sample Example and Interactive Session

6.6.2.1 Sample example.   As an example, an axially compressed column has been solved by the interactive microcomputer program. The support conditions are fixed-hinged. The column length is 20 ft. It is subjected to an axial compressive force of 300 K. The column is used as a main member. The yield stress of steel is 36 ksi. First the program was asked to design a W8 section. But no W8 section would carry the specified load. Thus, the program selected the lightest W10 shape, that is, W10 × 68 shown in Fig. 6.16.

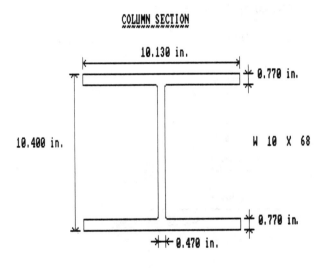

Figure 6.16

Next, using the redesign menu of Fig. 6.15 the program was asked to check the adequacy of an HP12 × 74, shown in Fig. 6.17. Then the program was asked to design a double channel ([ ]) with a width of 10 in. The answer is 2 C10 × 30, shown in Fig. 6.18. Subsequently, the program was asked to design a double channel (][) with a width of 10 in. The answer is 2 C15 × 33.9, which is shown in Fig. 6.19.

Using the redesign menu once more, this time the program was asked to find the lightest W shape. The answer is W12 × 65, which is shown in Fig. 6.20. Finally, the program was asked to find the lightest double W section with a width of 20 in. The answer is 2 W16 × 26, which is shown in Fig. 6.21.

COLUMN SECTION

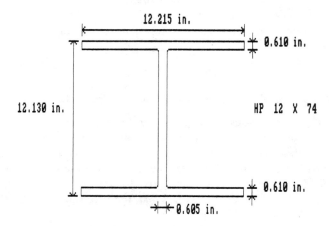

PRESS ANY KEY TO CONTINUE

Figure 6.17

COLUMN SECTION

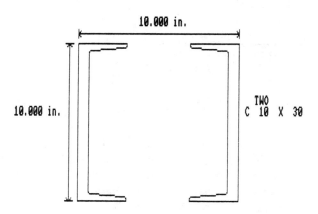

PRESS ANY KEY TO CONTINUE

Figure 6.18

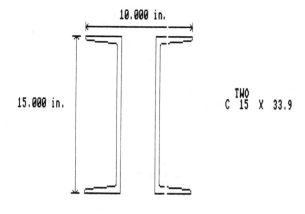

Figure 6.19

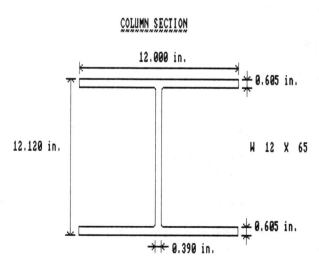

Figure 6.20

COLUMN SECTION

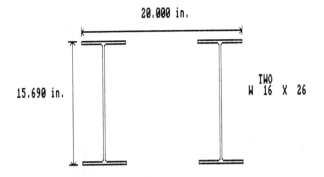

PRESS ANY KEY TO CONTINUE

Figure 6.21

6.6.2.2 Interactive session.

```
RUN
*COLUMN DESIGN-AISC*
*** MAIN MENU ***
^^^^^^^^^^^^^^^^^^^^
1. NEW PROBLEM
2. SUMMARY OF RESULTS
3. DISPLAY CROSS-SECTION
4. REDESIGN
5. SAVE DATA ON A DISK
6. QUIT
ENTER A NUMBER ==>>
? 1


        ***** DESIGN OF AXIALLY COMPRESSED COLUMNS *****
        ^^^^^^^^^^^^^^^^^^^^^^^^^^^^^^^^^^^^^^^^^^^^^^^^^^

            ((( Based on AISC Specification 1980 )))
            ^^^^^^^^^^^^^^^^^^^^^^^^^^^^^^^^^^^^^^



            *** BASIC INPUT DATA ***
            ^^^^^^^^^^^^^^^^^^^^^^^^

DO YOU DESIRE TO USE THE DATA FROM A DISK (Y/N) ? n

PROBLEM HEADING ? sample example

PROBLEM DESCRIPTION ?

UNBRACED LENGTH OF THE COLUMN (ft.) ? 20

IS THE SECTION USED AS A SECONDARY MEMEBER (Y/N) ? n
```

```
*** STEEL TYPE MENU ***
````````````````````````````
 1. A36 (Fy=32 ksi)   11. A529 (Fy=42 ksi)
 2. A36 (Fy=36 ksi)   12. A242 (Fy=42 ksi)
 3. A441 (Fy=40 ksi)   13. A242 (Fy=46 ksi)
 4. A441 (Fy=42 ksi)   14. A242 (Fy=50 ksi)
 5. A441 (Fy=46 ksi)   15. A588 (Fy=42 ksi)
 6. A441 (Fy=50 ksi)   16. A588 (Fy=46 ksi)
 7. A572 (Fy=42 ksi)   17. A588 (Fy=50 ksi)
 8. A572 (Fy=50 ksi)   18. A514 (Fy=90 ksi)
 9. A572 (Fy=60 ksi)   19. A514 (Fy=100 ksi)
10. A572 (Fy=65 ksi)   20. OTHERS
ENTER THE NUMBER PLEASE ==>>
? 2
```

```
*** COLUMN TYPE MENU ***
CASE  A    B    C    D    E    F    G
TYPEOTHER
K==) .65  .80  1.2  1.0  2.1  2.0
KT=) 0.5  0.7  1.0  1.0  2.0  2.0
PRESS ENTER FOR DESCRIPTIONS.
SELECT TYPE OF THE COLUMN (A-G).? b
```

```
                 ***  LOADS  ***
                 ```````````````````
```

```
AXIALLY COMPRESSED DEAD LOAD (Ksi) ? 150
```

```
AXIALLY COMPRESSED LIVE LOAD (Ksi) ? 150
```

```
DO YOU DESIRE TO SAVE THE DATA ON A DISK (Y/N) ? n
```

```
*** SECTION MENU ***
```````````````````````
1. W SHAPES
2. HP SHAPES
3. TWO C-SECTIONS ([])
4. TWO C-SECTIONS (][)
5. TWO W-SECTIONS (II)
6. TWO HP-SECTIONS (II)
ENTER A NUMBER ==>>
? 1
```

```
DO YOU WANT TO CHECK A SECTION OR DESIGN (CHK/DSN) ? dsn
```

```
             ***  DESIGN  OF  COLUMN  ***
             ``````````````````````````````
```

```
DO YOU WANT TO CHOOSE THE NOMINAL DEPTH?
(IF YOU DON'T CHOOSE, THE PROGRAM WILL FIND THE
LIGHTEST SECTION POSSIBLE FOR THE GIVEN SHAPE) (Y/N)? y
```

```
AVAILABLE NOMINAL DEPTHS FOR 'W' SECTIONS ARE :
 4  5  6  8  10  12  14  16  18  21  24  27  30  33  36
NOMINAL DEPTH OF THE SECTION ? 8
```

```
                  searching ...
```

```
NEXT LARGER SECTION DEPTH IS CHOSEN
```

```
SELECTED SECTION :
                   *****    W 10 X 68    *****

            A   =  20.000 in.^2       bf  = 10.130 in.
            d   =  10.400 in.         tf  =  0.770 in.
            tw  =   0.470 in.         rT  =  2.790 in.
            Ix  = 394.00 in.^4        Iy  = 134.00 in.^4
            Sx  =  75.70 in.^3        Sy  = 26.40 in.^3
            rx  =   4.44 in.          ry  =  2.59 in.
```

```
THE MAXIMUM RATIO OF THE ACTUAL STRESS TO THE ALLOWABLE STRESS IS  0.938
THE SECTION IS COMPACT
TOTAL WEIGHT OF THE COLUMN :  1.36 Kips.
PRESS ANY KEY TO CONTINUE


*COLUMN DESIGN-AISC*
*** MAIN MENU ***
```````````````````````
1. NEW PROBLEM
2. SUMMARY OF RESULTS
3. DISPLAY CROSS-SECTION
4. REDESIGN
5. SAVE DATA ON A DISK
6. QUIT
ENTER A NUMBER ==>>
? 4

           ****   REDESIGN MENU   ****
           ``````````````````````````````````
       1.  READ FROM A FILE
       2.  PROBLEM HEADING & DESCRIPTION
       3.  TYPE OF THE STEEL (FY= 36ksi)
       4.  TYPE OF THE COLUMN (K= 0.80)
       5.  LENGTH OF THE COLUMN (L= 20.00ft.)
       6.  DEAD LOAD ( 150.00kips)
       7.  LIVE LOAD ( 150.00kips)
  -->>8.  SECTION TYPE( W) & NOMINAL DEPTH(10)
  -->>9.  CHECK/DESIGN & NOMINAL DEPTH(10)
       M.  RETURN TO MAIN MENU


    IF YOU WANT TO UPDATE ANY OF THE PREVIOUS
    DATA, ENTER THE NUMBER(S) PLEASE.
    (PRESS RETURN KEY WHEN FINISHED)




*** SECTION MENU ***
````````````````````````
1. W SHAPES
2. HP SHAPES
3. TWO C-SECTIONS ([])
4. TWO C-SECTIONS (][)
5. TWO W-SECTIONS (II)
6. TWO HP-SECTIONS (II)
ENTER A NUMBER ==>>
? 2


DO YOU WANT TO CHECK A SECTION OR DESIGN (CHK/DSN) ? chk


SECTION DESIGNATION : DEPTH, WEIGHT
(e.g. W,36,300)        HP,? 12,74


                       searching ...


GIVEN SECTION :
                    *****   HP 12 X 74   *****

              A   =  21.800 in.^2        bf  =  12.215 in.
              d   =  12.130 in.          tf  =   0.610 in.
              tw  =   0.605 in.          rT  =   2.790 in.
              Ix  = 569.00 in.^4         Iy  =  186.00 in.^4
              Sx  =  93.80 in.^3         Sy  =   30.40 in.^3
              rx  =   5.11 in.           ry  =    2.92 in.
```
```

```
THE MAXIMUM RATIO OF THE ACTUAL STRESS TO THE ALLOWABLE STRESS IS  0.816
THE SECTION IS COMPACT
SECTION IS SAFE
TOTAL WEIGHT OF THE COLUMN :  1.48 Kips.
PRESS ANY KEY TO CONTINUE

            ####   REDESIGN MENU   ####
            ~~~~~~~~~~~~~~~~~~~~~~~~~~~~~

      1.  READ FROM A FILE
      2.  PROBLEM HEADING & DESCRIPTION
      3.  TYPE OF THE STEEL (FY= 36ksi)
      4.  TYPE OF THE COLUMN (K=  0.80)
      5.  LENGTH OF THE COLUMN (L= 20.00ft.)
      6.  DEAD LOAD ( 150.00kips)
      7.  LIVE LOAD ( 150.00kips)
   ->>8.  SECTION TYPE(HP) & NOMINAL DEPTH(12)
   ->>9.  CHECK/DESIGN & NOMINAL DEPTH(12)
      M.  RETURN TO MAIN MENU


      IF YOU WANT TO UPDATE ANY OF THE PREVIOUS
      DATA, ENTER THE NUMBER(S) PLEASE.
      (PRESS RETURN KEY WHEN FINISHED)




### SECTION MENU ###
~~~~~~~~~~~~~~~~~~~~
1.  W SHAPES
2.  HP SHAPES
3.  TWO C-SECTIONS ([])
4.  TWO C-SECTIONS (][)
5.  TWO W-SECTIONS (II)
6.  TWO HP-SECTIONS (II)
ENTER A NUMBER ==>>
? 3
                              <-->
ENTER THEN SECTION WIDTH (in.) ([  ]) ==> 10


DO YOU WANT TO CHECK A SECTION OR DESIGN (CHK/DSN) ? dsn



            ### DESIGN  OF COLUMN ###
            ~~~~~~~~~~~~~~~~~~~~~~~~~~

DO YOU WANT TO CHOOSE THE NOMINAL DEPTH?
(IF YOU DON'T CHOOSE, THE PROGRAM WILL FIND THE
LIGHTEST SECTION POSSIBLE FOR THE GIVEN SHAPE) (Y/N)? y

AVAILABLE NOMINAL DEPTHS FOR 'C' SECTIONS ARE :
 3  4  5  6  7  8  9  10  12  15
NOMINAL DEPTH OF THE SECTION ? 10


                    searching ...


SELECTED SECTION :
                #####   C 10 X 30   #####

            A  =   8.820 in.^2        bf  =   3.033 in.
            d  =  10.000 in.          tf  =   0.436 in.
            tw =   0.673 in.          rT  =   0.000 in.
            Ix =  103.00 in.^4        Iy  =   3.94 in.^4
            Sx =   20.70 in.^3        Sy  =   1.65 in.^3
            rx =   3.42 in.           ry  =   0.67 in.

THE MAXIMUM RATIO OF THE ACTUAL STRESS TO THE ALLOWABLE STRESS IS  0.956
THE SECTION IS COMPACT
TOTAL WEIGHT OF THE COLUMN :  1.20 Kips.
PRESS ANY KEY TO CONTINUE
```

```
####   REDESIGN MENU   ####
^^^^^^^^^^^^^^^^^^^^^^^^^^^^^

  1.  READ FROM A FILE
  2.  PROBLEM HEADING & DESCRIPTION
  3.  TYPE OF THE STEEL (FY= 36ksi)
  4.  TYPE OF THE COLUMN (K=  0.80)
  5.  LENGTH OF THE COLUMN (L= 20.00ft.)
  6.  DEAD LOAD ( 150.00kips)
  7.  LIVE LOAD ( 150.00kips)
→→8.  SECTION TYPE( C) & NOMINAL DEPTH(10)
  9.  CHECK/DESIGN & NOMINAL DEPTH(10)
  M.  RETURN TO MAIN MENU


    IF YOU WANT TO UPDATE ANY OF THE PREVIOUS
    DATA, ENTER THE NUMBER(S) PLEASE.
    (PRESS RETURN KEY WHEN FINISHED)
```

```
### SECTION MENU ###
^^^^^^^^^^^^^^^^^^^^^^^
1.  W SHAPES
2.  HP SHAPES
3.  TWO C-SECTIONS ([])
4.  TWO C-SECTIONS (][)
5.  TWO W-SECTIONS (II)
6.  TWO HP-SECTIONS (II)
ENTER A NUMBER ==>>
? 4
                         <-->
ENTER THEN SECTION WIDTH (in.) (] [) ==> 10
```

```
        ### DESIGN OF COLUMN ###
        ^^^^^^^^^^^^^^^^^^^^^^^^

DO YOU WANT TO CHOOSE THE NOMINAL DEPTH?
(IF YOU DON'T CHOOSE, THE PROGRAM WILL FIND THE
LIGHTEST SECTION POSSIBLE FOR THE GIVEN SHAPE) (Y/N)? y

AVAILABLE NOMINAL DEPTHS FOR 'C' SECTIONS ARE :
  3  4  5  6  7  8  9  10  12  15
NOMINAL DEPTH OF THE SECTION ? 10



                searching ...



NEXT LARGER SECTION DEPTH IS CHOSEN
NEXT LARGER SECTION DEPTH IS CHOSEN

SELECTED SECTION :
                  *****    C 15 X 33.9    *****

            A   =   9.960 in.^2       bf  =  3.400 in.
            d   =  15.000 in.         tf  =  0.650 in.
            tw  =   0.400 in.         rT  =  0.000 in.
            Ix  = 315.00 in.^4        Iy  =  8.13 in.^4
            Sx  =  42.00 in.^3        Sy  =  3.11 in.^3
            rx  =   5.62 in.          ry  =  0.90 in.

THE MAXIMUM RATIO OF THE ACTUAL STRESS TO THE ALLOWABLE STRESS IS  0.949
THE SECTION IS COMPACT
TOTAL WEIGHT OF THE COLUMN :  1.36 Kips.
PRESS ANY KEY TO CONTINUE
```

```
****   REDESIGN MENU   ****
``````````````````````````````````

1. READ FROM A FILE
2. PROBLEM HEADING & DESCRIPTION
3. TYPE OF THE STEEL (FY= 36ksi)
4. TYPE OF THE COLUMN (K= 0.80)
5. LENGTH OF THE COLUMN (L= 20.00ft.)
6. DEAD LOAD (150.00kips)
7. LIVE LOAD (150.00kips)
8. SECTION TYPE(C) & NOMINAL DEPTH(15)
9. CHECK/DESIGN & NOMINAL DEPTH(15)
M. RETURN TO MAIN MENU

IF YOU WANT TO UPDATE ANY OF THE PREVIOUS
DATA, ENTER THE NUMBER(S) PLEASE.
(PRESS RETURN KEY WHEN FINISHED)

*** SECTION MENU ***
```````````````````````
1.  W SHAPES
2.  HP SHAPES
3.  TWO C-SECTIONS ([])
4.  TWO C-SECTIONS ()[()
5.  TWO W-SECTIONS (II)
6.  TWO HP-SECTIONS (II)
ENTER A NUMBER ==))
? 1




        ***   DESIGN  OF COLUMN  ***
        ```````````````````````````

DO YOU WANT TO CHOOSE THE NOMINAL DEPTH?
(IF YOU DON'T CHOOSE, THE PROGRAM WILL FIND THE
LIGHTEST SECTION POSSIBLE FOR THE GIVEN SHAPE) (Y/N)? n

            searching ...

SELECTED SECTION :
                 *****   W 12 X 65   *****

        A = 19.100 in.^2   bf = 12.000 in.
        d = 12.120 in.   tf = 0.605 in.
        tw =   0.390 in.   rT = 3.280 in.
        Ix = 533.00 in.^4   Iy = 174.00 in.^4
        Sx =   87.90 in.^3   Sy = 29.10 in.^3
        rx =   5.28 in.   ry =   3.02 in.

THE MAXIMUM RATIO OF THE ACTUAL STRESS TO THE ALLOWABLE STRESS IS 0.920
THE SECTION IS COMPACT
TOTAL WEIGHT OF THE COLUMN : 1.30 Kips.
PRESS ANY KEY TO CONTINUE
```
```

```
####   REDESIGN MENU   ####
,,,,,,,,,,,,,,,,,,,,,,,,,,,,,,,

1.  READ FROM A FILE
2.  PROBLEM HEADING & DESCRIPTION
3.  TYPE OF THE STEEL (FY= 36ksi)
4.  TYPE OF THE COLUMN (K=  0.80)
5.  LENGTH OF THE COLUMN (L= 20.00ft.)
6.  DEAD LOAD ( 150.00kips)
7.  LIVE LOAD ( 150.00kips)
8.  SECTION TYPE( W) & NOMINAL DEPTH(12)
9.  CHECK/DESIGN & NOMINAL DEPTH(12)
M.  RETURN TO MAIN MENU

IF YOU WANT TO UPDATE ANY OF THE PREVIOUS
DATA, ENTER THE NUMBER(S) PLEASE.
(PRESS RETURN KEY WHEN FINISHED)


### SECTION MENU ###
,,,,,,,,,,,,,,,,,,,,,,
1.  W SHAPES
2.  HP SHAPES
3.  TWO C-SECTIONS ([])
4.  TWO C-SECTIONS (][)
5.  TWO W-SECTIONS (II)
6.  TWO HP-SECTIONS (II)
ENTER A NUMBER ==))
? 5
                         <-->
ENTER THEN SECTION WIDTH (in.) (I  I) ==) 20


        ### DESIGN OF COLUMN ###
        ,,,,,,,,,,,,,,,,,,,,,,,,,

DO YOU WANT TO CHOOSE THE NOMINAL DEPTH?
(IF YOU DON'T CHOOSE, THE PROGRAM WILL FIND THE
LIGHTEST SECTION POSSIBLE FOR THE GIVEN SHAPE) (Y/N)? n


            searching ...


SELECTED SECTION :
                #####   W 16 X 26   #####

        A   =   7.680 in.^2      bf  =  5.500 in.
        d   =  15.690 in.        tf  =  0.345 in.
        tw  =   0.250 in.        rT  =  1.360 in.
        Ix  = 301.00 in.^4       Iy  =   9.59 in.^4
        Sx  =  38.40 in.^3       Sy  =   3.49 in.^3
        rx  =   6.26 in.         ry  =   1.12 in.

THE MAXIMUM RATIO OF THE ACTUAL STRESS TO THE ALLOWABLE STRESS IS  0.982
THE SECTION IS COMPACT
TOTAL WEIGHT OF THE COLUMN : 1.04 Kips.
PRESS ANY KEY TO CONTINUE
```

6.7 LOAD AND RESISTANCE FACTOR DESIGN OF COLUMNS

According to LRFDS, the design strength of compression members excluding those with slender compression elements (see Sec. 5.11.1) is equal to $\phi_c P_n$, where

ϕ_c = resistance factor for compression = 0.85

P_n = nominal axial strength = AF_{cr}

A = gross cross-sectional area of member

$$F_{cr} = \begin{cases} F_y \exp\left(-0.419\lambda_c^2\right) & \text{for } \lambda_c \leq 1.5 \\ 0.877 F_y / \lambda_c^2 & \text{for } \lambda_c > 1.5 \end{cases} \tag{6.18}$$

$$\lambda_c = \frac{KL}{\pi r}\sqrt{\frac{F_y}{E}} \tag{6.19}$$

In Eq. (6.19), KL/r is the maximum slenderness ratio of the column. For LRFD of columns with slender compression elements, the reader should refer to the AISC LRFD publication [48].

6.8 INTERACTIVE MICROCOMPUTER-AIDED LOAD AND RESISTANCE FACTOR DESIGN OF COLUMNS

6.8.1 Program Capabilities and Limitations

In this section, an interactive BASIC program is presented for design of columns according to LRFDS. The following load combinations and the corresponding load factors are considered:

$$1.4D_n$$
$$1.2D_n + 1.6L_n$$

where

D_n = dead load acting on the column

L_n = live load acting on the column

The program can select a W, M, S, or HP shape from the AISC sections database. After having selected the cross-sectional shape of the column from the section menu (Fig. 5.28), two options are available to the user. He can select the nominal depth of the column section (for example, 12 for W12). In this case, the program will find the lightest section for the column within the specified shape and nominal depth. If the user does not choose the nominal depth, the program will find the lightest available section within the given shape. To find the lightest section, the program uses four data files for the four types of structural shapes (W, M, S, and HP). In these data files, sections are arranged in order of increasing cross-sectional area. This is different from the main AISC sections database, in which sections are arranged in the same order as the AISCM.

To start the iterative design process, a lower bound value for the required gross area is determined by using the equation:

$$A_{(\text{lower bound})} = \frac{F_{\max}}{\phi_c F_y} \tag{6.20}$$

where $F_{\max}$ is the maximum axial compressive force due to the critical load combination.

Using the lower bound value of A, a tentative section is chosen from the AISC sections database for a given shape and nominal depth (for example, W12), or from one of the four data files mentioned previously if only the cross-sectional shape is specified by the user. On the basis of the properties of this section, F_{cr} is calculated by using Eq. (6.18). The program also considers sections with slender compression elements. If one of the following conditions is not satisfied, the column section is called a section with slender compression elements:

$$\frac{b_f}{2t_f} \leq \frac{95}{\sqrt{F_y}} \tag{6.21}$$

$$\frac{h_c}{t_w} \leq \frac{253}{\sqrt{F_y}} \tag{6.22}$$

Next, the section is checked for the following requirements:

$$\phi_c P_n \geq F_{\max} \tag{6.23}$$

$$\frac{KL}{r} \leq 200 \tag{6.24}$$

If any one of the above checks is not satisfied, the next larger section within the given shape and nominal depth will be checked. The design will be complete when all the above checks are satisfied.

After the completion of a design, the user can use the redesign menu (Fig. 6.15) to change any of the input data. Different changes can be made at the same time. The interactive program will come up with a new design accommodating all the changes.

Similar to the program for interactive design of columns according to the AISCS presented in Sec. 6.6, this program uses five menus. The main menu is

```
*COLUMN DESIGN-LRFD*
*** MAIN MENU ***

1. NEW PROBLEM
2. SUMMARY OF RESULTS
3. DISPLAY CROSS-SECTION
4. REDESIGN
5. SAVE DATA ON A DISK
6. QUIT


ENTER A NUMBER ==>>? ■
```

Figure 6.22

shown in Fig. 6.22. The section menu is shown in Fig. 5.28. The remaining three menus—that is, the column type menu (Fig. 6.11), the steel type menu (Fig. 5.18), and the redesign menu (Fig. 6.15)—are the same as those in the program for interactive design of columns according to the AISCS.

6.8.2 Sample Example

As an example, the same axially compressed column solved in Sec. 6.6.2 according to the AISCS is solved by this program according to LRFDS. The support conditions are fixed-hinged. The column length is 20 ft. It is subjected to dead and live axial forces of 150 K each. The column is used as a main member. The yield stress of steel is 36 ksi.

Initially a W10 section was selected. The answer is W10 × 68, which is the same as that obtained by the interactive program for design of columns according to the AISCS (see Sec. 6.6.2). Next, the program was asked to find the lightest available W section. The answer is W12 × 65, which is the same as that obtained by the program for design of columns according to the AISCS.

By using the redesign menu, dead load was increased to 200 K and live load was decreased to 100 K while the total axial load was kept the same, 300 K. The program was asked to find the lightest W section. The answer is W10 × 60, which is lighter than the previous minimum weight design of W12 × 65. Finally, the dead load was further increased to 250 K and the live load was further decreased to 50 K while the total axial load was kept the same, 300 K. The program was again asked to find the lightest W section. The answer is W12 × 58. As the ratio of live load to dead load decreases, it appears that design on the basis of the new LRFD specification yields a lighter section than design on the basis of the AISC allowable stress design specification.

For the sake of brevity, the interactive session for this example is not presented in the book.

6.9 PROBLEMS

6.1 The column *AB* shown in Fig. 6.23 is made of a C10 × 30 and a C10 × 20. The two ends of the column are hinged with no lateral displacement in any direction. At point D,

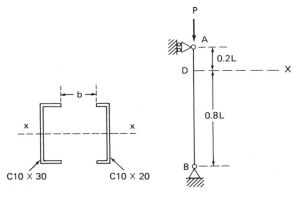

C10 × 30   C10 × 20

Figure 6.23

lateral displacement is prevented in the x direction only. The length of the column is $L = 15\,\text{ft}$. Channels are made of A36 steel with yield stress of 36 ksi. (a) What is the maximum possible allowable axial compressive load for this column? (b) What is the corresponding minimum distance b between the channels?

6.2 A column of length 24 ft is subjected to an axial load of 165 kips. Column ends are hinged. The column is supported at the midheight only in its weak direction. Using A36 ($F_y = 36\,\text{ksi}$) steel, select the least weight tube for the column.

6.3 In Problem 6.2, suppose you ought to use a section made of four L3.5 × 3.5 × $\frac{5}{16}$ with double lacings as shown in Fig. 6.24. Find the minimum dimensions a and b of the column cross section. Next, after rounding these minimum dimensions to the nearest inch, design the double lacing bars. Check all the necessary requirements.

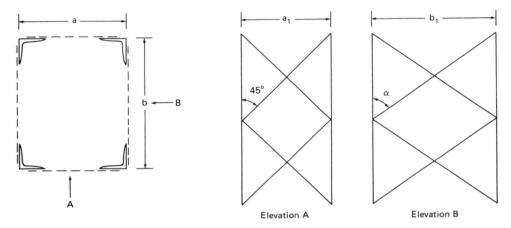

Elevation A   Elevation B

Figure 6.24

6.4 A column is hinged at one end and fixed at the other end. It is made of 2 L5 × 5 × $\frac{1}{2}$ and 2 PL5 × $\frac{1}{2}$ as shown in Fig. 6.25. The angles and plates are made of A36 steel with a yield stress of 36 ksi. Find the maximum axial compressive load capacity of this column, assuming that angles and plates are connected to each other properly so that none of them will buckle individually. Use the design value for the effective length factor.

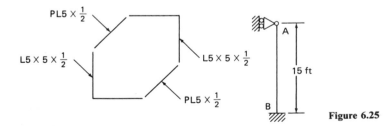

PL5 × $\frac{1}{2}$

L5 × 5 × $\frac{1}{2}$

L5 × 5 × $\frac{1}{2}$

PL5 × $\frac{1}{2}$

15 ft

Figure 6.25

6.5 A built-up column consists of an L8 × 6 × $\frac{1}{2}$ and a 0.5 in.-thick plate welded together as shown in Fig. 6.26. Both angle and plate are made of A36 steel with a yield stress of 36 ksi. The length of the column is 15 ft. The bottom end of the column is fixed and its top end is free. Find the maximum axial compressive load capacity of this column.

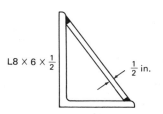

L8 × 6 × $\frac{1}{2}$ $\frac{1}{2}$ in.

Figure 6.26

6.6 Solve Problem 6.5, using A572 steel with a yield stress of 60 ksi.

6.7 Find the lightest column made of A36 steel ($F_y = 36$ ksi) with hinged ends and a length of 10 ft to carry an axial compressive force of 250 K. Use the following shapes:
a. W14
b. WT
c. square structural tube
d. pipe
e. angle with equal legs

6.8 Find the maximum axial compressive load capacity of a column with a length of 20 ft and the cross section shown in Fig. 6.27. The two ends of the columns are hinged.

C15 × 50

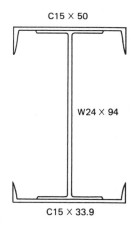

W24 × 94

C15 × 33.9   **Figure 6.27**

Assume the following types of steel:
a. A36 ($F_y = 36$ ksi)
b. A441 ($F_y = 50$ ksi)
c. A572 ($F_y = 65$ ksi)

6.9 Solve Problem 6.8, assuming that the column is an L8 × 6 × 1.

7

Design
of Beam-Columns

7.1 INTRODUCTION

Columns in steel buildings are often subjected to bending moments in addition to
axial compressive forces. Even when beams are connected to the column through
simple connections, such as framing angles shown in Fig. 7.1, they exert bending
moment on the column due to eccentrically applied support reactions. Columns in
moment-resisting frames, of course, are subjected to considerable bending. Members
acted on simultaneously by compressive axial forces and bending moments are
referred to as *beam-columns*.

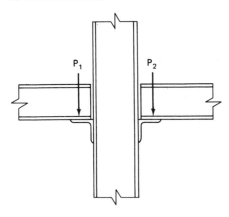

Figure 7.1

7.2 STRESSES IN BEAM-COLUMNS

When a section is subjected to an axial load P and bending moments M_x and M_y about its principal axes x and y, respectively (Fig. 7.2), stress at any point can be

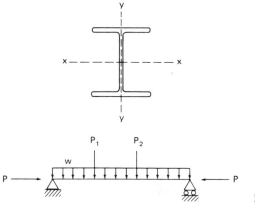

Figure 7.2

calculated approximately from the following expression:

$$f = \frac{P}{A} \pm \frac{M_x y}{I_x} \pm \frac{M_y x}{I_y} \tag{7.1}$$

where A is the cross-sectional area, and I_x and I_y are the moments of inertia with respect to the principal axes x and y, respectively. Equation (7.1) does not take into account the additional bending moment by the axial forces P due to the lateral deflection of the beam-column. Maximum normal stress from Eq. (7.1) is

$$f_{\max} = \frac{P}{A} + \frac{M_x c_x}{I_x} + \frac{M_y c_y}{I_y} = f_a + f_{bx} + f_{by} \tag{7.2}$$

where f_a is the compressive stress due to axial load P, and f_{bx} and f_{by} are the maximum bending stresses due to bending moments M_x and M_y acting on the section. Noting that we have three different allowable stresses F_a, F_{bx}, and F_{by} for axial compressive stress, bending about the major axis x, and bending about the minor axis y, respectively, the following relation must be satisfied:

$$\frac{f_a}{F_a} + \frac{f_{bx}}{F_{bx}} + \frac{f_{by}}{F_{by}} \leq 1.0 \tag{7.3}$$

Until 1963, this equation was used by AISCS for design of beam-columns. This equation is now limited to small axial compressive stresses, that is, $f_a/F_a \leq 0.15$. For larger axial compressive stresses, Eq. (7.3) has been modified to take into account the effect of additional moments produced by lateral deflections.

7.3 BEAM-COLUMN INTERACTION FORMULAS

According to AISCS 1.6.1, members subjected to combined axial force and bending shall be proportioned to satisfy the following equations.

For $\dfrac{f_a}{F_a} \leqslant 0.15$:

$$\frac{f_a}{F_a} + \frac{f_{bx}}{F_{bx}} + \frac{f_{by}}{F_{by}} \leqslant 1.0 \tag{7.4}$$

For $\dfrac{f_a}{F_a} > 0.15$:

$$\frac{f_a}{F_a} + \frac{C_{mx} f_{bx}}{\left(1 - \dfrac{f_a}{F'_{ex}}\right) F_{bx}} + \frac{C_{my} f_{by}}{\left(1 - \dfrac{f_a}{F'_{ey}}\right) F_{by}} \leqslant 1.0 \tag{7.5}$$

$$\frac{f_a}{0.60 F_y} + \frac{f_{bx}}{F_{bx}} + \frac{f_{by}}{F_{by}} \leqslant 1.0 \tag{7.6}$$

where

F_a = allowable axial compressive stress if only axial force existed
F_b = allowable compressive bending stress if only bending moment existed
f_a = actual axial compressive stress
f_b = actual compressive bending stress
$F'_e = \dfrac{12\pi^2 E}{23(KL_b/r_b)^2}$ = Euler stress divided by a factor of safety of $\frac{23}{12} = 1.92$
L_b = unbraced length in the *plane of bending*
r_b = radius of gyration in the plane of bending
K = effective length factor in the plane of bending

In Eqs. (7.4) through (7.6), subscripts x and y indicate the axis of bending. Coefficient $1./(1. - f_a/F'_e)$ is an amplification factor which takes care of the increased moments caused by lateral displacements. This amplification factor introduces greater conservatism compared to the older Eq. (7.4). Depending on the actual slenderness ratios, axial load, lateral loads, and end restraint conditions, this amplification factor may become excessively conservative. To offset this situation, a modification or reduction factor C_m is introduced in Eq. (7.5) whose value is less than or equal to one.

To find the value of C_m, AISCS divides beam-columns into these categories:

1. For compression members in frames subject to sidesway or joint translation [for example, member AB in moment-resisting frame of Fig. 7.3(a)]:

$$C_m = 0.85 \tag{7.7}$$

2. For compression members in frames braced against sidesway or joint translation and not subjected to transverse loading between their ends [for example,

member AB in the braced frame of Fig. 7.3(b)]:

$$C_m = 0.6 - 0.4 \frac{M_1}{M_2} \geqslant 0.4 \qquad (7.8)$$

where M_1/M_2 is the ratio of the smaller moment to the larger moment at the ends of the unbraced length in the plane of bending. The ratio M_1/M_2 is positive when the end moments M_1 and M_2 are in the same direction (reverse curvature) and negative otherwise (single curvature). A member in single curvature in general has larger lateral displacement than a corresponding member in double curvature and consequently is subjected to larger moments and bending stresses. When the two end moments have the same magnitude but opposite directions ($M_1 = -M_2$), C_m becomes equal to unity. The value of C_m in Eq. (7.8) is least when $M_1/M_2 = 0.5$ in reverse curvature.

3. For compression members in frames braced against sidesway and subjected to transverse loading [for example, member CD in the braced frame of Fig. 7.3(b)], the value of C_m is found from the following expression:

$$C_m = 1 + \psi \frac{f_a}{F_e'} \qquad (7.9)$$

where ψ is a dimensionless factor whose values for several end restraint and loading conditions are given in Table 7.1. For other cases, AISCS recommends the following values:

$$\text{For members with moment restraint at the ends: } C_m = 0.85 \qquad (7.10)$$

$$\text{For members with simply supported ends: } \qquad C_m = 1.0 \qquad (7.11)$$

In Eqs. (7.4)-(7.6), F_a is found on the basis of the maximum slenderness ratio without regard to the plane of bending. In contrast, F_e' is calculated on the basis of the slenderness ratio in the plane of bending.

Equations (7.6) and (7.5) may be considered as yield and stability criteria, respectively. In calculating f_b in Eq. (7.6) the larger of the two end moments is used. On the other hand, when the member is subject to intermediate transverse loads, the maximum bending moment between points of supports must be used to calculate f_b in Eq. (7.5).

When bending takes place about one axis, the term corresponding to the other axis in Eqs. (7.4)-(7.6) shall be deleted.

7.4 DESIGN OF BEAM-COLUMNS USING EQUIVALENT AXIAL COMPRESSIVE LOAD

Equations (7.4)-(7.6) cannot be readily used for design of beam-columns. These equations can be rearranged so that we need to design the beam-column for an equivalent axial compressive load. In other words, when a member is acted upon by an axial compressive force P and a bending moment M, the bending moment can be converted into an equivalent axial load P' and consequently the member needs to be designed for an equivalent axial load of $P_{eq} = P + P'$.

TABLE 7.1 REDUCTION FACTOR C_m (AISC COMMENTARY TABLE C1.6.1)

Case	ψ	C_m
a	0	1.0
b	-0.4	$1 - 0.4\,\dfrac{f_a}{F'_e}$
c	-0.4	$1 - 0.4\,\dfrac{f_a}{F'_e}$
d	-0.2	$1 - 0.2\,\dfrac{f_a}{F'_e}$
e	-0.3	$1 - 0.3\,\dfrac{f_a}{F'_e}$
f	-0.2	$1 - 0.2\,\dfrac{f_a}{F'_e}$

In the case of bending about the x-axis only, Eq. (7.5) becomes

$$\frac{f_a}{F_a} + \frac{C_{mx}f_{bx}}{(1 - f_a/F'_{ex})F_{bx}} \leq 1.0 \tag{7.12}$$

Substituting for $f_a = P/A$ and $f_{bx} = M_x/S_x$ and multiplying the two sides by AF_a, we find that Eq. (7.12) at the limit becomes

$$P + \frac{M_x A}{S_x}\left(\frac{F_a}{F_{bx}}\right)\left[\frac{C_{mx}}{1 - P/(AF'_{ex})}\right] = AF_a \tag{7.13}$$

Note that $P_{eq} = AF_a$ would be the design load if the member were axially loaded.

Defining

$$a_x = \frac{12\pi^2 EAr_x^2}{23} \tag{7.14}$$

we can write

$$AF'_{ex} = \frac{12\pi^2 EA}{23(KL/r_x)^2} = \frac{a_x}{(KL)^2} \tag{7.15}$$

$$\frac{C_{mx}}{1 - P/(AF'_{ex})} = \frac{a_x C_{mx}}{a_x - P(KL)^2} \tag{7.16}$$

If we denote $B_x = A/S_x$, called the bending factor, the equivalent axial compressive load will be

$$P_{eq} = P + B_x M_x C_{mx}\left(\frac{F_a}{F_{bx}}\right)\left[\frac{a_x}{a_x - P(KL)^2}\right] \tag{7.17}$$

Similarly, in general, the three Eqs. (7.4), (7.5), and (7.6) can be written in the following equivalent axial load form:

$$P_{eq} = P + P'_x + P'_y = P + B_x M_x\left(\frac{F_a}{F_{bx}}\right) + B_y M_y\left(\frac{F_a}{F_{by}}\right) \tag{7.18}$$

$$P_{eq} = P + P'_x + P'_y = P + B_x M_x C_{mx}\left(\frac{F_a}{F_{bx}}\right)\left[\frac{a_x}{a_x - P(KL)^2}\right]$$

$$+ B_y M_y C_{my}\left(\frac{F_a}{F_{by}}\right)\left[\frac{a_y}{a_y - P(KL)^2}\right] \tag{7.19}$$

$$P_{eq} = P + P'_x + P'_y = P\left(\frac{F_a}{0.60F_y}\right) + B_x M_x\left(\frac{F_a}{F_{bx}}\right) + B_y M_y\left(\frac{F_a}{F_{by}}\right) \tag{7.20}$$

In Eq. (7.19), K is the effective length factor and L is the actual unbraced length in the plane of bending. For I sections, the ranges of bending factors B_x and B_y are

$$0.168 \leqslant B_x = \frac{A}{S_x} \leqslant 1.133 \tag{7.21}$$

$$0.408 \leqslant B_y = \frac{A}{S_y} \leqslant 4.722 \tag{7.22}$$

Also,

$$a_x = 149,000 Ar_x^2 \tag{7.23}$$

$$a_y = 149,000 Ar_y^2 \tag{7.24}$$

In these equations, A is in square inches, r_x is in inches, and S_x and S_y are in cubic inches.

Design of beam-columns is an iterative process. To start this process, the following rough approximation may be used for the equivalent axial load:

$$P_{eq} = P + B_x M_x + B_y M_y \tag{7.25}$$

For the initial trial selection, average values may be used for B_x and B_y, as summarized in Table 7.2.

TABLE 7.2 AVERAGE VALUES FOR
BENDING FACTORS B_x AND B_y

Type of section	B_x (1/in.)	B_y (1/in.)
W14	0.18	0.50
W12	0.21	0.69
W10	0.26	0.83
W8	0.33	1.02
W6	0.46	1.83

7.5 EFFECTIVE LENGTH OF COLUMNS IN BRACED AND UNBRACED FRAMES

The end conditions for a column in a braced or unbraced frame is usually different from the end conditions for the columns shown in Fig. 6.2. For a practical method of finding the effective length or equivalent hinged-ends length for a column in a frame, the following assumptions are usually made [31]:

1. Columns have elastic behavior.
2. Columns are prismatic.
3. The frame is a rectangular and symmetrical structure.
4. Axial forces in girders are negligible.
5. Columns at a joint carry the end girder moments in proportion to their stiffnesses.
6. All columns attain their buckling loads simultaneously.
7. At the incipient buckling the rotations of the girder at its ends are equal and opposite.

On the basis of these assumptions, one can find the following transcendental equation for the effective length factor K in a braced frame such as the one shown in Fig. 7.3(b) [31]:

$$\frac{G_A G_B}{4}\left(\frac{\pi^2}{K^2}\right) + \left(\frac{G_A + G_B}{2}\right)\left(1 - \frac{\pi/K}{\tan \pi/K}\right) + \frac{2}{\pi/K}\tan\frac{\pi}{2K} = 1 \qquad (7.26)$$

In this equation, subscripts A and B refer to the ends of column AB and

$$G = \frac{\sum \dfrac{I_c}{L_c}}{\sum \dfrac{I_g}{L_g}} \qquad (7.27)$$

where I_c is the moment of inertia and L_c the unsupported length of a column section,

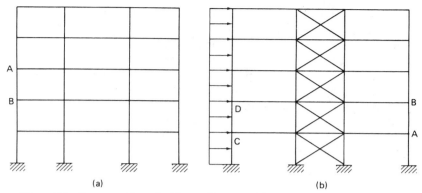

Figure 7.3 Frames with and without sidesway. (a) unbraced frame; (b) braced frame.

I_g is the moment of inertia and L_g the unsupported length of a girder, and $\sum$ indicates a summation for all members connected to the joint under consideration and lying in the plane of buckling.

Similarly, the transcendental equation for the effective length factor K in an unbraced frame is as follows [31]:

$$\frac{G_A G_B (\pi/K)^2 - 36}{6(G_A + G_B)} = \frac{\pi/K}{\tan(\pi/K)} \qquad (7.28)$$

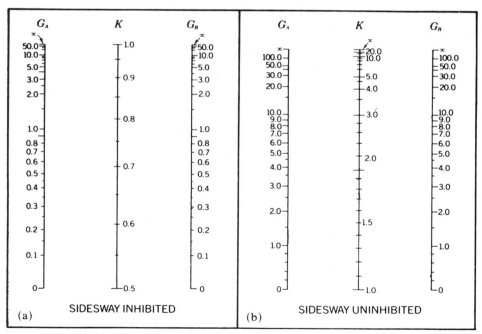

Figure 7.4 Alignment charts for the effective length factors in braced and unbraced frames [47]. Printed by permission of the AISC.

Equations (7.26) and (7.28) must be solved numerically. For manual design of frames, nomograms or alignment charts have been developed to be used in place of these equations. These alignment charts are presented in Fig. 7.4. To find the effective length factor for a column, first quantities G_A and G_B are found for the two ends of the column. By connecting the corresponding points on the alignment chart, a straight line is obtained. The intersection of this line with the K-line in the chart yields the value of the effective length factor.

For columns supported by a footing or foundation through a simple support, G is theoretically infinity, but AISC recommends a practical value of 10. For columns rigidly connected to a footing or foundation, a practical value of 1.0 is recommended for G [47].

7.6 EXAMPLES OF DESIGN OF BEAM-COLUMNS

Example 1

Design the W14 steel column shown in Fig. 7.5. The column must be designed for an axial load of 400 Kips, bending moment about the major axis of 200 K-ft, and bending

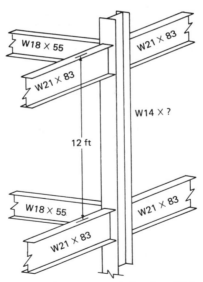

Spacing of columns in each direction: 25 ft.   **Figure 7.5**

moment about the minor axis of 100 K-ft. The column is a member of a moment-resisting space frame (with "rigid" connections). Use A36 steel and assume the column in the tier above and below to be the same as the column shown in the figure. The average values of bending coefficients B_x and B_y for W14 sections are 0.18 1/in. and 0.5 1/in., respectively.

Solution

$$P = 400 \text{ K} \qquad M_x = 200 \text{ K-ft} \qquad M_y = 100 \text{ K-ft}$$

From Eq. (7.25):

$$P_{eq} \simeq 400 + 0.18(200)(12) + 0.5(100)(12) = 1432 \text{ K}$$

In general, to design a beam-column, several trials are necessary. For the first trial one may assume a certain value for the allowable axial compressive stress F_a. In this example, let us assume

$$F_a = 0.50 F_y = 18 \text{ ksi}$$

In that case, a rough estimate for the cross-sectional area will be

$$A = \frac{P_{eq}}{F_a} = \frac{1432}{18} = 80 \text{ in.}^2$$

Note that Eq. (7.25) in general overestimates the design requirement. Try W14 × 233:

$$I_x = 3010 \text{ in.}^4 \qquad A = 68.5 \text{ in.}^2 \qquad b_f = 15.89 \text{ in.}$$

$$I_y = 1150 \text{ in.}^4 \qquad S_x = 375 \text{ in.}^3 \qquad r_x = 6.63 \text{ in.}$$

$$r_T = 4.40 \text{ in.} \qquad S_y = 145 \text{ in.}^3 \qquad r_y = 4.10 \text{ in.}$$

$$\frac{d}{A_f} = 0.59 \text{ 1/in.} \qquad \frac{b_f}{2t_f} = 4.6 \qquad \frac{d}{t_w} = 15.0$$

To find the effective length factors K_x and K_y we use the alignment charts [Fig. 7.4(b)].

For W21 × 83: $I_x = 1830 \text{ in.}^4$

For W18 × 55: $I_x = 890 \text{ in.}^4$

$$(G_A)_x = (G_B)_x = \frac{\sum I_c/L_c}{\sum I_g/L_g} = \frac{2(3010/12)}{2(1830/25)} = 3.4$$

$$(G_A)_y = (G_B)_y = \frac{2(1150/12)}{890/25} = 5.4$$

From the alignment chart [Fig. 7.4(b)]:

$$K_x = 1.9 \qquad K_y = 2.3$$

$$\left(\frac{KL}{r}\right)_x = \frac{(1.9)(12)(12)}{6.63} = 41.3$$

$$\left(\frac{KL}{r}\right)_y = \frac{(2.3)(12)(12)}{4.10} = 80.8 \qquad \text{(controls the value of allowable axial compressive stress, } F_a\text{)}$$

$$C_c = \sqrt{\frac{2\pi^2 E}{F_y}} = \sqrt{\frac{2\pi^2(29,000)}{36}} = 126.1 > \left(\frac{KL}{r}\right)_y$$

$$\text{F.S.} = \frac{5}{3} + \frac{3(KL/r)_y}{8C_c} - \frac{(KL/r)_y^3}{8C_c^3}$$

$$= \frac{5}{3} + \frac{3(80.8)}{8(126.1)} - \frac{(80.8)^3}{8(126.1)^3} = 1.874$$

$$F_a = \frac{\left[1 - \frac{(KL/r)_y^2}{2C_c^2}\right]F_y}{\text{F.S.}} = \frac{\left[1 - \frac{1}{2}\left(\frac{80.8}{126.1}\right)^2\right](36)}{1.874} = 15.26 \text{ ksi}$$

$$f_a = \frac{P}{A} = \frac{400}{68.5} = 5.84 \text{ ksi}$$

$$\frac{f_a}{F_a} = \frac{5.84}{15.26} > 0.15$$

∴ Equations (7.5) and (7.6) must be satisfied.

Check compactness (Sec. 5.5.1.1 and Table 5.1):

$$L_u = (12)(12) = 144 \text{ in.}$$

$$\frac{76b_f}{\sqrt{F_y}} = \frac{76(15.89)}{\sqrt{36}} = 201.3 \text{ in.} > L_u = 144 \text{ in.}$$

$$\frac{20,000}{(d/A_f)F_y} = \frac{20,000}{(0.59)(36)} = 941.6 > L_u = 144 \text{ in.}$$

$$\frac{b_f}{t_f} = 9.2 < \frac{130}{\sqrt{F_y}} = 21.7$$

$$\frac{f_a}{F_y} = \frac{5.84}{36} = 0.162 > 0.16$$

$$\frac{h}{t_w} = \frac{d - 2t_f}{t_w} = \frac{16.04 - 2(1.72)}{1.07} = 11.8 < \frac{257}{\sqrt{F_y}} = 42.8$$

∴ The section is compact.

$$F_{bx} = 0.66F_y = 24 \text{ ksi}$$

$$F_{by} = 0.75F_y = 27 \text{ ksi}$$

$$F'_{ex} = \frac{12\pi^2 E}{23(KL/r)_x^2} = \frac{12\pi^2(29,000)}{23(41.3)^2} = 87.55 \text{ ksi}$$

$$F'_{ey} = \frac{12\pi^2 E}{23(KL/r)_y^2} = \frac{12\pi^2(29,000)}{23(80.8)^2} = 22.87 \text{ ksi}$$

$$C_{mx} = C_{my} = 0.85$$

$$f_{bx} = \frac{M_x}{S_x} = \frac{(200)(12)}{375} = 6.40 \text{ ksi}$$

$$f_{by} = \frac{M_y}{S_y} = \frac{(100)(12)}{145} = 8.28 \text{ ksi}$$

Check Eq. (7.5):

$$\frac{f_a}{F_a} + \frac{C_{mx}f_{bx}}{\left(1 - \dfrac{f_a}{F'_{ex}}\right)F_{bx}} + \frac{C_{my}f_{by}}{\left(1 - \dfrac{f_a}{F'_{ey}}\right)F_{by}}$$

$$= \frac{5.84}{15.26} + \frac{0.85(6.4)}{\left(1 - \dfrac{5.84}{87.55}\right)24} + \frac{0.85(8.28)}{\left(1 - \dfrac{5.84}{22.87}\right)27} = 0.98 < 1$$

Check Eq. (7.6):

$$\frac{f_a}{0.60F_y} + \frac{f_{bx}}{F_{bx}} + \frac{f_{by}}{F_{by}} = \frac{5.84}{22} + \frac{6.4}{24} + \frac{8.28}{27} = 0.84 < 1$$

$$\boxed{\text{USE W14} \times 233}$$

Example 2

Design a 10 × 10 square tube from the AISCM for a column with a length of 20 ft and subjected to an axial load of 400 Kips and a uniformly distributed load of intensity 0.2 Kips/ft acting on one of its sides (Fig. 7.6). The column is used in a braced frame. Assume that the column ends are pinned. Use A572 steel with a yield stress of 60 ksi.

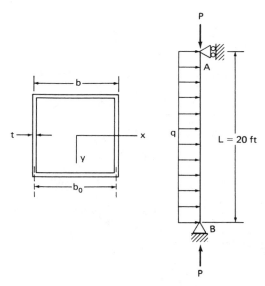

Figure 7.6

Solution

$$P = 400 \text{ Kips} \qquad q = 0.2 \text{ Kips/ft} \qquad K = 1 \qquad L = 20 \text{ ft}$$

$$M_y = \frac{qL^2}{8} = \frac{0.2(20)^2}{8} = 10 \text{ K-ft} = 120 \text{ K-in.}$$

The properties of a square box section can be found in terms of its thickness t and width of the section b_o (distance between the midplanes of the opposite plates) as follows:

$$A = 4b_o t$$

$$S_x = S_y = \frac{4b_o t(b_o^2 + t^2)}{3(b_o + t)}$$

$$r_x = r_y = \sqrt{\frac{b_o^2 + t^2}{6}}$$

$$B_x = B_y = \frac{3(b_o + t)}{b_o^2 + t^2}$$

For the first trial section, considering that t is small compared with b, we can write

$$B_y \simeq \frac{3}{b_o} \simeq \frac{3}{b} = \frac{3}{10} = 0.3 \ 1/\text{in.}$$

$$r_x = r_y \simeq \frac{b_o}{\sqrt{6}} \simeq 0.4b = 0.4(10) = 4 \text{ in.}$$

Now, the equivalent axial load from Eq. (7.25) is equal to

$$P_{eq} = P + B_y M_y = 400 + 0.3(120) = 436 \text{ Kips}$$

$$\frac{KL}{r} = \frac{(20)(12)}{4} = 60$$

$$C_c = \sqrt{\frac{2\pi^2 E}{F_y}} = \sqrt{\frac{2\pi^2(29,000)}{60}} = 97.7 > \frac{KL}{r}$$

Thus, the allowable axial compressive stress from Eq. (6.14) becomes $F_a = 26.03$ ksi.

$$\text{Required } A \simeq \frac{P_{eq}}{F_a} = \frac{436}{26.03} = 16.75 \text{ in.}^2$$

Try TUBE 10×10 with wall thickness $t = 0.5$ in.

$$A = 18.4 \text{ in.}^2 \qquad S_x = S_y = 54.2 \text{ in.}^3 \qquad r_x = r_y = 3.84 \text{ in.}$$

Check for local flange buckling (line 5 and column 4 of Table 5.1):

$$\frac{b_f}{t_f} = \frac{10}{0.5} = 20 < \frac{238}{\sqrt{F_y}} = \frac{238}{\sqrt{60}} = 30.7$$

From Table 7.1 (Case a): $C_{my} = 1.0$

$$\frac{KL}{r} = \frac{(20)(12)}{3.84} = 62.5 < C_c = 97.7$$

From Eq. (6.14): $F_a = 25.52$ ksi

$$f_a = \frac{P}{A} = \frac{400}{18.4} = 21.74 \text{ ksi}$$

$$\frac{f_a}{F_a} = \frac{21.74}{25.52} = 0.852 > 0.15$$

$$f_{by} = \frac{M_y}{S_y} = \frac{120}{54.2} = 2.21 \text{ ksi}$$

$$F'_{ey} = \frac{12\pi^2 E}{23(KL/r)^2} = \frac{12\pi^2(29,000)}{23(62.5)^2} = 38.23 \text{ ksi}$$

Check for compactness [Eqs. (5.11), (5.12), and (5.13) and Table 5.1].

$$\frac{d}{b} = 1 < 6$$

$$\frac{t_f}{t_w} = 1 < 2$$

$$L_u = 240 \text{ in.} < \frac{1950b}{F_y} = \frac{(1950)(10)}{60} = 325 \text{ in.}$$

$$\frac{b_f}{t_f} = 20 < \frac{190}{\sqrt{F_y}} = \frac{190}{\sqrt{60}} = 24.5$$

$$\frac{f_a}{F_y} = \frac{21.74}{60} = 0.36 > 0.16$$

$$\frac{d}{t_w} = \frac{10}{0.5} = 20 < \frac{257}{\sqrt{F_y}} = \frac{257}{\sqrt{60}} = 33.2$$

$\therefore$ The section is compact.

$$F_{by} = 0.66F_y = 0.66(60) = 40 \text{ ksi}$$

Check Eq. (7.5):

$$\frac{f_a}{F_a} + \frac{C_{my}f_{by}}{\left(1 - \frac{f_a}{F'_{ey}}\right)F_{by}} = 0.852 + \frac{(1.0)(2.21)}{\left(1 - \frac{21.74}{38.23}\right)(40)} = 0.980 < 1.0$$

USE TUBE 10 × 10, $t = 0.5$ in.

7.7 INTERACTIVE MICROCOMPUTER-AIDED DESIGN OF BEAM-COLUMNS ACCORDING TO AISCS

This is an interactive BASIC program for design of beam-columns according to AISCS. The input data may be read from an existing data file or may be given by the user interactively. The program can either design a section or check the adequacy of a given section for given boundary and loading conditions. It can display the cross section of the member.

The user can select one of six different types of cross sections from the section menu, the same as the one in the program for design of columns as shown in Fig. 6.14. They are W, HP, double C, double W, and double HP. The program has the following limitations:

1. The member is subjected to bending about major or minor axis only.
2. The member can either have full lateral support or no lateral support at all. Intermediate supports are not allowed.
3. The effective length factor is the same in the direction of both principal axes.
4. The end conditions and loads acting on the member are limited to those cases shown in the beam-column type menu of Fig. 7.7.

TYPE	END CONDITIONS AND LOADING	K	Psi
1		1.0	0
2		0.80	-0.4
3		0.65	-0.4
4		1.0	-0.2
5		0.80	-0.3
6		0.65	-0.2
7		TO BE GIVEN BY USER	N.A

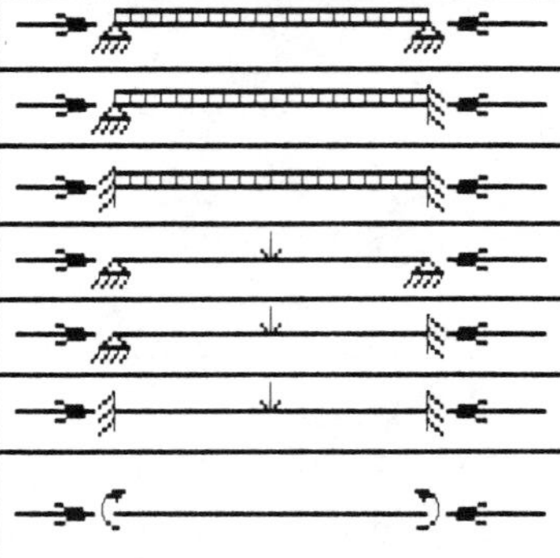

*** BEAM-COLUMN TYPE MENU ***

K- EFFECTIVE LENGTH FACTOR
SELECT TYPE OF BEAM-COLUMN (1-7)   ? ▮

Figure 7.7

To start the iterative process of design of beam-columns, an approximate equivalent axial load is calculated by using the equation

$$P_{eq} = P + M_{max}B$$

where

P = axial compressive load acting on the member
M_{max} = maximum bending moment acting on the member
B = average value of bending coefficient for the selected shape and nominal depth

Thus, an approximate value of cross-sectional area is calculated by using the equation

$$A_{req} = P_{eq}/(0.6F_y)$$

Next, a tentative section is chosen from the AISC sections database for the given shape and nominal depth (for example, W12). Using the properties of the section, we can estimate the equivalent axial load more accurately using either Eq. (7.18) or Eqs. (7.19) and (7.20) with applicable terms for major or minor axis of bending. Now, the required cross-sectional area of the member is computed by using the equation

$$A_{req} = P_{eq} / F_a$$

This iterative process is repeated until the same section is obtained in two consecutive iterations. Finally, the appropriate interaction Eq(s). (7.4) or (7.5) and (7.6) are checked. The value of the left-hand side of interaction Eq. (7.4) or the larger of the left-hand-side values of the two interaction Eqs. (7.5) and (7.6) is displayed for the user along with the properties of the section selected.

After the completion of a design, the user has the flexibility of changing any portion(s) of input data by using the redesign menu shown in Fig. 7.8.

There are five menus in the program: main menu, column type menu (Fig. 7.7), steel type menu (Fig. 5.18), section menu (Fig. 6.14), and the redesign menu (Fig. 7.8).

```
          *** REDESIGN MENU ***
          NNNNNNNNNNNNNNNNNNNNNNNN

   1.  MAIN MENU          2.  READ FROM A FILE

   3.  SAVE ON A FILE     4.  YIELD STRESS

   5.  SPAN LENGTH        6.  AXIS OF BENDING

   7.  LATERAL SUPPORT    8.  BEAM COLUMN TYPE

   9.  AXIAL LOAD        10.  SECTION & WIDTH

  11.  CHECK/DESIGN      12.  NOTHING

  IF YOU WANT TO UPDATE ANY OF THE PREVIOUS DATA,

  ENTER THE NUMBER PLEASE ==>>    ? ▮
```

Figure 7.8

7.8 LOAD AND RESISTANCE FACTOR DESIGN OF BEAM-COLUMNS

7.8.1 Method One for Members in Braced and Unbraced Frames

According to LRFDS, members subjected to combined axial force and bending shall be proportioned to satisfy the following interaction equations:

For $\dfrac{P_u}{\phi_c P_n} \geq 0.2$: $\dfrac{P_u}{\phi_c P_n} + \dfrac{8}{9}\left(\dfrac{M_{ux}}{\phi_b M_{nx}} + \dfrac{M_{uy}}{\phi_b M_{ny}}\right) \leq 1.0$ (7.29)

For $\dfrac{P_u}{\phi_c P_n} < 0.2$: $\dfrac{P_u}{2\phi_c P_n} + \dfrac{M_{ux}}{\phi_b M_{nx}} + \dfrac{M_{uy}}{\phi_b M_{ny}} \leq 1.0$ (7.30) where

P_n = nominal axial load strength
P_u = required axial load strength
M_n = nominal flexural strength determined according to Sec. 5.11
M_u = required flexural strength

$$M_u = B_1 M_{NT} + B_2 M_{LT} \qquad (7.31)$$

M_{NT} = required flexural strength of member assuming there is no lateral translation (in.-Kips)
M_{LT} = required flexural strength of member as a result of lateral translation of the frame only (in.-Kips)
B_1 = the member instability amplification factor
B_2 = the frame instability amplification factor

$$B_1 = \frac{C_m}{1 - \dfrac{P_u}{P_e}} \geqslant 1.0 \qquad (7.32)$$

$$B_2 = \frac{1}{1 - \dfrac{\sum P_u \Delta_{OH}}{\sum HL}} \qquad (7.33)$$

or

$$B_2 = \frac{1}{1 - \dfrac{\sum P_u}{\sum P_e}} \qquad (7.34)$$

$\sum P_u$ = required axial load strength of all columns in a story (Kips)

$$P_e = \text{Euler load} = \frac{\pi^2 EI}{(L)^2} = \frac{AF_y}{\lambda_c^2} \, (\text{Kips}) \qquad (7.35)$$

λ_c is the slenderness parameter defined by Eq. (6.19) but with $K = 1.0$

Δ_{OH} = lateral displacement of the story under consideration
$\sum H$ = sum of the story horizontal forces producing Δ_{OH} (Kips)
L = story height

The coefficient C_m is the same as that defined in Sec. 7.3. The axial force times the deflection produced by primary moments (due to transverse loads or end moments acting on the member) causes additional moments referred to as secondary or P-δ moment. The B_1 factor in Eq. (7.31) takes care of this secondary moment.

In unbraced frames, vertical gravity loads times the drift (lateral displacement) produces overturning moment and additional drift. The B_2 factor in Eq. (7.31) takes care of this instability effect, referred to as the $P\Delta$ effect. The $P\Delta$ effect in braced frames is negligible ($B_2 = 0$).

7.8.2 Method Two for Members in Braced Frames

For I shapes in braced frames subjected to combined axial force and biaxial bending, LRFDS provides alternate nonlinear interaction equations as follows:

$$\left(\frac{M_{ux}}{\phi_b M'_{px}}\right)^{\xi} + \left(\frac{M_{uy}}{\phi_b M'_{py}}\right)^{\xi} \leq 1.0 \tag{7.36}$$

$$\left(\frac{C_{mx} M_{ux}}{\phi_b M'_{nx}}\right)^{\eta} + \left(\frac{C_{my} M_{uy}}{\phi_b M'_{ny}}\right)^{\eta} \leq 1.0 \tag{7.37}$$

where

$$\xi = 1.6 - \frac{P_u/P_y}{2[\text{Ln}\,(P_u/P_y)]} \qquad \text{for } 0.5 \leq b_f/d \leq 1.0 \tag{7.38}$$

$$\eta = \begin{cases} 0.4 + \dfrac{P_u}{P_y} + \dfrac{b_f}{d} \geq 1.0 & \text{for } b_f/d \geq 0.3 \\[2mm] 1 & \text{for } b_f/d < 0.3 \end{cases} \tag{7.39}$$

$$b_f = \text{flange width, in.}$$

$$d = \text{member depth, in.}$$

$$M'_{px} = 1.2 M_{px}[1 - (P_u/P_y)] \leq M_{px} \tag{7.40}$$

$$M'_{py} = 1.2 M_{py}[1 - (P_u/P_y)^2] \leq M_{py} \tag{7.41}$$

$$M'_{nx} = M_{nx}[1 - (P_u/\phi_c P_n)][1 - (P_u/P_{ex})] \tag{7.42}$$

$$M'_{ny} = M_{ny}[1 - (P_u/\phi_c P_n)][1 - (P_u/P_{ey})] \tag{7.43}$$

$$P_y = A F_y \tag{7.44}$$

$$M_{px} = \text{plastic moment about the } x\text{-axis}$$

$$M_{py} = \text{plastic moment about the } y\text{-axis}$$

Chen and Lui [26] have compared the linear and nonlinear interaction equations. Except when b_f/d is small and the ratio of moment to axial force is large, they concluded, the nonlinear interaction equations yield more economical sections than the linear interaction equations.

7.9 PROBLEMS

7.1 Design a W14 section made of A36 steel ($F_y = 36$ ksi) for the beam-column shown in Fig. 7.9. The member is subjected to an axial load of 150 Kips. Bending about the major axis is due to a uniformly distributed load of intensity $q = 1$ K/ft. Bending about the minor axis is due to a couple of magnitude 12 K-ft applied at the midpoint C of the member. Assume lateral support at the two ends A and B only.

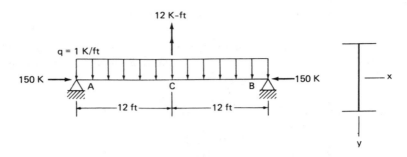

Figure 7.9

7.2 The W-shape column shown in Fig. 7.10 is part of a laterally braced frame. It is subjected to two centric axial compressive forces of P_1 and P_3 at the two ends and the force P_2 with an eccentricity of 12 in. with respect to the center of the column. The three forces are applied in the plane of the web. In addition to the two hinged ends A and B, the column also has lateral support in the x-direction at point D at a distance of 8 ft from the top end A. The column is made of A572 steel with a yield stress of 50 ksi. Check the adequacy of a W14 × 43 section for the column. Neglect the weight of the column.

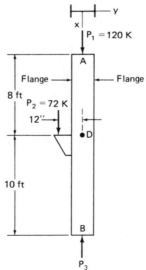

Figure 7.10

7.3 Design the moment-resisting frame shown in Fig. 7.11 using A36 steel ($F_y = 36$ ksi). Lateral support is provided at points B, C, E (midpoint of column AB), F (midpoint of column CD), and G (midpoint of beam BC). Select the lightest W14 for the column and the lightest W section for the beam.

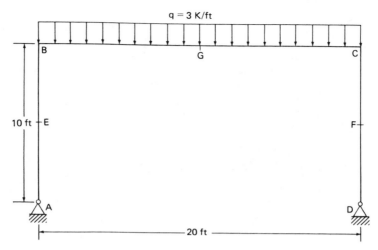

Figure 7.11

7.4 Find the lightest W12 section for the column in Example 1 of this chapter.

7.5 Design a 12 × 12 tube for Example 2 of this chapter.

7.6 Solve Problem 5.17 as a beam-column subjected to axial compressive forces of 20 K at the two ends in addition to the lateral vertical and horizontal loads.

7.7 Solve Problem 5.18 as a beam-column subjected to axial compressive forces of 25 K at the two ends in addition to the lateral vertical and horizontal loads.

Design
of Connections

8.1 RIVETED CONNECTIONS

Nowadays, riveted connections are rarely used. A designer may encounter riveted connections when analyzing an existing riveted structure for increased loading, for example. A single rivet is shown in Fig. 8.1. It consists of a cylindrical shank and a rounded head. The other rounded head is usually formed by heating the rivet and using a rivet gun operated with compressed air.

Figure 8.1

According to ASTM (American Society for Testing and Materials) classifications, three types of rivets may be used in structural steel design:

A502, Grade 1. These rivets have a low carbon content and are more ductile than the ordinary structural steel and thus easier to drive.

A502, Grade 2. They are carbon-manganese rivets and stronger than Grade 1.

A502, Grade 3. These rivets have the same strength as the Grade 2 but have better corrosion resistance.

TABLE 8.1   ALLOWABLE TENSILE AND SHEAR STRESSES FOR RIVETS (AISCS TABLE 1.5.2.1)

Type of rivet	Allowable tensile stress (F_t)	Allowable shear stress (F_v) bearing-type connection
A502, Grade 1 hot-driven rivets	23 ksi	17.5 ksi
A502, Grades 2 and 3 hot-driven rivets	29 ksi	22 ksi

Allowable tensile and shear stresses are given in Table 8.1. Rivets may not be used as friction-type connections, because the amount of friction between the rivet and the parts being connected is not dependable.

The allowable bearing stress for rivets is the same as for bolts and equal to (AISCS 1.5.1.5.3)

$$F_p = 1.5F_u \tag{8.1}$$

where F_u is the specified minimum tensile strength of the steel used in the connected parts. To simplify the design calculations, the bearing stress is assumed uniform over the projected area of bolts and rivets, that is, a rectangular area equal to the diameter of the rivet or bolt times the thickness of the plate.

8.2 BOLTED CONNECTIONS

Today, bolting has practically superseded riveting. Bolting is fast and does not require as much skilled labor as riveting or welding does.

8.2.1 High-Strength Bolts

Two types of high-strength bolts are commonly used for bolted connections in steel structures. They are designated as A325 and A490. These bolts have hexagon heads and use hexagon nuts. The common sizes for these bolts are in the range $\frac{1}{2}$ to $1\frac{1}{2}$ in. In steel buildings, the most common sizes appear to be $\frac{3}{4}$ in. and $\frac{7}{8}$ in.

The allowable tensile and shear stresses are summarized in Table 8.2. The letter N in the high-strength bolt designation indicates that threads are included in the shear plane, while the letter X indicates that threads are excluded from the shear plane.

The AISCS provides two types of connections for high-strength bolts: friction type and bearing type. In the friction-type connection sufficient slip resistance is provided under service conditions. In other words, a high factor of safety is used to prevent bearing of the bolt shank against the side of the hole. The bearing-type connection is used whenever occurrence of slippage under occasional overloads is tolerated. Evidently, allowable shear stresses for bearing-type connections are larger than the corresponding values for friction-type connections. In Table 8.2, the allowable shear stress values for the friction-type high-strength bolts are for

TABLE 8.2   ALLOWABLE STRESSES FOR BOLTS (AISCS TABLE 1.5.2.1)

| Designation | Allowable tensile stress (F_t), ksi | Allowable shear stress (F_v), ksi | |
		Friction type (standard holes)	Bearing type
A307	20	—	10.0
A325N	44	17.5	21.0
A325X	44	17.5	30.0
A490N	54	22	28.0
A490X	54	22	40.0

standard-size holes. For oversized, short-slotted, and long-slotted holes, the allowable shear stress is decreased as given in AISCS Table 1.5.2.1.

The allowable bearing stress for bolts is the same as for rivets, as given by Eq. (8.1). It should be noted that experimental research on bolted connection has shown that neither the bolts nor the connected parts fail in bearing. The magnitude of the bearing stress has an influence on the efficiency of the connection, however. Thus, the AISCS requirement for checking bearing stresses is to take into account the decreased efficiency of the connection due to high bearing stresses.

Due to the larger allowable shear stresses, the bearing-type connection normally yields the more economical solution. The friction-type connection is used in structures subjected to impact and vibration resulting in considerable stress variations or reversals.

8.2.2 Unfinished Bolts

Unfinished bolts, also called common bolts or rough bolts, are made of low-carbon steel and designated by A307. They are cheaper than the high-strength bolts and can be used for static loading only. They are usually used in light structures and secondary or bracing members. They usually have square heads and nuts to reduce the cost. A connection made of unfinished bolts may not necessarily be less expensive than an equivalent connection made of high-strength bolts, because the required number of bolts is usually larger for unfinished bolts. These bolts are used in sizes varying from $\frac{1}{4}$ in. to 4 in. The allowable tensile and shear stresses for A307 bolts are given in Table 8.2.

8.2.3 Bolts Subjected to Eccentric Shear

Consider the group of bolts shown in Fig. 8.2(a) subjected to a force F having an eccentricity of e from the centroid C of the bolt group. To find the magnitude of shear force in each bolt, it is convenient to decompose the loading in Fig. 8.2(a) to the statically equivalent loading in Fig. 8.2(b), which in turn can be considered as

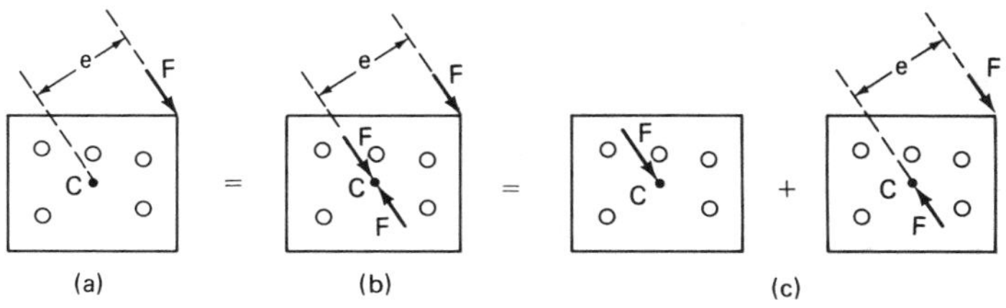

Figure 8.2

the sum of the force F passing through the centroid C and a couple of magnitude Fe as shown in Fig. 8.2(c). The shear force in each bolt due to force F passing through the centroid is equal to F divided by the number of bolts N. Now we calculate the shear forces in the bolts due to the couple of magnitude Fe, using the following assumptions:

1. The gusset plate is perfectly rigid.
2. The bolts are elastic.

The couple $M_c = Fe$ causes the plate to rotate about the centroid of the bolt group. This rotation produces strains in the bolts which are proportional to their distances from the centroid. Since bolts are assumed to behave elastically, shear forces $R_1, R_2, \ldots, R_5$ developed in the bolts are also proportional to their corresponding distances $r_1, r_2, \ldots, r_5$ from the centroid (Fig. 8.3). In other words, if we denote the small angle of rotation of the plate by θ (in radians) and displacement of bolt j in the direction of shear force R_j by d_j, we can write

$$d_j = r_j\theta \tag{8.2}$$

$$R_j = Kd_j = K\theta r_j \tag{8.3}$$

$$\frac{R_1}{r_1} = \frac{R_2}{r_2} = \ldots = \frac{R_j}{r_j} = K\theta \tag{8.4}$$

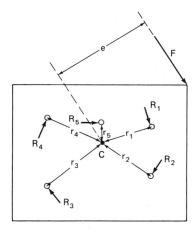

Figure 8.3

In Eqs. (8.3) and (8.4), K is a proportionality factor. By writing the moment equilibrium we obtain

$$M_c = \sum_{i=1}^{N} r_i R_i = K\theta \sum_{i=1}^{N} r_i^2 \tag{8.5}$$

Deleting $K\theta$ between Eqs. (8.3) and (8.5), we will have

$$R_j = \frac{M_c r_j}{\sum_{i=1}^{N} r_i^2} \tag{8.6}$$

This shear force R_j is normal to the line drawn from the centroid C to the bolt j. It is often easier to use the vertical and horizontal components of the shear R_j. By selecting the origin of coordinates at the centroid C (Fig. 8.4) we can write

$$R_{jx} = R_j \sin \alpha = R_j \frac{-y_j}{r_j} = -\frac{M_c y_j}{\sum_{i=1}^{N} r_i^2}$$

$$R_{jy} = R_j \cos \alpha = R_j \frac{x_j}{r_j} = \frac{M_c x_j}{\sum_{i=1}^{N} r_i^2}$$

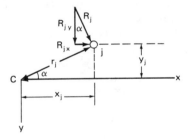

Figure 8.4

Finally, taking into account the contribution of the concentric force, we have

$$R_{jx} = \frac{F_x}{N} - \frac{M_c y_j}{\sum_{i=1}^{N} r_i^2} \tag{8.7}$$

$$R_{jy} = \frac{F_y}{N} + \frac{M_c x_j}{\sum_{i=1}^{N} r_i^2} \tag{8.8}$$

where F_x and F_y are the x and y components of the force F, respectively. Note that in deriving Eqs. (8.7) and (8.8), it is assumed that the positive x-axis is to the right and the positive y-axis is downward.

8.2.4 Bolts Subjected to Combined Shear and Tension

Experiments on bolts subjected to combined shear and tension have demonstrated that their strength can be represented by an elliptical interaction curve with the major axis half-length equal to the allowable stress F_t and the minor axis half-length equal to the allowable stress F_v given in Table 8.2 (AISCS commentary 1.6.3) as shown in Fig. 8.5. In the AISCS, this curve has been replaced by three straight

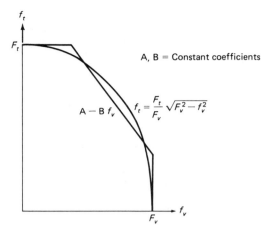

Figure 8.5

lines, as shown in Fig. 8.5. Thus, according to AISCS 1.6.3 the allowable tensile stress F_t for bolts subjected to combined shear and tension is given in terms of the actual shear stress f_v in Table 8.3.

TABLE 8.3   ALLOWABLE TENSILE STRESS FOR BOLTS IN COMBINED SHEAR AND TENSION

Type	F_t
A325N	$55 - 1.8f_v \leq 44$
A325X	$55 - 1.4f_v \leq 44$
A490N	$68 - 1.8f_v \leq 54$
A490X	$68 - 1.4f_v \leq 54$

When a friction-type bolted connection is subjected to tension in addition to shear, the clamping force is reduced and therefore the allowable shear stress F_v must also be reduced to take into account the loss of pretension. The reduction factor given by AISCS 1.6.3 is $1 - f_t A_b / T_b$ and the allowable shear stress in the friction-type connection is given by

$$F_v = \left(1 - \frac{f_t A_b}{T_b}\right)(17.5) \qquad \text{for A325 bolts} \tag{8.9}$$

$$F_v = \left(1 - \frac{f_t A_b}{T_b}\right)(22) \qquad \text{for A490 bolts} \tag{8.10}$$

where f_t is the actual tensile stress in the bolt, A_b is the cross-sectional area of the bolt, and T_b is the specified pretension load of the bolt given in Table 8.4. According to AISCS 1.23.5, all A325 and A490 bolts should be tightened by the minimum pretension load given in Table 8.4.

TABLE 8.4 MINIMUM BOLT PRETENSIONS (EQUAL
TO 70 PERCENT OF THE MINIMUM TENSILE STRENGTH
OF BOLTS ROUNDED OFF TO NEAREST KIP) (AISCS
1.23.5)

Bolt size (in.)	A325 Bolts	A490 Bolts
$\frac{1}{2}$	12	15
$\frac{5}{8}$	19	24
$\frac{3}{4}$	28	35
$\frac{7}{8}$	39	49
1	51	64
$1\frac{1}{8}$	56	80
$1\frac{1}{4}$	71	102
$1\frac{3}{8}$	85	121
$1\frac{1}{2}$	103	148

In the case of combined dead and live loads and wind (or seismic) load, the constants in Table 8.3 shall be increased 33 percent, but the coefficients applied to f_v shall be kept the same. The reduced allowable shear stress given by Eq. (8.9) or (8.10) shall also be increased 33 percent.

8.2.5 Examples

Example 1

For the connection shown in Fig. 8.6, find the number of $\frac{7}{8}$-in. A490 bearing-type bolts, assuming that threads are excluded from the shear planes. Plates are made of A36 steel with an ultimate stress of $F_u = 58$ ksi.

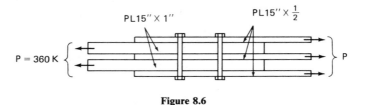

PL15″ × 1″   PL15″ × $\frac{1}{2}$

P = 360 K   P

Figure 8.6

Solution The free-body diagram for one bolt is shown in Fig. 8.7. Denoting the number of bolts by N, maximum shear in the bolt is $P/3N$. The allowable shear stress for the bolts is $F_v = 40$ ksi (Table 8.2). For finding the cross-sectional area of one bolt A_b, the table of square and round bars in the AISCM may be used.

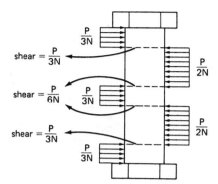

Figure 8.7

$$A_b = 0.60 \text{ in.}^2$$

$$A_b F_v = (0.60)(40) = 24 \text{ Kips} = \frac{P}{3N} = \frac{360}{3N} \Rightarrow N = 5$$

The allowable bearing stress from Eq. (8.1) is

$$F_p = 1.5 F_u = 1.5(58) = 87 \text{ ksi}$$

The bearing stress on the plate 15 in. × $\frac{1}{2}$ in. is larger than that on the plate 15 in. × 1 in. Hence, check bearing on PL 15 in. × $\frac{1}{2}$ in.

$$f_p = \frac{P/3N}{dt} = \frac{P}{3Ndt} = \frac{360}{3(5)(\frac{7}{8})(\frac{1}{2})} = 54.9 \text{ ksi} < F_p \qquad \underline{\underline{\text{O.K.}}}$$

Therefore, use $\boxed{N = 5}$.

Example 2

A built-up girder is made of a W21 × 93 section and two C12 × 30 sections, as shown in Fig. 8.8. In each plane of connection, the three sections are connected by four bearing-type A490 high-strength bolts of 0.5-in. diameter. Determine the pitch (longitudinal spacing of the bolts) where the shear force on the girder is $V = 100$ Kips. Assume that bolt threads are excluded from the shear planes.

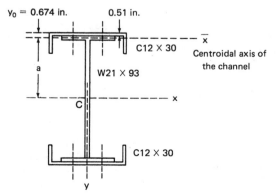

Figure 8.8

Solution   Properties of W21 × 93: $d = 21.62$ in.

$$I_x = 2070 \text{ in.}^4 \qquad t_f = 0.93 \text{ in.} \qquad b_f = 8.42 \text{ in.}$$

Properties of C12 × 30: $A = 8.82$ in.^{2}

$$y_o = 0.674 \text{ in.} \qquad I_{\bar{x}} = 5.14 \text{ in.}^4 \qquad t_w = 0.51 \text{ in.}$$

The bolts must carry the longitudinal shear on the plane between the channel and the W shape. The total shear force between the channel and the flange of the W shape for a unit length of the beam is equal to

$$q = \frac{VQ}{I}$$

where Q is the first moment of the area of the channel about the centroidal axis of the built-up section and I is the moment of inertia of the built-up section.

$$a = \frac{21.62}{2} + 0.51 - 0.674 = 10.646 \text{ in.}$$

$$I = 2070 + 2[5.14 + 8.82(10.646)^2] = 4079.55 \text{ in.}^4$$

$$Q = Aa = (8.82)(10.646) = 93.90 \text{ in.}^3$$

$$q = \frac{VQ}{I} = \frac{(93.90)(100)}{4079.55} = 2.30 \text{ Kips/in.}$$

Allowable shear stress for bearing-type high-strength A490 bolts is (Table 8.2)

$$F_v = 40 \text{ ksi}$$

Cross-sectional area of one bolt is

$$A_b = \frac{\pi d^2}{4} = \frac{\pi (0.5)^2}{4} = 0.1963 \text{ in.}^2$$

Denoting the longitudinal spacing of the bolts by p, we find that the total horizontal shear between the channel and the flange of the W shape over a length of p is qp. The shear capacity of the bolts over the same length is $2A_bF_v$. Therefore,

$$2A_bF_v = qp$$

$$p = \frac{2A_bF_v}{q} = \frac{2(0.1963)(40)}{2.30} = 6.83 \text{ in.}$$

According to AISCS 1.18.2.3, the maximum longitudinal spacing of bolts connecting two rolled shapes in contact with each other shall not be greater than 24 in. Therefore, use $\boxed{P = 6\frac{3}{4} \text{ in.}}$.

Example 3

Using A490 bearing-type connection, find the bolt size for the connection shown in Fig. 8.9. The column is a W14 × 99, and the thickness of the gusset plate is one inch. Both column and gusset plate are made of A36 steel. Bolt threads are excluded from the shear plane.

Solution   First, we must find the centroid of the bolt group (point C in Fig. 8.9). Denoting the cross-sectional area of one bolt by A_b, the location of the centroid is determined by finding the distances $\bar{x}$ and $\bar{y}$ shown in Fig. 8.9.

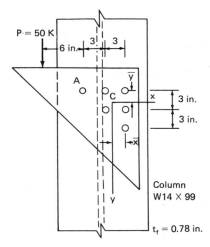

Figure 8.9

$$\bar{x} = \frac{6A_b + 2(3)A_b}{6A_b} = 2 \text{ in.}$$

$$\bar{y} = 2 \text{ in.}$$

The eccentricity of the load from the centroid is

$$e = -(6 + 3 + 3 - 2) = -10 \text{ in.}$$

and the magnitude of the corresponding couple is

$$M = Pe = -(50)(10) = -500 \text{ K-in.}$$

The horizontal and vertical components of the shear force R_j acting on the bolt j can be calculated from Eqs. (8.7) and (8.8).

$$P_x = 0 \qquad P_y = 50 \text{ K} \qquad N = 6$$

$$\sum_{i=1}^{N} r_i^2 = 2(4^2 + 2^2) + (1^2 + 1^2) + 2(2^2 + 1^2) + (2^2 + 2^2) = 60 \text{ in.}^2$$

$$R_{jx} = \frac{P_x}{N} - \frac{My_j}{\sum_{i=1}^{N} r_i^2} = \frac{500}{60} y_j = \frac{25}{3} y_j$$

$$R_{jy} = \frac{P_y}{N} + \frac{Mx_j}{\sum_{i=1}^{N} r_i^2} = \frac{50}{6} - \frac{500}{60} x_j = \frac{25}{3}(1 - x_j)$$

The most stressed bolt is bolt A, for which $x_A = -4$ in. and $y_A = -2$ in. Therefore, the resultant shear force acting on this bolt is

$$R_A = \tfrac{25}{3}\sqrt{y_A^2 + (1 - x_A)^2} = \tfrac{25}{3}\sqrt{2^2 + 5^2} = 44.88 \text{ Kips}$$

The design of bolts is commonly based on the most stressed bolt. The allowable shear stress for A490X bearing-type connections is $F_v = 40$ ksi (Table 8.2). Thus, the required cross-sectional area for one bolt is

$$\text{Required } A_b = \frac{R_A}{F_v} = \frac{44.88}{40} = 1.12 \text{ in.}^2$$

Try $1\frac{1}{4}$-in. bolts, $A_b = 1.23$ in.^{2} Noting that the bolts are in single shear and the thickness of the column flange ($t_f = 0.78$ in.) is less than the thickness of the gusset plate, we should check the bearing for $t = 0.78$ in.

$$f_p = \frac{R_A}{dt} = \frac{44.88}{(1.25)(0.78)} = 46.03 \text{ ksi} < F_p = 1.5F_u = 1.5(58) = 87 \text{ ksi} \qquad \underline{\underline{\text{O.K.}}}$$

$$\boxed{\text{USE } 1\frac{1}{4}\text{-in. bolts}}$$

8.3 WELDED CONNECTIONS

8.3.1 Types of Welding

In welded connections, different elements are connected by heating their surfaces to a plastic or fluid state. Notwithstanding the availability of both gas and arc welding, welded connections in steel structures are ordinarily done by arc welding. To obtain satisfactory connections, additional metal is used for joining different elements. In electric arc welding, the additional material is a metallic rod, which is used as the electrode. In this type of welding, the electric arc produced between the elements being welded and the electrode heats the elements and the electrode to the melting point. This transformation of electrical energy into thermal energy and the resulting high temperature (up to 10,000°F) causes the metallic electrode to melt off into the joint. Small droplets of the molten metallic electrode are in fact driven onward to the joint. Thus, overhead welding is possible by electric arc welding.

Molten steel must be protected from the surrounding air; otherwise, gases contained in the molten steel can combine chemically with oxygen and nitrogen in the air. This chemical reaction leaves small pockets of gases in the weld after it has cooled down, making it porous. The resulting weld will be brittle with very little resistance to corrosion. To prevent this undesirable brittleness of the weld, two types of arc welding are commonly used. One is called Shielded Metal Arc Welding (with acronym SMAW) and the other is Submerged (or hidden) Arc Welding (with acronym SAW).

In SMAW, the weld is protected by using an electrode covered with a layer of mineral compounds. Melting of this layer during the welding produces an inert gas encompassing the weld area. This inert gas shields the weld by preventing the

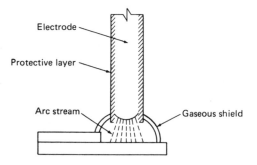

Figure 8.10 Shielded metal arc welding (SMAW).

molten metal from having contact with the surrounding air (Fig. 8.10). The protecting layer of the electrode leaves a slag after the mold has cooled down. The slag can be removed by peening and brushing.

In SAW, the surface of the weld and the electric arc is covered by some granular fusible material and thus is protected from the surrounding air. In this method, a bare metal electrode is used as filler material. Compared with SMAW, SAW welds provide deeper penetration, and this fact is reflected in the allowable shear stress values recommended by AISCS (see Sec. 8.3.3). Also, SAW welds show good ductility and corrosion resistance and high impact strength.

8.3.2 Advantages of Welding

1. In welded connections, in general, fewer pieces are used. This will speed up the detailing and fabrication process.
2. In welded connections, gusset and splice plates may be eliminated. Bolts or rivets are not needed either. Thus, the total weight of a welded steel structure is somewhat less than that of the corresponding bolted structure.
3. Connecting unusual members (such as pipes) is easier by welding than by bolting.
4. Welding provides truly rigid joints and continuous structures.

One possible drawback of welding is the need for careful execution and supervision. For this reason, welding is sometimes done in the shop and bolting in the field. In other words, shop-welding is complemented by the bolting in the field.

8.3.3 Types of Welds

The two common types of welds in welded steel structures are *groove welds* and *fillet welds*. Fillet welds are much more popular in structural steel design than groove welds. Two different types of groove welds are shown in Fig. 8.11. They are the partial penetration (single vee) groove weld and full penetration (double vee) groove weld. Groove welds can be used when the pieces to be connected can be lined up in the same plane.

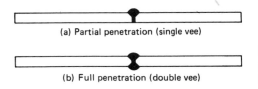

(a) Partial penetration (single vee)

(b) Full penetration (double vee)

Figure 8.11 Groove welds. (a) Partial penetration (single vee); (b) full penetration (double vee).

Fillet welds are shown in Fig. 8.12. Depending on the direction of the applied load and the line of the fillet weld, fillet welds are classified as longitudinal fillet weld or transverse fillet weld. In the former, the shear force to be transferred is parallel to the weld line; in the latter, the force to be transmitted is perpendicular to the weld line.

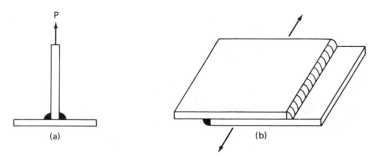

Figure 8.12 Fillet welds. (a) Transverse fillet; (b) longitudinal fillet.

Fillet welds can be either equal-leg or unequal-leg, as shown in Fig. 8.13. The intersection point of the original faces of the steel elements being connected is called the *root* of the weld. The surface of the weld should have a slight convexity. In computation of the strength of the weld, however, this convexity is not taken into

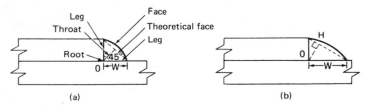

Figure 8.13 Fillet welds. (a) Equal leg; (b) unequal leg.

account and the theoretical flat surface is used. A convex surface for a fillet weld is clearly superior to a concave surface. When the weld cools down, it shrinks. This shrinkage causes surface tension in concave welds and surface compression in convex welds (Fig. 8.14). The concave surface in tension tends to crack, causing the separation of the weld from the faces of the pieces being connected. The normal distance from the root to the theoretical face of the weld is called the *throat* of the weld.

Figure 8.14 Fillet welds with convex and concave surfaces. (a) Convex surface under compression; (b) concave surface under tension.

Experiments performed on fillet welds indicate that they are weaker in shear than in tension and compression. Also, equal-leg fillet welds fail in shear through the throat (at angles of about 45 deg with the legs of the weld). For equal-leg fillet welds, the relation between the dimensions of the leg w and the throat t is

$$t = 0.707w \tag{8.11}$$

Thus, shear stress is the controlling factor in the design of fillet welds; it is customarily calculated by dividing the force P acting on the weld by the effective throat area of the weld. The effective throat area is computed by multiplying the throat thickness by the length of the fillet weld. This method of finding average shear stress is used for both longitudinal and transverse fillet welds.

The AISCS does not recognize the fact that transverse fillet welds are stronger (about one-third) than the longitudinal fillet welds. Experiments indicate that transverse fillet welds fail in planes somewhat different from the 45-deg plane. The size of a fillet weld is indicated by the size of its leg. For example, a $\frac{7}{8}$-in. fillet weld means a fillet weld with a leg size of $w = \frac{7}{8}$ in.

As pointed out previously, SAW welds provide a deeper penetration than the SMAW welds. This fact is recognized in the AISCS by allowing a larger throat area for SAW welds. According to AISCS 1.14.6.2, for SAW welds with sizes $\frac{3}{8}$ in. or smaller, the effective throat thickness is taken as equal to leg size, and for fillet welds greater than $\frac{3}{8}$ in., the effective throat is taken equal to the theoretical throat thickness plus 0.11 in.

$$t = \begin{cases} w & \text{for } w \leqslant \frac{3}{8} \text{ in.} \\ 0.707w + 0.11 & \text{for } w > \frac{3}{8} \text{ in.} \end{cases} \qquad (8.12)$$

8.3.4 Allowable Stress on Welds

Allowable stresses for different types of welds are given in Table 1.5.3 of the AISCS. The allowable shear stress on the effective area of fillet welds is 0.30 times the nominal tensile strength of the weld metal, but the shear stress in the base metal may not be larger than 0.40 times the yield stress of base metal. Electrodes are designated as E60, E70, and so on, where E stands for electrode and the number following the letter E is the minimum tensile strength of the weld, in ksi.

8.3.5 Minimum and Maximum Sizes of Fillet Welds

The minimum size of fillet weld is determined on the basis of the thicker of the two pieces connected, as given in Table 8.5 (AISCS 1.17.2).

The maximum size of fillet welds along edges of an element less than $\frac{1}{4}$ in. thick is equal to the thickness of the element. Along edges of an element with thickness of $\frac{1}{4}$ in. or more, the maximum size of the fillet weld is equal to the thickness of the element minus $\frac{1}{16}$ in. (AISCS 1.17.3).

TABLE 8.5   MINIMUM SIZE OF FILLET WELDS (AISCS TABLE 1.17.2A)

Thickness of thicker part connected (in.)	Minimum leg size of fillet weld (in.)
To $\frac{1}{4}$ inclusive	$\frac{1}{8}$
Over $\frac{1}{4}$ to $\frac{1}{2}$	$\frac{3}{16}$
Over $\frac{1}{2}$ to $\frac{3}{4}$	$\frac{1}{4}$
Over $\frac{3}{4}$	$\frac{5}{16}$

8.3.6 Minimum Length of Fillet Weld

The minimum effective length of a fillet weld shall be at least equal to four times its nominal size; otherwise weld size shall be limited to $\frac{1}{4}$ of its effective length (AISCS 1.17.4).

The effective length of any segment of intermittent fillet weld shall be at least 1.5 in. and four times the weld size (AISCS 1.17.5).

8.3.7 Eccentrically Loaded Welded Connections

The analysis of eccentrically loaded welded connections is similar to the analysis of eccentrically loaded bolts, as covered in Sec. 8.2.3. Consider a bracket welded to the flange of a column, as shown in Fig. 8.15. The size of the fillet weld is assumed

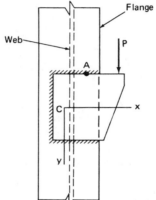

Figure 8.15

to be the same on the three edges of the bracket. Suppose that the connection is subjected to horizontal and vertical shears of F_x and F_y that produce a moment M_c about the centroid C of weld lines. Thus, the total components of shear force per unit length at point A with coordinates x_A and y_A are

$$q_x = q_{xp} + q_{xm} = \frac{F_x}{L} - \frac{M_c y_A}{J} \tag{8.13}$$

$$q_y = q_{yp} + q_{ym} = \frac{F_y}{L} + \frac{M_c x_A}{J} \tag{8.14}$$

Finally, the resultant shear force per unit length of weld at point A is

$$q_A = \sqrt{q_x^2 + q_y^2} \tag{8.15}$$

Note that in Eqs. (8.13) and (8.14) it is assumed that the positive x-axis points to the right and the positive y-axis points downward.

Denoting the total length of fillet weld by L, the horizontal shear force per unit length of the weld at any point A due to the direct shear F_x is

$$q_{xp} = \frac{F_x}{L}$$

and the vertical shear force per unit length of the weld due to the direct shear F_y is

$$q_{yp} = \frac{F_y}{L}$$

The shear forces per unit length at point A (with coordinates x_A and y_A) due to couple M_c are

$$q_{xm} = -\frac{M_c y_A}{J}$$

$$q_{ym} = \frac{M_c x_A}{J}$$

where

$J = I_x + I_y$
J = polar moment of inertia of the weld of unit width about point C
I_x = moment of inertia of the weld of unit width about the x-axis
I_y = moment of inertia of the weld of unit width about the y-axis

8.3.8 Examples

Example 4

A welded built-up girder is made of a W24 × 94 and a C12 × 25 section, as shown in Fig. 8.16. The maximum shear force in the girder is $V = 150$ Kips. Design the intermittent fillet weld for connecting the two sections, as shown in Fig. 8.16. Use submerged arc welding (SAW) and E70 electrodes.

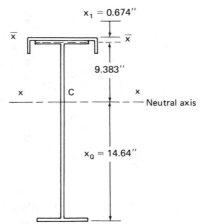

Figure 8.16

Solution   Properties of W24 × 94:

$$A = 27.7 \text{ in.}^2 \qquad b_f = 9.065 \text{ in.} \qquad t_f = 0.875 \text{ in.}$$

$$I_x = 2700 \text{ in.}^4 \qquad d = 24.31 \text{ in.}$$

Properties of C12 × 25:

$$A = 7.35 \text{ in.}^2 \qquad d = 12.0 \text{ in.} \qquad t_w = 0.387 \text{ in.} \qquad I_{\bar{x}} = 4.47 \text{ in.}^4$$

We first locate the neutral (centroidal) axis of the cross section of the built-up girder (x-axis in Fig. 8.16).

$$x_o = \frac{\sum yA}{\sum A} = \frac{(7.35)(24.31 + 0.387 - 0.674) + 27.7(24.31)/2}{7.35 + 27.7} = 14.64 \text{ in.}$$

The moment of inertia of the built-up section about the x-axis:

$$I_x = 2700 + 27.7(14.64 - 24.31/2)^2 + 4.47 + 7.35(9.383)^2 = 3522.62 \text{ in.}^4$$

The two lines of fillet weld must carry the longitudinal shear on the plane between the channel and the W shape. The total shear force between the channel and the flange of the W shape for a unit length of the beam is equal to

$$q = \frac{VQ}{I_x}$$

where Q is the first moment of the area of the channel about the centroidal axis of the built-up section.

$$Q = (7.35)(9.383) = 68.97 \text{ in.}^3$$

$$q = \frac{(150)(68.97)}{3522.62} = 2.94 \text{ K/in.}$$

- Minimum size of the fillet weld (Sec. 8.3.5): $\frac{5}{16}$ in.

Use $\boxed{w = \frac{5}{16} \text{ in.}}$

- Maximum size of the fillet weld along the flange edge of the W shape (Sec. 8.3.5):

$$0.875 - \tfrac{1}{16} = 0.813 \text{ in.}$$

- The minimum length of a segment of intermittent weld (Sec. 8.3.6): the larger of $4w = 1.25$ and 1.5 in.

Use $\boxed{L_1 = 1.5 \text{ in.}}$ length of a segment of weld

The effective throat thickness (Eq. 8.12):

$$t = w = \tfrac{5}{16} \text{ in.}$$

The allowable shear stress of the weld is

$$F_y = 0.30(70) = 21 \text{ ksi}$$

   If the longitudinal spacing of intermittent welds is denoted by a, the total horizontal shear between the channel and the flange of the W shape over a length of a is aq. The shear capacity of the two lines of fillet weld over the same length is

$2L_1F_vw$. Therefore,

$$aq = 2L_1F_vw$$

$$a = \frac{2L_1F_vw}{q} = \frac{2(1.5)(21)(5/16)}{2.94} = 6.7 \text{ in.}$$

The maximum longitudinal spacing of intermittent welds connecting two rolled shapes in contact (AISCS 1.18.2.3): $a_{max} = 24$ in.

Use $\boxed{a = 6.5 \text{ in.}}$

Example 5

Determine the size of the submerged arc fillet weld for the connection of a C12 × 25 beam to a W14 × 120 column as shown in Fig. 8.17. Note that the beam is connected to the column at its end and at the edge of the column flange. Use E70 electrodes.

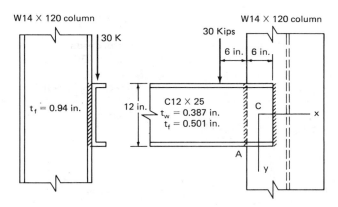

Figure 8.17

Solution   We design the weld on the basis of the maximum shear stress, which is at point A (Fig. 8.17).

$$x_A = -3 \text{ in.} \qquad y_A = 6 \text{ in.}$$

$$F_x = 0 \qquad F_y = 30 \text{ Kips} \qquad M_c = -30(9) = -270 \text{ K-in.}$$

Total length of the weld: $L = 2(12) = 24$ in.

$$I_x = 2(\tfrac{1}{12})(1)(12)^3 = 288 \text{ in.}^4/\text{in.}$$

$$I_y = 2(12)(3)^2 = 216 \text{ in.}^4/\text{in.}$$

$$J = I_x + I_y = 504 \text{ in.}^4/\text{in.}$$

$$q_x = \frac{F_x}{L} - \frac{M_c y_A}{J} = \frac{(270)(6)}{504} = 3.21 \text{ K/in.}$$

$$q_y = \frac{F_y}{L} + \frac{M_c x_A}{J} = \frac{30}{24} + \frac{(270)(3)}{504} = 2.86 \text{ K/in.}$$

$$q_A = \sqrt{q_x^2 + q_y^2} = \sqrt{3.21^2 + 2.86^2} = 4.30 \text{ K/in.}$$

For SAW and E70 electrode, the allowable shear stress is

$$F_v = 0.30(70) = 21 \text{ ksi}$$

The required effective thickness of the weld throat is

$$t = \frac{q_A}{F_v} = \frac{4.30}{21} = 0.205 \text{ in.} = \frac{3.3}{16} \text{ in.} < \frac{3}{8} \text{ in.}$$

Therefore,

$$w = t = \tfrac{1}{4} \text{ in.}$$

- The minimum size of the fillet weld for 0.94-in.-thick plate (column flange) (Sec. 8.3.5): $\frac{5}{16}$ in.
- The maximum size of the fillet weld along the edge of the beam web (Sec. 8.3.5): $0.387 - \frac{1}{16} = 0.325$ in.

Use $\boxed{\tfrac{5}{16} \text{ in.}}$ SAW fillet weld.

Example 6

To make a moment-resisting joint in a rigid frame, a W21 × 147 girder is directly welded to a W14 × 145 column as shown in Fig. 8.18. The beam, made of A36 steel, has an end reaction of 194 Kips and an end bending moment of 210 K-ft. Using E70 electrodes, find the size of the submerged arc fillet weld (SAW) for the connection of beam flanges and web to the column flange.

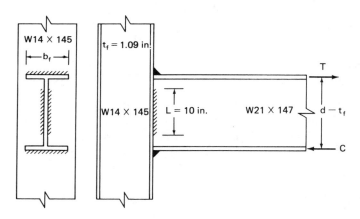

Figure 8.18

Solution Assume that the web welds carry all of the shear and the flange welds carry all of the moment.

$$V = 194 \text{ Kips} \qquad M = 210 \text{ K-ft}$$

Properties of the beam:

$$d = 22.06 \text{ in.} \qquad t_f = 1.15 \text{ in.} \qquad t_w = 0.72 \text{ in.} \qquad b_f = 12.51 \text{ in.}$$

The allowable shear stress of the weld: $F_v = 0.30(70) = 21$ ksi.

(a) Web Welds
If the effective throat thickness is denoted by t, the shear stress in the weld is

$$f_v = \frac{V}{2tL} \leq F_v$$

The required throat thickness:

$$t = \frac{V}{2LF_v} = \frac{194}{2(10)(21)} = 0.462 \text{ in.} > \tfrac{3}{8} \text{ in.}$$

For $w > \tfrac{3}{8}$ in.:

$$t = 0.707w + 0.11 = 0.462 \text{ in.}$$

$$w = 0.5 \text{ in.}$$

Minimum weld size (Sec. 8.3.5): $w_{min} = \tfrac{5}{16}$ in.

Use $\boxed{w = \tfrac{1}{2} \text{ in.}}$

(b) Flange welds
The force T to be carried by the weld is found by dividing the beam end bending moment by the center-to-center distance of flanges:

$$T = \frac{M}{d - t_f}$$

$$f_v = \frac{T}{tb_f} = \frac{M}{tb_f(d - t_f)} \leq F_v$$

$$t = \frac{M}{b_f(d - t_f)F_v} = \frac{(210)(12)}{(12.51)(22.06 - 1.15)(21)} = 0.459 \text{ in.} > \frac{3}{8} \text{ in.}$$

$$t = 0.707w + 0.11 = 0.459 \text{ in.}$$

$$w = 0.49 \text{ in.} > w_{min} = \tfrac{5}{16} \text{ in.}$$

Use $\boxed{w = \tfrac{1}{2} \text{ in.}}$

8.4 MICROCOMPUTER-AIDED DESIGN OF BUILDING CONNECTIONS

8.4.1 Introduction

In this section, an interactive MicroCAD system is presented for design of beam-column connections in steel building, called STEELCON [13–17]. In addition to designing the connections, STEELCON can display and plot any isometric views as well as three orthographic views, that is, front face, side face, and top face views of the connection. The orthographic views allow for all pertinent dimensions to be displayed on the screen or plotted on a printer.

At present, STEELCON can design eight different types of simple (Type 2) and three different types of moment-resisting (Type 1) connections. Design of different types of connections is based on the AISC specification. Under no circumstances does the program violate AISCS or allow the user to enter data in violation of the AISC code.

STEELCON is designed to be highly interactive. It often allows the user to correct his mistakes without causing the program to terminate. Wherever the user is prompted to input something, a check is made to see whether the inputted data is within the permissible range. Some of the prompts to the user require a yes or no answer. For this type of prompt, if the user types either y or Y, the answer is assumed to be yes; otherwise, the answer is assumed to be no. For all other prompts, if the data fall within the allowable range, the program will continue. If the data are outside the allowable range, however, an error message will be given and an audible signal will be heard. The user is then asked to reenter the data which caused the problem.

8.4.2 Program Structure

STEELCON presently occupies two floppy disks and consists of a series of subprograms linked together using BASIC's CHAIN command. The limited random access memory (RAM) of the computer necessitates the chaining of the subprograms.

The first subprogram used in the program is START.BAS. This is the starting subprogram, which determines what type of connection is desired, the sections used, and the load applied at the connection. The START.BAS subprogram CHAIN's to the subprogram that actually does the connection design. There are several different design subprograms due to the wide variety of connection types. For example, one subprogram designs all the beam-column connections that use angles. Another subprogram designs a simple type of connection using a single plate welded to the column flange and bolted to the beam web. After the connection design subprogram performs its task, the ARRAY.BAS subprogram is loaded into the computer's RAM. This subprogram fills the main graphic arrays, to be described in the following section. After this subprogram performs its task, control is passed to subprogram GRAPH.BAS, which is used to display the connection graphically onto the terminal screen or plot it on a printer. A final subprogram that may be referenced is DIM.BAS. This subprogram is accessed by GRAPH.BAS when the user wishes to see the dimensions and designations of the connection displayed. The DIM.BAS and GRAPH.BAS subprograms are often toggled back and forth as the user requests different displays and dimensions.

Another important concern with the computer's RAM is trying to minimize the memory required for all variables and arrays. When the computer loads a program into its RAM, the program itself takes up a considerable amount of the available RAM. The remaining RAM is available for variable and array storage. Measures must be taken to limit the memory requirements of the arrays and variables. One way to do this is to use integer variables and arrays whenever possible. In BASIC, integer variables use two bytes of memory while real variables use four bytes. By trying to use integers as much as possible, substantial memory savings can be gained. The memory savings, however, is achieved sometimes at the expense of a slightly longer code. For example, one array used in the program is the IS array. The IS array, whose significance will be discussed later, should contain the following possible values; -1, -0.5, 0, 0.5, and 1. Since to store 0.5 and -0.5 requires the array to be real, the total number of bytes required to store the 30 by

4 array would be 480. However, if instead of storing the above numbers, the values -10, -5, 0, 5, and 10 are stored, the array can be stored as an integer array. This allows a savings of 240 bytes of memory. The only modification required in the code is to multiply the values in IS by 0.1 whenever the array is referenced.

8.4.3 Graphics Data Structures

8.4.3.1 Face frame representation of the connection.

The graphic portion of the program enables the user to view the designed connection on the terminal screen. The graphic display of the connection is a three-dimensional face frame representation. Setting up the graphics in a three-dimensional fashion allows for the connection to be viewed from any angle by referencing the same data base. This allows for a more efficient storage of information than if each set of viewing angle data is contained in a separate array. It also allows the user to choose any viewing angle desired for observing the connection.

The data structure used for the graphics subprogram is based on three main arrays. The CR array is a two-dimensional array with three columns containing information about X, Y, and Z coordinates of the nodal points in the connection. The CRF array is simply the CR array after it has been transformed by a combination of rotations about the X-, Y-, and Z-axes. The last main graphic array is the ICS array. The ICS array is a one-dimensional integer array that contains information concerning the order of connecting nodes in the CRF array.

The CR and ICS arrays are filled in the START.BAS and ARRAY.BAS subprograms. In START.BAS, the arrays obtain the appropriate values for all I sections selected for design. The remainder of the array values, pertaining to all other parts of the connection, are calculated and stored in the ARRAY.BAS subprogram.

8.4.3.2 Graphic representation of rolled sections.

The calculation of the values to be stored in the CR array for each I section, T section, angle, and plate is based on the same principle. After an element has been designed by the program and okayed by the user, the nodal points describing that element are placed in the CR array. Every element used in the design is initially set up about the origin. Every I section, T section, angle, or plate is described with five parameters. The five parameters are their depth (D), width (B), flange thickness (TF), web thickness (TW), and length (L), as shown in Fig. 8.19. With these five parameters, any of the above-mentioned sections can be described in terms of the X, Y, and Z coordinates of its nodal points.

The process used to describe a plate is slightly different, because there is no need for flange thickness and web thickness parameters. However, the program still considers these parameters in the plate description. For calculation of nodal point coordinates, to be discussed in the following paragraphs, these two parameters are set to zero for every nodal point in the plate description. This in effect eliminates their contribution to the final coordinate location while still keeping the same data structure form as all the other sections. By doing this, the same array can be used for all sections, rather than having to define an array specifically for plates.

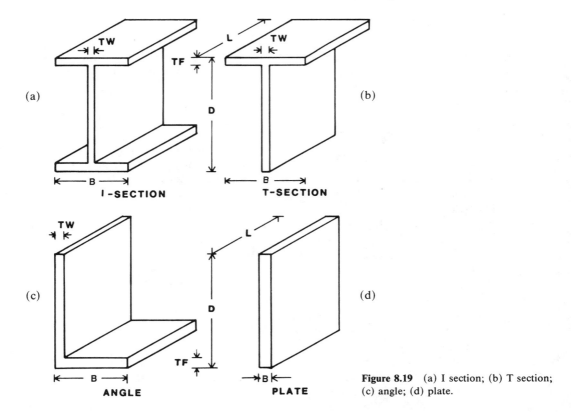

Figure 8.19 (a) I section; (b) T section; (c) angle; (d) plate.

In order to calculate the X and Y coordinates of a node (the first two columns of the CR array), a two-dimensional integer array, IS, is used. The IS array contains information concerning the relative magnitudes of the B, TW, D, and TF parameters. These values are used to find the X and Y coordinates of various nodal points for the four different sections, as seen in Table 8.6. By determining beforehand what type of section is being set up, the proper starting location in the IS array is referenced. The number of nodal points needed for each description is 24 for I sections, 16 for T sections, 12 for angles, and 8 for plates, as shown in Figs. 8.20 through 8.23.

By simply multiplying the values in the first two columns of the IS array by B and TW of the section, respectively, and dividing the sum by 10, the X coordinates of the nodal points of the section are found. Similarly, by multiplying the values in the last two columns of the IS array by D and TF of the section, respectively, and dividing the sum by 10, the Y coordinates of the nodal points of the section are obtained. The reason for the division by 10 is explained in the previous section. The Z coordinates of the nodal points of each section, that is, CR(, 3), are either a positive or a negative half of the total length of the section, depending on which end of the section is being set up.

TABLE 8.6 DIVISION OF THE "IS" ARRAY

Section	Point no.	Multipliers for finding the X coordinate		Multipliers for finding the Y coordinate	
		B	TW	D	TF
I section	1	−5	0	−5	0
	2	−5	0	−5	10
	3	0	−5	−5	10
	4	0	−5	5	−10
	5	−5	0	5	−10
	6	−5	0	5	0
	7	5	0	5	0
	8	5	0	5	−10
	9	0	5	5	−10
	10	0	5	−5	10
	11	5	0	−5	10
	12	5	0	−5	0
Angle	1	0	0	0	0
	2	0	0	10	0
	3	0	10	10	0
	4	0	10	0	10
	5	10	0	0	10
	6	10	0	0	0
T section	1	0	−5	−10	5
	2	0	−5	0	−5
	3	−5	0	0	−5
	4	−5	0	0	5
	5	5	0	0	5
	6	5	0	0	−5
	7	0	5	0	−5
	8	0	5	−10	5
Plate	1	−5	0	−5	0
	2	−5	0	5	0
	3	5	0	5	0
	4	5	0	−5	0

Note: Refer to Figs. 8.20–8.23 for section descriptions.

The length of all T sections, angles, and plates is calculated in the appropriate design subprogram before the CR array is set up. The actual length of the I section is immaterial to the connection display. However, for display purposes, a determination of the length of I sections must be made. The length selected for each I section depends on the type of structural member: 36 in. for columns and 12 in. for beams. The selected length for beams and columns is based on their appearance in the final connection display. The initial unmagnified drawing of the front, side, and top views each has its lower window corner at −18 in. in the Y direction and −24 in.

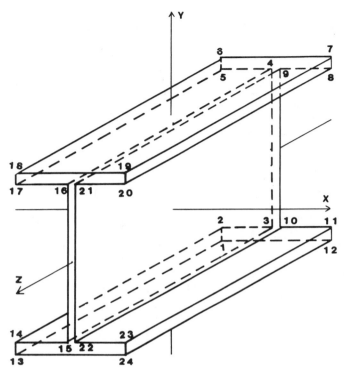

Figure 8.20

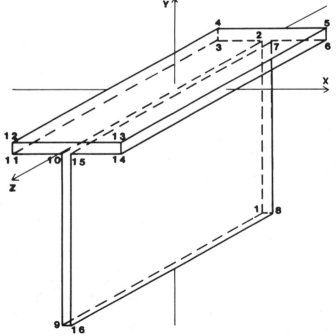

Figure 8.21

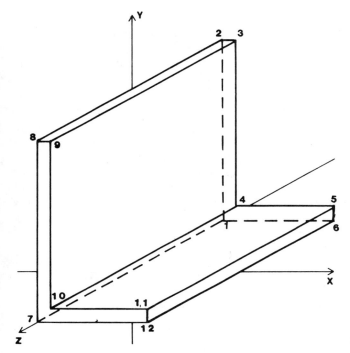

Figure 8.22

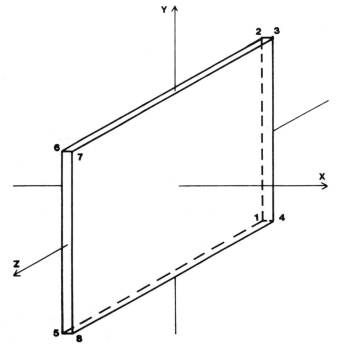

Figure 8.23

in the X direction. The upper right window corner is selected at 18 in. in the Y direction and 24 in. in the X direction. The above-mentioned selection of the lengths for the sections allows all I sections to fit within the limits of the window without being clipped. In addition, sufficient room will be left for displaying dimensions and designations without infringing upon the connection drawing.

As stated before, initially all the sections are set up about the origin. In order to actually display the connection, however, these sections have to be moved and rotated into their final positions. These transformations are accomplished by using the various three-dimensional transformation matrices given in Table 8.7. By multiplying the original points in the CR array, set up about the origin, by the final transformation matrix, the final values for the CR array are obtained. The transformations used are X, Y, and Z rotations about their respective axes, and X, Y, and Z translations along their respective axes. Each of the rotational directions has a separate transformation matrix handling that transformation. The translation along the X-, Y-, and Z-axes can be handled by one matrix only.

TABLE 8.7   TRANSFORMATION MATRICES

Rotation about X-axis by α Rotation about Y-axis by β

$$\begin{bmatrix} 1 & 0 & 0 & 0 \\ 0 & \cos(\alpha) & -\sin(\alpha) & 0 \\ 0 & \sin(\alpha) & \cos(\alpha) & 0 \\ 0 & 0 & 0 & 1 \end{bmatrix} \qquad \begin{bmatrix} \cos(\beta) & 0 & \sin(\beta) & 0 \\ 0 & 1 & 0 & 0 \\ -\sin(\beta) & 0 & \cos(\beta) & 0 \\ 0 & 0 & 0 & 1 \end{bmatrix}$$

Rotation about Z-axis by γ X, Y, and Z translation

$$\begin{bmatrix} \cos(\gamma) & -\sin(\gamma) & 0 & 0 \\ \sin(\gamma) & \cos(\gamma) & 0 & 0 \\ 0 & 0 & 1 & 0 \\ 0 & 0 & 0 & 1 \end{bmatrix} \qquad \begin{bmatrix} 1 & 0 & 0 & 0 \\ 0 & 1 & 0 & 0 \\ 0 & 0 & 1 & 0 \\ TX & TY & TZ & 1 \end{bmatrix}$$

For example, to obtain a final transformation of 90 degrees rotation about the X-axis, 90 degrees rotation about the Y-axis, and +5 units translation in the X direction, three matrices must be set up and subsequently multiplied. Then, the points in the CR array corresponding to the transformed section are multiplied by the final transformation matrix. It is important to consider the order in which the transformations are performed. Because multiplication of matrices is not commutative, if the transformation order is incorrect, the section will end up in the wrong final position, as shown in the example of Fig. 8.24.

The final connection display is based on the CRF array, not the CR array. The only difference between the CR array and the CRF array is that the latter array contains the final locations of the nodal points before they are displayed on the screen. The CRF array is set up by multiplying the final CR array by a transformation matrix, depending on the isometric or orthographic view desired. The same transformation matrices as before are involved, except for the translation transformation

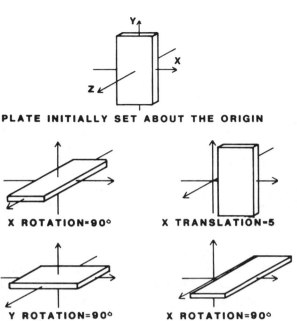

PLATE INITIALLY SET ABOUT THE ORIGIN

X ROTATION=90°   **X TRANSLATION=5**

Y ROTATION=90°   **X ROTATION=90°**

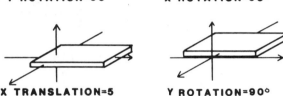

X TRANSLATION=5   **Y ROTATION=90°**   Figure 8.24

matrix, which is not needed. In the portion of the program where CRF is calculated, all of the transformations are strictly rotational. For example, if a connection is designed and a side face view is desired, the only transformation needed is a -90-degree rotation about the Y-axis. To obtain a top view, the CR array is modified to the CRF array by multiplying it by an X-transformation matrix, that is, a rotation of 90 degrees about the X-axis. It is important to note that in the formulation of the CRF array, the whole CR array is multiplied by the final transformation matrix to obtain the CRF array. This differs from the formulation of the CR array, where one portion of the array, representing a rolled section or a connector, is set up at a time using the final transformation matrix for that portion.

For an efficient algorithm on microcomputers, the multiplication of the CR array by the transformation matrices to obtain the CRF array is too time-consuming and thus should be avoided. For the three different orthographic views, i.e., front, side, and top view, it is much easier and faster to forgo all of the matrix multiplications and simply set up a three-condition branch for each of the orthographic views. For a front face view, the CR and CRF arrays are identical. For a side face view, the Y values do not change, but the new X coordinates are observed to be the negative Z coordinates, and the new Z coordinates are observed to be the positive X

coordinates. For a top face view, the X coordinates remain unchanged, but the new Y coordinates are taken from the negative Z coordinates, and the new Z coordinates are taken from the positive Y coordinates. Setting up the three-part branch reduces the computation time involved in calculating the CRF array. For isometric views, however, this shortcut is not possible. Because the program allows the user the option to input any X and Y rotational angles desired, the multiplication of transformation matrices is unavoidable.

The CRF array may not be needed. To do any displaying, all that needs to be done is to transform the CR array to the new directional angle and then record what transformations have been done. Next, with any subsequent transformations, all that has to be done is to reverse the order and sign of the previous transformation and apply the current set of transformations. If this approach is taken, the branching previously mentioned for the orthographic views is not necessary. Therefore, there is a tradeoff involved. Extra storage is required when using the CRF array; however, computation time is reduced.

The final main graphic array is the ICS array. Up to now, we presented the arrays necessary for locating the nodal points in the connection. If just points are plotted, there will be a screen full of dots indecipherable to the user. There has to be some means of indicating which nodal points are connected to each other. A connectivity list is provided through the one-dimensional integer array, ICS. Before all the coordinates for a particular section are set up, the ICS array is filled with a list of points to connect. The order of connecting nodes in the CRF array is specified in the ICS array.

TABLE 8.8   ICS CONNECTIVITY ARRAY FOR AN I-SECTION

−1	12	11	10	9	8	7	6	5	4	3	2	1	−13	14	cont.
15	16	17	18	19	20	21	22	23	24	13	−1	2			cont.
14	13	1	−2	3	15	14	2	−3	4	16	15	3	−4		cont.
5	17	16	4	−5	6	18	17	5	−6	7	19	18	6		cont.
−7	8	20	19	7	−8	9	21	20	8	−9	10	22	21		cont.
9	−10	11	23	22	10	−11	12	24	23	11	−12				cont.
1	13	24	12												

Note: Refer to Fig. 8.20 for locations of nodal points.

The graphic display is set up so that the connection will be displayed one face at a time. This face frame representation of the connection requires that the ICS array have some means of distinguishing when one face ends in the storage of the sections and another one begins. This is accomplished by starting each face connectivity list with a negative value for the CRF position it is pointing to, as indicated in Table 8.8. In this approach, it is necessary to check whether the value in the ICS array is negative. If it is negative, the nodal point is the beginning of a new face in the storage; rather than drawing to the new location using BASIC's LINE command, a move is made to the new location using BASIC's PSET command.

The ICS array is set up so that back faces, those not facing the viewer, are connected in a counterclockwise direction, while those facing the viewer are set up in a clockwise direction, as seen in the example of Fig. 8.25(a). This back-face, front-face distinction is important for displaying the connection, as explained later. The counterclockwise direction for back faces and clockwise direction for front

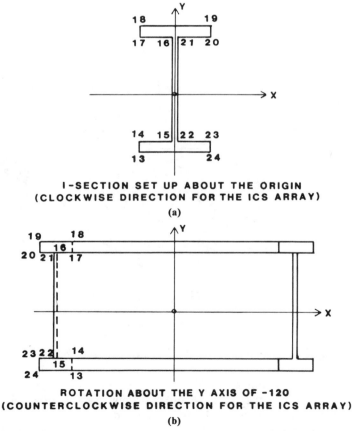

Figure 8.25 (a) I section set up about the origin (clockwise direction for the ICS array); (b) rotation about the Y-axis of −120° (counterclockwise direction for the ICS array).

faces hold true even if transformations are involved to find the final location of the section. Therefore, even if a face is set up clockwise at the beginning, it may eventually become counterclockwise after a transformation, indicating that it is no longer a front face, as shown in the example of Fig. 8.25(b).

8.4.3.3 Graphic representation of bolts. The discussion of the graphic arrays up to now has dealt strictly with steel rolled sections. Included in any graphic output are the connectors, that is, welds and bolts. The manner in which the aforementioned graphic arrays are set up for the connectors is quite similar to the steel rolled sections, but there are a few differences.

For bolts the CR array is set up to contain information about the bolt head on one side and the nut on the other side, as shown in Fig. 8.26. The dimensions of each bolt from $\frac{5}{8}$-in. to $1\frac{1}{2}$-in. size are contained in a two-dimensional, three-column array, BOLTDIM. This array has eight rows for the eight different sizes of bolts within the aforementioned range. The first column of the BOLTDIM array contains the values for the width across the flats for each bolt size, as shown in Fig. 8.26.

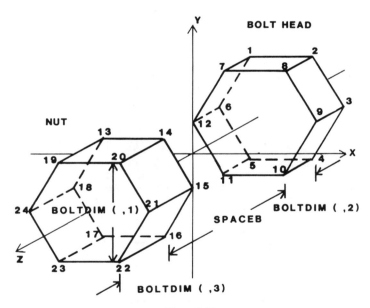

Figure 8.26

This value is all that is needed to find the X and Y coordinates for the bolt to be stored in the CR array. By multiplying the BOLTDIM(, 1) value corresponding to the selected bolt diameter by the appropriate sine and cosine values, while circumnavigating the bolt head and nut clockwise, the values for the first two columns of CR for the particular bolt are calculated.

The heights of the bolt head and the nut are stored in BOLTDIM(, 2) and BOLTDIM(, 3), respectively, as seen in Fig. 8.26. An additional item is also needed to obtain the Z coordinates, the gap between the back sides of the nut and the bolt head, SPACEB. This gap is equal to the thickness of the parts connected, which varies for each connection, plus the thickness of the washer, taken as 0.134 inch for all bolts.

The storage of bolts in the CR array does not include information about the bolt shank or washer. Instead of storing the shank and washer as additional parts in the face frame representation of bolts, these two parts are drawn in the graphic subprogram independent of all graphic arrays set up. The washer and bolt shank are drawn only for orthographic views and not for the isometric views; the reason for this is that the orthographic views have the option to be magnified, but the isometric views do not. With the ability to magnify, all the connection details such as washers and bolt shanks can be displayed more clearly. The exclusion of washers and bolt shanks from the database leads to a significant reduction in both computation time and storage at the expense of slightly more verbose code in the graphic subprogram. Also, with this method, the drawing of the washer and bolt shank does not have to be a face frame idealization of a curved surface. Instead, BASIC's CIRCLE statement can be used effectively to produce a better-looking representation.

Filling the ICS array for bolts is similar to that for steel sections. To indicate whether a face being drawn is a bolt face or not, the beginning point for a bolt face in the ICS array is selected to have a negative value of the CRF position it is pointing to minus 10,000, as seen in Table 8.9. Therefore, any value in the ICS array less than −10,000 and greater than −20,000 is the starting point for a bolt face. Because of the way bolted connections are constructed, it is always known that the back

TABLE 8.9 ICS CONNECTIVITY ARRAY FOR BOLTS

−10,001	6	5	4	3	2	1	−10,001	2	8	7	1	−10,002	cont.
3	9	8	2	−10,003	4	10	9	3	−10,004	5	11	10	cont.
4	−10,005	6	12	11	5	−10,006	1	7	12	6	−15,019		cont.
20	21	22	23	24	19	−15,013	14	20	19	13			cont.
−15,014	15	21	20	14	−15,015	16	22	21	15				cont.
−15,016	17	23	22	16	−15,017	18	24	23	17				cont.
−15,018	13	19	24	18									

Note: Refer to Fig. 8.26 for the locations of nodal points.

faces of the nut and bolt head are against the parts being connected and are never visible. Thus, it is possible to skip over these two faces in the formulation of the ICS array for the bolt (Table 8.9). It should be noted that the savings incorporated in the ICS array cannot be included in the CR and CRF arrays. The locations of the hidden face points are vital in the description of the sides of bolt heads and nuts.

Because bolts have a repetitive nature in their arrangement, this repetition can be used to advantage in the graphic arrays' storage. If every bolt in a connection is stored individually in all three main graphic arrays, the required amount of memory and the computation time would increase dramatically.

To circumvent this problem, the storage of bolts in the graphic arrays takes on a few additional features. First, rather than storing all the bolts in the three main graphic arrays, only one bolt in a sequence or row of bolts is stored. The total number of bolts in the Ith sequence is calculated in the design subprogram and denoted by NBOLTF(I). A two-dimensional array, BOLTREP, is used to store information needed to display the remaining bolts in the sequence. The BOLTREP array has four columns with each row representing a row of bolts. The first column stores the starting point in the CRF array for the nodal points of the bolt. The second through fourth columns contain the spacing of the bolt group in the X, Y, and Z directions.

The BOLTREP array works in conjunction with the drawing of a group of bolts by first determining whether the face being drawn is a bolt face as discussed previously. If the face is part of a bolt, then a determination is made to find which bolt row the face is in. This is done by comparing the value at the starting location in the ICS array for that face with the values in the first column of the BOLTREP array. If the absolute value of the number in the ICS array minus 10,000 is less than or equal to the BOLTREP array's first column plus 23 and greater than or equal to the value in the first column of BOLTREP, then the bolt face corresponds to the BOLTREP sequence in the row where the above test is true. The number 23 in the above test is used because there are 24 nodal points needed to store a bolt.

By setting up the bolt storage in this manner, a FOR-NEXT loop is included in the graphic subprogram to draw the remaining bolts in the sequence without referring to an individual set of data for each bolt. This loop goes from 1 to NBOLTF(I) minus one. The last three columns in the BOLTREP array enable the remaining bolts in the sequence to be displayed in any of the three directions.

Because bolt shanks and washers are not stored in the graphic arrays but instead are drawn whenever a bolt is drawn, an indication of when to draw them must be included. The washer and bolt shank can be seen on the nut end of any bolt display. Because of this, an additional piece of information must be included in the ICS array to indicate if the bolt face being drawn is on the bolt head side or the nut side. Whenever a face on the nut side is drawn, the bolt shank and washer must also be drawn. This is accomplished by having the beginning point of a nut face in the ICS array subtracted by 15,000, rather than 10,000, as was done for the bolt head side as shown in Table 8.9.

8.4.3.4 Graphic representation of welds.

The graphic description of a line of fillet weld or simply a weld is a right isosceles triangular prism as seen in Fig.

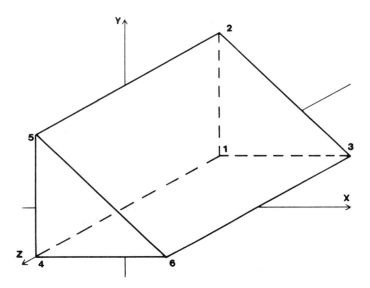

Figure 8.27

8.27. All X, Y, and Z coordinates in a weld are determined by the weld leg size and its length selected previously in one of the design subprograms.

The storage of welds in the ICS array has a few unique aspects. A means of indicating that the face stored is a weld face must be included. This indication is quite similar to the one used for bolts. With welds, all beginning nodal points in the nodal point connectivity list are indicated by the negative CRF location it is pointing to minus 20,000, as shown in Table 8.10. Therefore, any ICS array value

TABLE 8.10   ICS CONNECTIVITY ARRAY FOR A WELD

−20,001	3	2	1	−20,004	5	6	4	−20,002	3	6	5	2

Note: Refer to Fig. 8.27 for location of nodal points.

less than −20,000 is considered to be the starting point in a weld face. The reason for having weld faces specified uniquely is different from that for bolts. In bolts, the reason is to save computation time and storage. A different color is used to represent the welds in order to make them easier to see on the final connection display. Without this distinction it would be difficult to distinguish a line weld from the sections it is connecting.

As with bolts, the ICS array storage for a weld need not include all the faces. The two leg faces in a weld are never visible. The face frame description of a weld thus encompasses only three faces, the two end faces and the face created by extending the hypotenuse of the triangle (Fig. 8.27 and Table 8.10).

8.4.4 Connection Display Algorithms

The graphic subprogram, GRAPH.BAS, takes all of the graphic arrays described previously and translates them into a physical drawing of the connection. Four different views are provided for the user to select from. The first choice, an isometric view, in fact provides an infinite choice of possible viewing angles. The second through fourth choices draw the most common views used to display a connection, that is, front face, side face, and top face views, henceforth referred to as the *orthographic views.*

8.4.4.1 Orthographic views.

The orthographic views offer all the detailed information needed by the user. After the user selects one of the orthographic views, the graphic subprogram goes to the three-part branch to set up the CRF array needed for the final drawing, as previously discussed. The drawing of the orthographic views involves a priority list algorithm. The idea behind a priority list algorithm is to sort all of the faces from those farthest to those closest to the viewer. The priority list algorithm can be quite extensive and time-consuming, so any lessening of the data to be sorted and drawn will reduce the time involved to do the algorithm considerably. One way to reduce the data is to employ back-face elimination before starting the algorithm.

Back-face elimination, as its name implies, involves determining which faces of the connection cannot be seen by the viewer because they face away from him or her. The faces are not removed from any of the main graphic arrays. Instead, they are ignored in any subsequent testing and drawing of the selected view. By using back-face elimination, approximately half of all the faces in the connection can be eliminated from the priority sort algorithm. This can lead to significant savings in computation time during the bubble sorting of the faces.

Back-face elimination is accomplished by calculating the area of each face using the following equation:

$$\text{Area} = \tfrac{1}{2} \sum_{I=1}^{N} X(I)[\, Y(I-1) - Y(I+1)\,]$$

where

N = number of sides in the polygon
$X(I)$ = coordinate of the nodal point corresponding to $CR(I, 1)$
$Y(I)$ = coordinate of the nodal point corresponding to $CR(I, 2)$

This equation will determine the area for any closed polygon. The area calculation has a unique characteristic. If the path used in going around the face is counterclockwise, the final area will be negative, but if the path used is clockwise, the area will be positive. The ICS array is set up originally with all back-face nodal points ordered counterclockwise. Therefore, if the area calculation returns a negative value, then the face is facing away from the viewer and will not be used in any subsequent calculations. The back-face, front-face distinction can change if any rotational transformation is employed. Just because a face is initially set up as a back face does not mean it cannot become a front face later on or vice versa, as shown in Fig. 8.25.

Back-face elimination is done one face at a time. If a face is not a back face, then three arrays are referenced and filled with the appropriate values for that face. The first two arrays, NSTART and NEND, are used in parallel. These one-dimensional integer arrays contain the starting and end locations, respectively, in the ICS array for each face. The starting location is determined by checking for a negative number in the ICS array. The end location is one less than the starting point of the next face. These starting and end arrays are later reordered in order to have the faces from farthest to closest to the viewer. The number of rows in the NSTART and NEND arrays is equal to the number of faces to be drawn. The main graphic arrays—CR, CRF, and ICS—are never changed in GRAPH.BAS. The procedure used to draw the faces in order is to use the sorted NSTART and NEND arrays to indicate the location in the ICS array of the next face to be drawn.

The next array to be referenced is XYZSL, which is a two-dimensional array with six columns and the number of rows equal to the number of faces to be drawn. The first three columns contain the smallest values for the face in the CRF array in the X, Y, and Z directions, respectively. The last three columns contain the corresponding largest values for the three directions. In effect, what this array contains is the boundaries of the smallest cube that will contain the face entirely within. For orthographic views only the third column of XYZSL, that is, the smallest value of the face in the Z direction, is needed. This same array is set up for isometric views, but in those views all six columns need to be filled for each face. The smallest Z value in the XYZSL array is the basis for sorting of the faces in the orthographic views.

After the NSTART, NEND, and XYZSL arrays are filled for all visible faces, the sorting is done based on the XYZSL(,3) value. The face with the smallest value is placed at the top of the NSTART and NEND arrays with all faces after that ordered to the largest value. The smaller the minimum Z value, the farther away from the viewer is the face.

Orthographic views of standard types of connections have the advantage that nearly all the faces are parallel or perpendicular to the viewing plane. All perpendicular faces and some parallel faces fail the area test by having either zero or negative area. Each remaining face, which is parallel to the viewing plane, has its minimum Z value equal to its maximum Z value. Because of this characteristic of orthographic views, it is only necessary to sort the faces based on the Z value and then a far-to-near drawing of the faces.

If the faces are drawn with no hidden line removal, the resulting wire frame drawing will be difficult to comprehend. Because of this, a technique to remove hidden lines is used. The displaying of the faces uses the painter's algorithm for hidden line removal. The painter's algorithm uses BASIC's PAINT statement to eliminate hidden lines. An example of how the painter's algorithm works is shown in Fig. 8.28. The first step in the algorithm is to outline the face in one color (color 1) using BASIC's PSET and LINE commands, as shown in Fig. 8.28. The PAINT command is then used to fill all the pixels within the outlined face by another color (color 2). After the initial drawing and painting of the face, the whole process is repeated one more time, except this time the drawing is done in another color (color 3) and the painting of the interior of the face is done in a final color

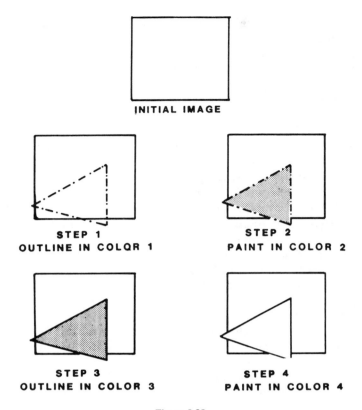

INITIAL IMAGE

STEP 1
OUTLINE IN COLOR 1

STEP 2
PAINT IN COLOR 2

STEP 3
OUTLINE IN COLOR 3

STEP 4
PAINT IN COLOR 4

Figure 8.28

(color 4, usually the background color), while still leaving an outline of the shape in color.

The painter's algorithm works well on the IBM PC in medium resolution, because four colors is the limit at any one time on the screen. By using this algorithm the effect achieved is like a painter using oils. The background is laid first and then the painter works to the foreground, painting over selected parts of the background.

All the drawing of the faces mentioned above is done by using BASIC's LINE and PSET commands. After a face has been sorted and a determination has been made that it should be drawn, the ICS and CRF arrays are referenced. The reference position in the ICS array is determined by the NSTART value for that face. With the NSTART value pointing to the start of the face in the ICS array, the PSET command is used to situate the beginning point's X and Y coordinates, that is,

$$CRF(ICS(NSTART(I)),1) \quad and \quad CRF(ICS(NSTART(I)),2)$$

where I is the position in the NSTART array of the face. After the initial point plotting is done, a loop is set up from $NSTART(I) + 1$ to $NEND(I)$ for the face to connect all the remaining points in the face by using the LINE command.

Because all four-sided faces in orthographic views have their lines either perpendicular or parallel to the screen outline, the efficiency of the painter's algorithm can be improved. This is accomplished by using the box option following Advanced BASIC's LINE command, and also the fill option, which does the painting inside the rectangle.

All the drawing described so far has been done in medium resolution. Before the program continues any further, the drawing is saved and redrawn in high resolution. This change of resolutions is needed because display/print in medium resolution has the drawback that only 40 characters can fit in each row, compared to 80 characters per row in high resolution. This leads to large numbers and letters that consume too much room on the screen, making it difficult to print the dimensions and designations on the screen without interfering with the connection display.

When the connection has been completely drawn, the user has the option to display the connection from a different view, magnify the current display, or have the dimensions and designations for the connection shown on the screen.

Writing the dimensions requires that the graphic subprogram chain to another subprogram, DIM.BAS. The dimension subprogram divides the task of writing the dimensions into three distinct parts. First, there are the I-section dimensions. These dimensions apply to all beams and columns used in the connection. Dimensions such as depth, flange width, flange and web thicknesses, and length are shown on the display. The second choice is the dimensions for the connecting parts. These dimensions apply to angles, T sections, and plates. Finally, there are the connector dimensions for the bolts and welds. The bolt dimensions do not contain the physical dimensions of the bolts, but rather the dimensions relative to the whole connection such as spacing, end distances, and the length of the bolt shank between the connecting parts. The separation of three different dimension outputs is needed because of the limited resolution of the microcomputer's display monitors.

Another choice available to the user is to zoom in on the drawing displayed. With this option, any detail of the connection can be enlarged. The user is prompted to enter the desired magnification value. Magnification of the connection is accomplished by changing the window's extreme coordinates, that is, $(-24, -18)$ and $(24, 18)$. In addition, the user has the option to select the focal point—that is, the centroid—of the window.

8.4.4.2 Isometric views.

The isometric views of the connection are drawn in a similar manner to the way the orthographic views are drawn. Again, a priority list algorithm is used, but for isometric views the algorithm is much more involved. The algorithm used in this work is a modification of an algorithm developed by Newell, et al. [34]. If an isometric view is desired, the user is prompted for both the X and Y rotation angles in degrees. No Z rotation is allowed because with Z rotation, the verticality of the connection becomes confused. With just X and Y rotations, all vertical lines remain vertical, thereby making the drawing much easier to comprehend.

In addition to determining the XYZSL values for each face, two small arrays, SWIND and BWIND, are needed for the isometric view. BWIND and SWIND are both one-dimensional arrays with two rows. BWIND contains the largest X

and Y coordinates in the entire set of faces which make up the connection in its two rows, respectively. Similarly, SWIND stores the smallest X and Y coordinates for all faces. These two arrays establish the largest window size possible to draw the isometric view without clipping any lines in the connection. Because the isometric view cannot be magnified by the user, it is important that the drawing be as large as possible while still displaying the entire connection.

Using the BWIND and SWIND values, the size of the window is determined. It is not possible to simply use the values in the two arrays as the corner points of the window, because the final drawing would be distorted. Instead, the following test is made:

$$(BWIND(1) - SWIND(1)) > (\tfrac{4}{3}(BWIND(2) - SWIND(2)))$$

If the above test is true, the X window coordinates in both SWIND and BWIND remain unchanged and the Y window coordinates are $\tfrac{3}{4}$ times the two X window coordinates. If the test is false, the Y window coordinates remain unchanged and the X window coordinates are $\tfrac{4}{3}$ times the size of the Y window coordinates. The $\tfrac{3}{4}$ and $\tfrac{4}{3}$ modifiers are needed because the aspect ratio of the screen must be considered in any graphic output. The terminal screen is not square, but instead is rectangular with its width $\tfrac{4}{3}$ times its height. This aspect ratio must be considered when one is setting up the isometric view's window.

The four parameters describing the plane of a front face are stored in the two-dimensional array PLANE. The equation used to describe a plane and thus a particular face is

$$Ax + By + Cz + D = 0$$

The four constants of this equation are stored in the four columns of array PLANE.

In orthographic views it is necessary only to sort the faces based on their minimum Z values and then draw the sorted faces in order. In an isometric drawing, faces to be drawn are not necessarily parallel to the viewing plane. Thus, drawing the faces based on their minimum Z values does not necessarily produce the correct drawing.

The first step involved in the drawing of an isometric view is again the sorting of faces from the smallest $XYZSL(, 3)$ value to the largest $XYZSL(, 3)$ value, just as in the orthographic views. After this sorting is accomplished, five tests are performed on the faces to determine the proper order in which they should be drawn. In order of difficulty these tests are: Z minimax test, bounding box test, P-behind-Q test, Q-in-front-of-P test, and the extensive edge check. With these five tests the final ordering of the faces in any isometric view can be accomplished.

To begin with, a loop is set up from one to the number of front faces to determine the face order. The face at the top of the list (farthest away from the viewer) is called the P face. The next face in the list is considered the Q face. With these two faces, up to five tests are performed to determine the order of drawing. The tests are set up from the shortest to the longest in computation time. This ordering of the tests yields the most efficient algorithm. If a face is drawn, a new P face is selected from the next position in the face list, and the process starts anew. This algorithm is slow at first in drawing the faces, but becomes increasingly faster as the number of faces remaining in the face list decreases. For more details about the algorithm, see Ref. 14.

8.4.5 General Design Considerations

At the beginning of the program execution, a brief introduction to the program is presented to the user. Following the introduction, the user is provided with a menu for the selection of moment-resisting or simple connections, as shown in Fig. 8.29. The two menus for selection of simple and moment-resisting connections are shown

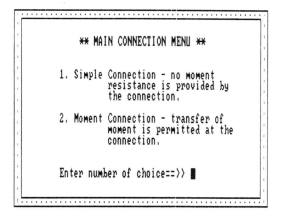

Figure 8.29

in Figs. 8.30 and 8.31, respectively. Design of connections varies from one to another. Some elements of design are common to all types of connections, however.

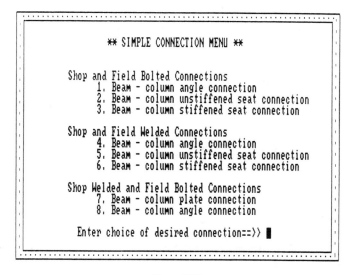

Figure 8.30

Figure 8.31

8.4.5.1 I Sections for Beams and Columns. STEELCON is designed to handle standard steel rolled I sections for all columns and beams. The user simply has to input the section desired by its standard designation, say W18 × 76. The inputted designation is then checked against a list of available I sections in the AISCS Manual stored in an external random access data file, called AISC.DAT. In addition to specifying the I section desired, the user also specifies the type of steel used for the section. There are three common designations of steel offered for the user to select from: A36, A572-grade50, and A529. If the user does not desire any of these, he or she has the opportunity to enter the properties that characterize each type of steel, that is, the ultimate strength F_u and the yield stress F_y. In all cases, a $\frac{1}{2}$-in. nominal gap between beams and columns is used in the design of the connection. This distance is needed in case there is an overrun in the length of a section being connected.

At the completion of the selection of I sections for beam and column, the program prompts the user to enter the load, in Kips, that the connection will be designed to carry. Any value less than 6 Kips is rejected, and the user is asked to enter a valid load (AISCS 1.15.1).

8.4.5.2 Angles. The angle selection process references an external random access data file, ANG.DAT, containing data for standard angles. The criteria for initial selection of the angles are ordered as follows. First, the angle thickness should be as thin as possible. Second, the short leg width should be as short as possible. Finally, the long leg width should be as short as possible. The angles in ANG.DAT are ordered in the same way as the above criteria.

The ordering of the criteria in the angle selection has the thickness as the most important factor. This is done for two reasons. First, by requiring that the angle thickness be as thin as possible, better ductility is achieved in the connection. Second, the angle thickness in general has the most considerable influence on the weight of the angle among the three parameters.

After the initial angle selection, the user has the option to select an angle other than the one selected by the program. Any change in the angle section is checked to make sure that the two widths and the thickness of the new angle are adequate.

If the angle is not adequate and thus is rejected by the program, the user will be told why the angle is rejected and given another opportunity to select an adequate angle.

8.4.5.3 Bolts.

There are six possible bolt types that can be selected. The selection of the bolt types is a two-step process. First, the designer selects either ASTM A325 or ASTM A490 bolts, which are the most frequently used types of bolts. After the first selection, the designer must specify the fastener type used, that is, friction-type connection (F), bearing-type connection with threads included in the shear plane (N), or bearing-type connection with threads excluded from the shear plane (X).

Diameters of the bolts range from $\frac{5}{8}$ in. to $1\frac{1}{2}$ in. in $\frac{1}{8}$-in. increments. Every connection type, unless noted otherwise, uses the above range of bolt sizes. The minimum number of bolts used in any connection part is two. Every design uses standard holes $\frac{1}{16}$ in. larger than the bolt diameter (AISCS 1.23.4.1). The design of the bolts has three parameters to be considered: the number of bolts needed, the diameter of the bolts, and the center-to-center spacing, or *pitch*, of the bolts.

The initial design returns the smallest bolt diameter capable of carrying the design load. The initial pitch is three inches or three times the bolt diameter, whichever is larger. The spacing selection is not the minimum allowed by AISCS, but instead is the preferred spacing used in most connection designs. If the user wishes to try a smaller pitch, he or she has the opportunity to do so after the initial bolt design. An initial edge distance of $1\frac{1}{2}$ in. for 1-inch or smaller bolt diameters, or $1\frac{3}{4}$ times the bolt diameter for bolts greater than 1-inch in diameter is used in the design process. This is a modification of the requirement of AISCS 1.16.5.1.

If the bolt type selected by the user requires a bolt diameter greater than $1\frac{1}{2}$ in., a message indicating that the bolt type selected does not work for the connection will be given. The program suggests that the user specify a bolt type with a higher shear strength. The design process is then repeated from the bolt-type selection. If the user selects the bolt type with the highest shear strength, that is, A490-X, and the connection still cannot be designed with $1\frac{1}{2}$-in. or smaller bolts, then another message is given indicating that the connection cannot be designed by using standard bolt sizes.

After the initial design for the bolts, the user has the opportunity to change several design parameters. The calculations are quite fast, so the user can explore many different possibilities within a short time to reach a final selection for the bolts.

8.4.5.4 Welds.

The welding process assumed for all welding is Shielded Metal Arc Welding (SMAW). This is the most commonly used welding process in design of steel buildings. The electrode selection is left for the user. Grades ranging from E-60 to E-110 are available to the user.

The possible fillet weld leg sizes range from $\frac{1}{8}$ in. to 1 in. A $\frac{1}{8}$-in. weld is the minimum size allowed by the specification (AISCS 1.17.2). The one-inch upper limit is set because there are few angles manufactured with a thickness greater than one inch. Weld sizes greater than one inch are impractical in most standard connections. The increments of possible weld sizes between the two extreme limits are $\frac{1}{16}$ in. for $\frac{1}{8}$-in. to $\frac{1}{2}$-in. welds and $\frac{1}{8}$ in. for weld legs greater than $\frac{1}{2}$ in.

The upper bound for the length of the weld in beam-column angle connections is determined by taking the total beam depth and subtracting the distance from the beam flange to the toe of the sections fillet (k) on each end and then subtracting one inch. The subtraction of the final inch allows the welder sufficient clearance to lay the weld.

After the initial selection of a weld, the user has the option to change the weld size. Changing the weld size may lead to a more economical design. However, two factors must be considered with any increase in the weld size. First, any weld leg size greater than $\frac{5}{16}$ in. requires more than one pass by the welder laying the weld. This will lead to cost increases due to the increased labor involved. Second, if the weld size is increased, the angle's required thickness must also be increased. With the ability to select many different weld sizes and observe the effect on the weld length and angle length and thickness, the user will be able to arrive at an economical solution.

8.4.5.5 Conclusion of the design. At the conclusion of the connection design, the user is given the opportunity to return to the beginning of the connection subprogram to redesign the connection. If the user is satisfied with the connection design, the program continues on to display the connection on the terminal screen. Before any display can be made, though, many graphic arrays must be filled for the designed connection. Setting up the arrays takes some time, so to entertain the user during the wait, the program provides an opportunity to play background music while the arrays are being set up. If music is desired by the user, a stirring rendition of "I Left My Heart in San Francisco" is played during the wait [40]. In addition, a summary of the connection design is displayed at this time. In the following sections, interactive designs of three types of simple (Type 2) connection and one type of moment-resisting (Type 1) connection are presented.

**8.4.6 Design of Simple Shop- and Field-Bolted Angle
Connections**

8.4.6.1 Comments. The example presented in this section is the same as that given in the AISC Manual (p. 4–17). The interactive microcomputer session for this design is presented in Sec. 8.4.6.2.

Following the selection of the bolts used for the shop connection, that is, the angle-to-beam-web connection, the program makes several design checks based on the beam and bolt selection. These checks include beam bearing capacity, minimum distance from the beam end (edge) to the center of the bolt holes, and the shear capacity of the beam. Any violation of the AISC code is brought to the user's attention for correction.

The field bolts—that is, the bolts connecting the angle to the column flange—are usually selected to be the same as the shop bolts. If the user does desire different field bolts, however, the program selects them in the same way as the shop bolts. In this type of connection, the beam is not allowed to frame into the column web.

This is because there is not sufficient room to enter an impact wrench to tighten the bolts. If the beam must frame to the column web, a seated connection will solve the above-mentioned problem.

In the selection of an angle, a check is made to ensure that the outstanding angle leg does not overrun the column flange. If this occurs, the outstanding angle leg width is reduced as much as possible. If the selection of the angle after this change still has the outstanding leg overrunning the column flange, a message indicating that using more bolts with a smaller diameter may solve the problem is given.

Figure 8.32

The main graphics menu, which is displayed after all arrays are set up, is shown in Fig. 8.32. The first choice, an isometric view, allows the user to select any viewing rotation desired about the X- and Y-axes. An example is shown in Fig. 8.33. In orthographic views, after the connection is displayed on the screen a second

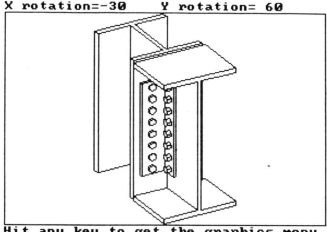

Figure 8.33

Side Face View

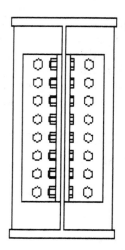

[1)Graphics menu 2)Zoom in 3)Dimensions] Enter your choice==>> █

Figure 8.34

Side Face View

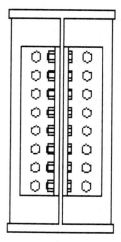

[1)I-sections 2)Connecting pieces 3)Connectors] Enter dimension type==>> █

Figure 8.35

graphic menu appears at the bottom of the screen, as shown in the side-face view of Fig. 8.34.

The option to show dimensions and designations on the screen is quite useful to the user. If this choice is selected by the user, he is immediately given another menu at the bottom of the screen, as seen in Figure 8.35. This dimension menu allows the user to specify which dimensions and designations he would like to see.

For example, Figs. 8.36 through 8.38 show the side-face view with dimensions and designations for I sections, angles, and bolts, respectively. Figure 8.39 shows the front-face view with dimensions and designations for the I sections. Figures 8.40 and 8.41 show the top view with dimensions and designations for I sections and bolts, respectively. In the top-face view of Fig. 8.41, I sections are not included. This enables the display to show parts of the connections that are obscured by the beam's top flange. The option to display the top view without the I sections is provided immediately after the user selects the top view from the main graphics menu.

If the user wishes to magnify the current display, he is prompted to enter the desired magnification value. Values under unity are not accepted. In addition to the magnification value, the user is prompted to enter the X and Y coordinates desired for the center point, that is, the focus of the window. Selecting the focal point may be difficult for the user. Therefore, the user is asked if he or she would like a ten-inch grid in each direction (in global coordinates) drawn on the screen, as shown in Fig. 8.42. Figures 8.43 and 8.44 show a front-face view and a top-face view for the connection after they have been magnified. Note that in these figures all the details of the connection, such as washers and bolt shanks, are clearly presented.

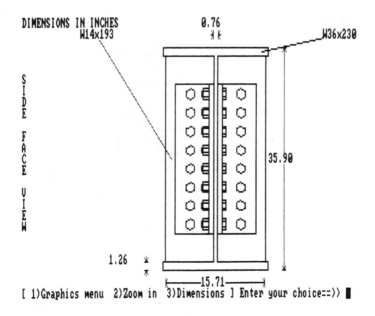

Figure 8.36

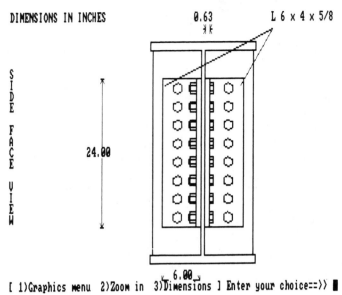

Figure 8.37

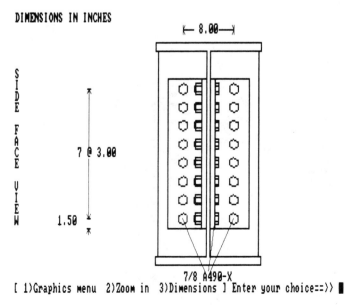

Figure 8.38

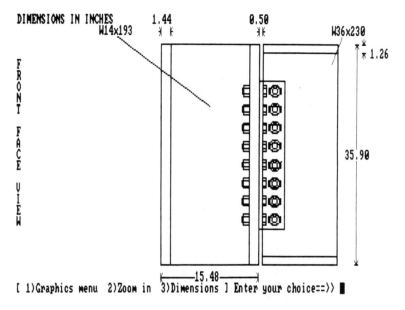

Figure 8.39

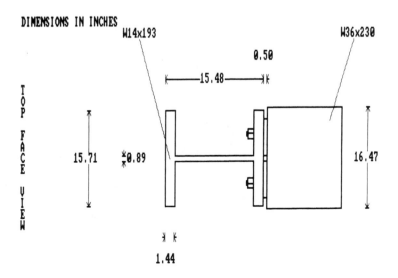

Figure 8.40

DIMENSIONS IN INCHES

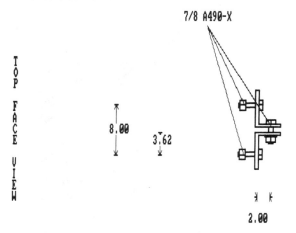

[1)Graphics menu 2)Zoom in 3)Dimensions] Enter your choice==)) █

Figure 8.41

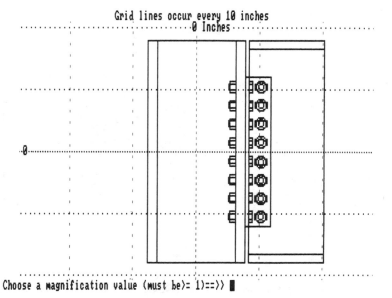

Choose a magnification value (must be)= 1)==)) █

Figure 8.42

Front Face View

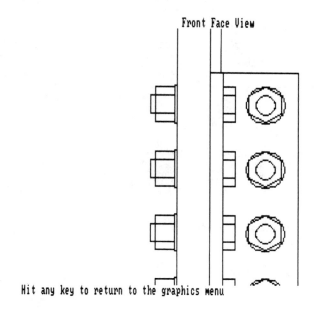

Hit any key to return to the graphics menu

Figure 8.43

Top Face View

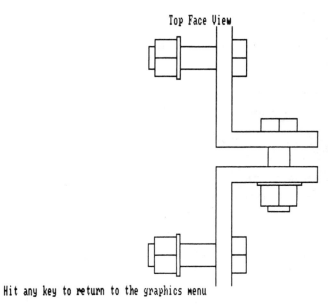

Hit any key to return to the graphics menu

Figure 8.44

8.4.6.2 Interactive session.

                        COLUMN INPUT

Enter I-section in the following format: I##x###
(ex. W36x300, S24x100, HP14x102, M14x 18) ==>> W14x193

                Data for W14x193 steel section

                Area, A= 56.80 in.^2
                Major Moment of Inertia, Ix= 2400.000 in.^4

    Depth, d= 15.480 in.   Web Thickness, tw= 0.890 in.
    Flange Width, b= 15.710 in.   Flange Thickness, tf= 1.440 in.

Is this section O.k.? (y or n)==>> Y

Setting up the connection and coordinate arrays for the section

            Possible types of steel

    1) A36 (Fy=36ksi , Fu=58ksi)
    2) A572-grade50 (Fy=50ksi , Fu=65ksi)
    3) A529 (Fy=42ksi , Fu=60ksi)
    4) Different than above types, Fy and Fu specified by user

Enter steel type desired for this section (1-4)==>> 1

                        BEAM INPUT

Enter I-section in the following format: I##x###
(ex. W36x300, S24x100, HP14x102, M14x 18) ==>> W36x230

                Data for W36x230 steel section

                Area, A= 67.60 in.^2
                Major Moment of Inertia, Ix=15000.000 in.^4

    Depth, d= 35.900 in.   Web Thickness, tw= 0.760 in.
    Flange Width, b= 16.470 in.   Flange Thickness, tf= 1.260 in.

Is this section O.k.? (y or n)==>> y

Setting up the connection and coordinate arrays for the section

            Possible types of steel

    1) A36 (Fy=36ksi , Fu=58ksi)
    2) A572-grade50 (Fy=50ksi , Fu=65ksi)
    3) A529 (Fy=42ksi , Fu=60ksi)
    4) Different than above types, Fy and Fu specified by user

Enter steel type desired for this section (1-4)==>> 1

What is the beam end reaction? (in kips)==>> 340

                        BOLTS

        1) A325 bolts
        2) A490 bolts
Which type of bolt do you want to use? Specify by number==>> 1

                    CONNECTION TYPE

        1) F-friction type
        2) N-bearing type, threads in shearing plane
        3) X-bearing type, threads not in shearing plane
Which type of connection do you want to use? Specify by number==>> 2

```
        ** DESIGN BOLTS FOR ANGLE TO BEAM WEB CONNECTION **

    Checking Shear Capacity of the Bolts

    Beam Web Connection Bolt Information:
       Number of bolts= 9 A325-N
       Diameter of bolts= 1.125 inches
       Spacing of bolts (center-center)= 3.375 inches
       Capacity of bolts= 375.7402 kips versus a reaction= 340 kips

   The above design is based on using the smallest diameter bolt that will work
   using a 3 inch center to center spacing or 3 times the bolt diameter,
   whichever is larger

    What would you like to change in the bolt design?
       1) Type of bolt
       2) Bolt diameter
       3) Spacing
       4) Number of bolts
       5) Nothing

Enter choice by number ==>> 4

  Absolute maximum number of  9 /8 inch bolts that will fit= 10

  Absolute minimum number of  9 /8 inch bolts for a safe connection= 9

Enter desired number of bolts to be used==>> 10


    Beam Web Connection Bolt Information:
       Number of bolts= 10 A325-N
       Diameter of bolts= 1.125 inches
       Spacing of bolts (center-center)= 3.023611 inches
       Capacity of bolts= 417.4892 kips versus a reaction= 340 kips

    What would you like to change in the bolt design?
       1) Type of bolt
       2) Bolt diameter
       3) Spacing
       4) Number of bolts
       5) Nothing

Enter choice by number ==>> 1

                 BOLTS

             1) A325 bolts
             2) A490 bolts
Which type of bolt do you want to use? Specify by number==>> 2

                 CONNECTION TYPE

             1) F-friction type
             2) N-bearing type, threads in shearing plane
             3) X-bearing type, threads not in shearing plane
Which type of connection do you want to use? Specify by number==>> 3

    Beam Web Connection Bolt Information:
       Number of bolts= 10 A490-X
       Diameter of bolts= .75 inches
       Spacing of bolts (center-center)= 3 inches
       Capacity of bolts= 353.43 kips versus a reaction= 340 kips

   The above design is based on using the smallest diameter bolt that will work
   using a 3 inch center to center spacing or 3 times the bolt diameter,
   whichever is larger

    What would you like to change in the bolt design?
       1) Type of bolt
       2) Bolt diameter
       3) Spacing
       4) Number of bolts
       5) Nothing
Enter choice by number ==>> 2
```

Enter desired bolt diameter (from .625 inches to 1.5 inches in .125 inch
increments)==>> .86

  Increment must be by .125

Enter desired bolt diameter (from .625 inches to 1.5 inches in .125 inch
increments)==>> .875

    Beam Web Connection Bolt Information:
      Number of bolts= 8 A490-X
      Diameter of bolts= .875 inches
      Spacing of bolts (center-center)= 2.33625 inches
      Capacity of bolts= 384.846 kips versus a reaction= 340 kips

    What would you like to change in the bolt design?
      1) Type of bolt
      2) Bolt diameter
      3) Spacing
      4) Number of bolts
      5) Nothing

Enter choice by number ==>> 3

  Absolute minimum bolt spacing= 2.365812 inches

  Absolute maximum bolt spacing (due to space limitations)= 4.071429 inches

Enter desired new bolt spacing (center to center) in inches==>> 3

    Beam Web Connection Bolt Information:
      Number of bolts= 8 A490-X
      Diameter of bolts= .875 inches
      Spacing of bolts (center-center)= 3 inches
      Capacity of bolts= 384.846 kips versus a reaction= 340 kips

    What would you like to change in the bolt design?
      1) Type of bolt
      2) Bolt diameter
      3) Spacing
      4) Number of bolts
      5) Nothing

Enter choice by number ==>> 5

        ** CHECK FOR BEAM BEARING CAPACITY **

Allowable bearing capacity in the beam web= 451.82 kips   versus a reaction=
340 kips

Meets requirements

        ** CALCULATING DISTANCE BETWEEN WEB END AND CENTER OF BOLT HOLES **

Minimum distance from end of the beam to center of holes in the
web= 2 inches

Would you like to increase this distance? (y or n)==>> n

        ** CHECK FOR BEAM WEB SHEAR CAPACITY **

  Shear limit for beam= 392.8896 kips

  Beam has sufficient shear strength

Would you like to specify a different design for the field bolts then the
shop bolts? (y or n)==>> n

   !! DETERMINE MINIMUM REQUIRED ANGLE THICKNESS !!

Tentative angle length= 24 inches

  Required thickness for bearing= .2859546 inches
  Required thickness for shear on gross area= .4918982 inches
  Required thickness for shear on net area= .5921281 inches   **!! Governs !!**

   !! CHECK ANGLE BEARING CAPACITY !!

Angle bearing capacity is sufficient

   !! FIND REQUIRED LEG WIDTHS !!

  Minimum angle leg widths in inches, Xmin is along the column, Ymin is along
the beam

  Ymin 3.625 inches
  Xmin 4.625 inches

Angle selected: L 6 x 4 x 5/8

  Angle thickness= .625 inches
  Leg width along beam web= 4 inches
  Leg width along column flange= 6 inches

Would you like to specify a different angle than the one
selected? (y or n) ==>> n

Would you like to increase the angle length? (y or n)==>> n

Would you like to change what you have designed? (y or n)==>> n

  The computer now has to set up the designed connection for later graphic
output. This takes a while so be patient. While you're waiting you can listen
to music, when this stops you may continue on to the graphics output

Would you like to listen to music while you wait? (y or n)==>> n

            !!!
            ! Summary of results for a simple !
            !   shop and field bolted   !
            !   beam to column angle connection !
            !!

Design end reaction= 340 kips

  !! I-Sections !!
Column section: W14x193   Fy= 36 ksi   Fu== 58 ksi
Beam section: W36x230   Fy= 36 ksi   Fu== 58 ksi

  !! T's, Angles and Plates !!
Angles connecting beam web to column flange: 2 L 6 x 4 x 5/8
Length of the angle= 24 inches

  !! Bolts !!
8 7 /8 inch A490-X bolts for angle to beam web connection
8 7 /8 inch A490-X bolts for each angle to column flange connection

HIT ANY KEY TO CONTINUE ON TO THE GRAPHICS MENU

8.4.7 Design of Simple Shop and Field Welded Angle Connections

8.4.7.1 Comments. The example presented in this section is the same as that given in the AISC Manual (p. 4–33). The interactive microcomputer session for this design is presented in Sec. 8.4.7.2.

The size of the field weld for the angle-to-column connection generally is as large or larger than the size of the shop weld, so it is designed first in this connection type. This is because the shop weld—that is, the weld connecting the angle to the beam web—has two horizontal components that contribute to the strength of the weld, as seen in Fig. 8.47. The field weld has only a small return at the top, which is not used in calculating the size of the weld needed. All field welds are designed with a return at the top of the angle with a length equal to twice the weld leg size.

All welds are checked to make sure that they are less than or equal to the angle thickness minus $\frac{1}{16}$ in. for angles with a thickness greater than or equal to $\frac{1}{4}$ in. or less than or equal to the angle thickness for angles less than $\frac{1}{4}$ in. thick (AISCS 1.17.3).

After a satisfactory design has been obtained, the user has the option to display the connection. Figure 8.45 shows the designed connection when seen from an isometric view. It should be noted that the welds are displayed in the color blue to make them easier to distinguish. A side-face view of the connection with dimensions and designations for both shop and field welds is shown in Fig. 8.46. The distinction between field and shop welds is shown with the darkened flag symbol. Figure 8.47 shows a front-face view of the design with all applicable dimensions for the angle. Finally, Fig. 8.48 shows a top-face view of the connection with designations and dimensions for the selected beam and column.

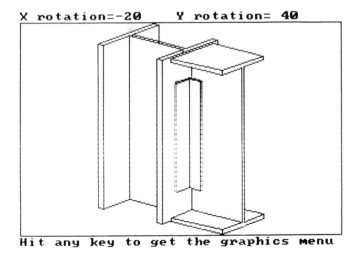

Figure 8.45

DIMENSIONS IN INCHES

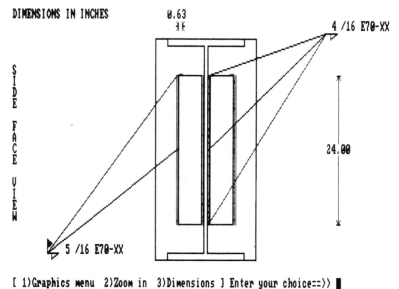

[1)Graphics menu 2)Zoom in 3)Dimensions] Enter your choice==>> ▉

Figure 8.46

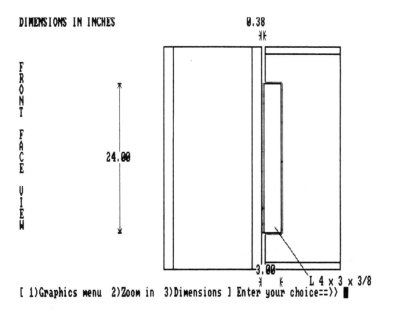

[1)Graphics menu 2)Zoom in 3)Dimensions] Enter your choice==>> ▉

Figure 8.47

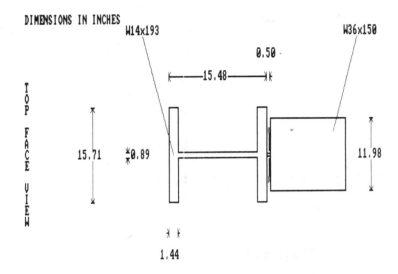

DIMENSIONS IN INCHES

[1)Graphics menu 2)Zoom in 3)Dimensions] Enter your choice==>> █

Figure 8.48

8.4.7.2 Interactive session.

```
                              COLUMN INPUT


Enter I-section in the following format: I##x###
(ex. W36x300, S24x100, HP14x102, M14x 18) ==>> W14x193


                    Data for W14x193 steel section

                    Area,   A= 56.80 in.^2
                    Major Moment of Inertia, Ix= 2400.000 in.^4

      Depth,   d= 15.480 in.          Web Thickness, tw=  0.890 in.
      Flange Width, b= 15.710 in.     Flange Thickness, tf=  1.440 in.

Is this section O.k.? (y,or n)==>> Y


  Setting up the connection and coordinate arrays for the section

            Possible types of steel

      1) A36 (Fy=36ksi , Fu=58ksi)
      2) A572-grade50 (Fy=50ksi , Fu=65ksi)
      3) A529 (Fy=42ksi , Fu=60ksi)
      4) Different than above types, Fy and Fu specified by user

Enter steel type desired for this section (1-4)==>> 1

                              BEAM INPUT


Enter I-section in the following format: I##x###
(ex. W36x300, S24x100, HP14x102, M14x 18) ==>> W36x150
```

Data for W36x150 steel section

Area, A= 44.20 in.^2
Major Moment of Inertia, Ix= 9040.000 in.^4

Depth,   d= 35.850 in.   Web Thickness, tw= 0.625 in.
Flange Width, b= 11.975 in.   Flange Thickness, tf= 0.940 in.

Is this section O.k.? (y or n)==>> y

Setting up the connection and coordinate arrays for the section

Possible types of steel

1) A36 (Fy=36ksi , Fu=58ksi)
2) A572-grade50 (Fy=50ksi , Fu=65ksi)
3) A529 (Fy=42ksi , Fu=60ksi)
4) Different than above types, Fy and Fu specified by user

Enter steel type desired for this section (1-4)==>> 4

Enter desired yield stength (Fy) for steel section in ksi==>> 36

Enter desired ultimate strength (Fu) for the steel section in ksi==>> 58

What is the beam end reaction? (in kips)==>> 1

 Minimum reaction you should design for=6 kips

What is the beam end reaction? (in kips)==>> 170

      ELECTRODE SELECTION
         1) E60
         2) E70 (most common)
         3) E80
         4) E90
         5) E100
         6) E110

Enter desired electrode type (1-6) for welding==>> 2

     ‡‡ DETERMINE PRELIMINARY ANGLE DIMENSIONS ‡‡

Maximum allowable angle length= 31 inches
Minimum allowable angle length= 16 inches

Select the width for the angle leg on the beam web in 1/2 in. increments
(3 or 3.5 in. is the most common width used)==>> 3

Select the width for the angle leg on the column flange in 1/2 in. increments
(4 in. is the most common width used)==>> 4

     ‡‡ FIELD WELD ‡‡

Design based on the minimum required size for the field weld:

Angle length required= 27 inches
Weld leg size= 4 /16 inches
Minimum angle thickness= .3125 inches

A larger weld size will require a smaller angle length which may be desireable
for the design. However, keep in mind that welds over 5/16 inches require more
than one pass by the welder which you want to avoid if possible.
Also, consider that an increased weld size may require a thicker angle

Would you like to change the weld size? (y or n)==>> y

Specify by a positive or negative integer number how much you would like to
increase or decrease the weld size in 16th's of an inch.==>> 1

The weld size you are now trying= 5 /16 inches

Angle length required= 22 inches
Weld leg size= 5 /16 inches
Minimum angle thickness= .375 inches

A larger weld size will require a smaller angle length which may be desireable
for the design. However, keep in mind that welds over 5/16 inches require more
than one pass by the welder which you want to avoid if possible.
Also, consider that an increased weld size may require a thicker angle

Would you like to change the weld size? (y or n)==>> n

‡‡ ANGLE SELECTION ‡‡

Angle selected: L 4 x 3 x 3/8

Angle thickness= .375 inches
Leg width along beam web= 3 inches
Leg width along column flange= 4 inches

Would you like to increase the angle length? (y or n)==>> n

‡‡ SHOP WELD ‡‡

Weld size needed for shop weld = 5 /16 inches

A larger weld size will require a smaller angle length which may be desireable
for the design. However, keep in mind that welds over 5/16 inches require more
than one pass by the welder which you want to avoid if possible.
Also, consider that an increased weld size may require a thicker angle

Would you like to change the weld size? (y or n)==>> y

Specify by a positive or negative integer number how much you would like to
increase or decrease the weld size in 16th's of an inch.==>> -1

The weld size you are now trying= 4 /16 inches

Angle length required= 24 inches
Weld leg size= 4 /16 inches
Minimum angle thickness= .3125 inches

A larger weld size will require a smaller angle length which may be desireable
for the design. However, keep in mind that welds over 5/16 inches require more
than one pass by the welder which you want to avoid if possible.
Also, consider that an increased weld size may require a thicker angle

Would you like to change the weld size? (y or n)==>> n

No reduction for the shop weld strength is needed because the beam web is
large enough

Would you like to change what you have designed? (y or n)==>> n

The computer now has to set up the designed connection for later graphic
output. This takes a while so be patient. While you're waiting you can listen
to music, when this stops you may continue on to the graphics output

Would you like to listen to music while you wait? (y or n)==>> n

```
!!!!!!!!!!!!!!!!!!!!!!!!!!!!!!!!!!!!!!!!!!!
!  Summary of results for a simple  !
!        shop and field welded      !
!  beam to column angle connection  !
!!!!!!!!!!!!!!!!!!!!!!!!!!!!!!!!!!!!!!!!!!!
```

Design end reaction= 170 kips

!! I-Sections !!
Column section: W14x193   Fy= 36 ksi   Fu== 58 ksi
Beam section: W36x150   Fy= 36 ksi   Fu== 58 ksi

!! T's, Angles and Plates !!
Angles connecting beam web to column flange: 2 L 4 x 3 x 3/8
Length of the angle= 24 inches

!! Welds !!
Shop weld information (angle to beam web)
    Weld size= 4 /16 inches using an E70-XX electrode
    Length of vertical weld= 24 inches
    Length of horizontal welds at top and bottom= 2.5 inches

Field weld information (angle to column flange)
    Weld size= 5 /16 inches using an E70-XX electrode
    Length of vertical weld= 24 inches
    Length of return at top= .625 inches

Erection bolts as needed

HIT ANY KEY TO CONTINUE ON TO THE GRAPHICS MENU

8.4.8 Design of Simple Shop-Welded and Field-Bolted Single Plate Connections

8.4.8.1 Comments.   Single plate connections are a relatively new development in connection design. The design procedure for this type of connection is described by Young and Disque [42] and is also presented in a recent AISC publication [46].

The procedure for design of a single plate connection begins with the designer entering the span of the beam used in the connection. The span length is needed later to determine the eccentricity of the applied load on the bolts. Design of the bolts used to connect the plate to the beam is done next. This connection type can only be designed with $\frac{3}{4}$-in.- to 1-in.-diameter bolts. This limitation is placed on the bolt selection because the research on this connection type has been done only for those bolt sizes.

After the designer is satisfied with the design of the field bolts, the eccentricity of the load must be computed. Note that the eccentricity is computed after the field bolt design is completed, because this calculation is influenced by the bolt design. The eccentricity of the loading is needed to determine the moment applied on the shop weld and plate. Even though this connection is classified as simple, there is a moment acting on the weld and plate that must be taken into account. The moment is assumed not to be transferred to the column; therefore, this moment is not ordinarily considered in the column design.

To find the moment applied on the plate and weld, the above eccentricity must be added to the distance from the center of the bolt holes to the column. This distance (ADIST) is entered by the user. The prompt to the designer for this distance indicates that a 3-in. distance is the most commonly used value. However, the designer may enter any distance desired, with a stipulation. Any distance less than twice the bolt diameter plus 0.5 in. will not be allowed, because it will not provide a sufficient end distance for the beam web end. This situation will be brought to the attention of the user. With ADIST known, the applied moment is then calculated.

For design of the plate, the thickness is determined first. A minimum plate thickness is calculated to resist the shear force. This minimum plate thickness is then compared to the minimum thickness necessary to resist the moment. The larger of the two is used as the basis for the minimum plate thickness selection. The thickness is rounded to $\frac{1}{16}$-in. increments. The combined effects of shear and moment forces are not considered in selecting the plate thickness.

A thick plate may inhibit the ductile behavior of the connection. Thus, a check is performed to see whether the plate thickness exceeds the specified limits. If the plate thickness does exceed the limit, a message indicating possible remedies is given. The remedies are using a stronger bolt designation or a larger bolt diameter. Both possible remedies return the user to the bolt selection portion of the program to start the design process anew.

One additional check for plate thickness is also included. The thickness of the plate is compared to the beam web thickness. If the plate is thicker than the web, a message is given indicating this and the user is asked if this is acceptable to him. If the design is not acceptable to the user, he has the opportunity to redesign the connection in order to find a thinner but adequate plate.

After the plate selection is made, the welds will be designed. Welds are placed on both sides of the plate when it is connected to the column flange. The welds are placed along the entire length of each side of the plate. No return is used for these welds. A return at the top would cause a notch in the plate, due to melting, at its most highly stressed fibers.

A sample interactive design session for this connection type is given in Sec. 8.4.8.2. This example is taken from Ref. 46 (pp. 3–54). Graphic output for the above example is shown in Figs. 8.49 through 8.55. Figure 8.49 shows an isometric

X rotation=-20   Y rotation= 45

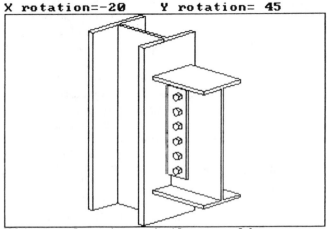

Hit any key to get the graphics menu   Figure 8.49

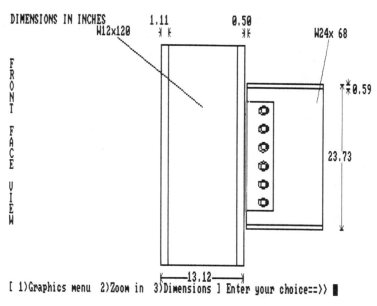

Figure 8.50

view of the connection. The front-face view of the connection is shown in Fig. 8.50 with dimensions and designations for I sections. The side-face view of the connection is shown in Fig. 8.51. Figure 8.52 shows the top view of the connection. Figure 8.53 shows the top view when I sections are excluded. A magnified version of this

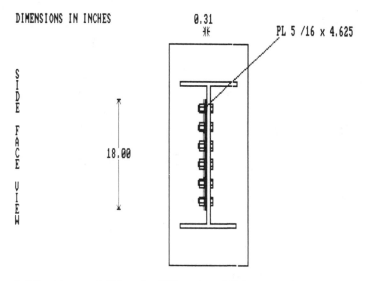

Figure 8.51

DIMENSIONS IN INCHES

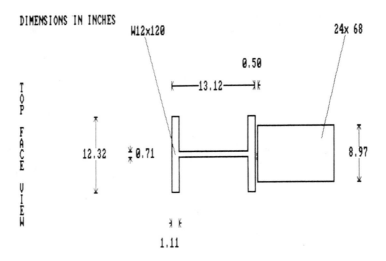

[1)Graphics menu 2)Zoom in 3)Dimensions] Enter your choice==>> ▮

Figure 8.52

figure is shown in Fig. 8.54. Finally, Fig. 8.55 displays the side-face view after magnification. Notice that the plate is used only on one side of the beam web. Welds are placed on both sides of the plate, but in this view one of the weld lines is obscured by the beam web.

DIMENSIONS IN INCHES

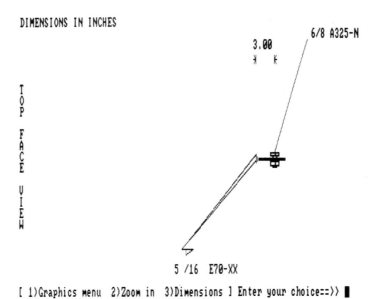

[1)Graphics menu 2)Zoom in 3)Dimensions] Enter your choice==>> ▮

Figure 8.53

Top Face View

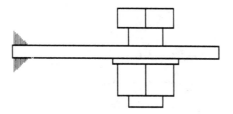

Hit any key to return to the graphics menu

Figure 8.54

Side Face View

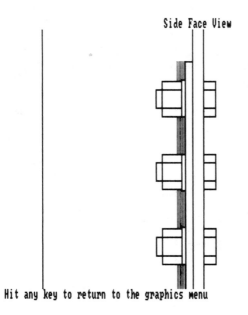

Hit any key to return to the graphics menu

Figure 8.55

8.4.8.2 Interactive session.

COLUMN INPUT

Enter I-section in the following format: I##x###
(ex. W36x300, S24x100, HP14x102, M14x 18)==>> W12x120

Data for W12x120 steel section

Area, A= 35.30 in.^2
Major Moment of Inertia, Ix= 1070.000 in.^4

Depth, d= 13.120 in.   Web Thickness, tw= 0.710 in.
Flange Width, b= 12.320 in.   Flange Thickness, tf= 1.105 in.

Is this section O.k.? (y or n)==>> y

Setting up the connection and coordinate arrays for the section

Possible types of steel

1) A36 (Fy=36ksi , Fu=58ksi)
2) A572-grade50 (Fy=50ksi , Fu=65ksi)
3) A529 (Fy=42ksi , Fu=60ksi)
4) Different than above types, Fy and Fu specified by user

Enter steel type desired for this section (1-4)==>> 1

BEAM INPUT

Enter I-section in the following format: I##x###
(ex. W36x300, S24x100, HP14x102, M14x 18)==>> W24x 68

Data for W24x 68 steel section

Area, A= 20.10 in.^2
Major Moment of Inertia, Ix= 1830.000 in.^4

Depth, d= 23.730 in.   Web Thickness, tw= 0.415 in.
Flange Width, b= 8.965 in.   Flange Thickness, tf= 0.585 in.

Is this section O.k.? (y or n)==>> y

Setting up the connection and coordinate arrays for the section

Possible types of steel

1) A36 (Fy=36ksi , Fu=58ksi)
2) A572-grade50 (Fy=50ksi , Fu=65ksi)
3) A529 (Fy=42ksi , Fu=60ksi)
4) Different than above types, Fy and Fu specified by user

Enter steel type desired for this section (1-4)==>> 1

What is the beam end reaction? (in kips)==>> 51.25

What is the length of the beam? (in feet)==>> 24

DESIGN THE BOLTS

BOLTS

1) A325 bolts
2) A490 bolts
Which type of bolt do you want to use? Specify by number==>> 1

CONNECTION TYPE

    1) F-friction type
    2) N-bearing type, threads in shearing plane
    3) X-bearing type, threads not in shearing plane
Which type of connection do you want to use? Specify by number==>> 2

  Checking the Shear Capacity of the Bolts

  Beam Web Connection Bolt Information:
    Number of bolts= 6 A325-N
    Diameter of bolts= .75 inches
    Spacing of bolts (center-center)= 3 inches
    Capacity of bolts= 55.66522 kips versus a reaction= 51.25 kips

The above design of the bolts is based on the smallest diameter bolt
that will work, from .75 in. to 1 in., with a 3 inch center to center spacing

  What would you like to change in the bolt design?
    1) Type of bolt used
    2) Bolt diameter
    3) Spacing
    4) Number of bolts
    5) Nothing

Enter choice by number==>> 5

        ** FIND MOMENT ACTING ON THE SHOP WELD AND THE PLATE **

The computed eccentricity of the loading from the center of the
bolt holes= 8.756619 inches

What is the distance from the center of the bolt holes to the edge of the
column,(in inches)? A distance of 3 inches is the most common==>> 3

  Moment connection is designed for= 602.5267 kip-inches

        ** SELECT A PLATE **

Minimum plate thickness for bending= .3100037 inches
Minimum plate thickness for shear= .1977238 inches

Try an initial plate thickness= 5 /16 inches

Would you like to increase this thickness? (y or n)==>> y

Specify increase in 1/16 in. increments==>> .125

 New plate thickness= 7 /16 inches

Is this thickness satisfactory to you? (y or n)==>> y

        ** DESIGN THE WELDS **

  Electrode Selection
        1)E60
        2)E70 (most common)
        3)E80
        4)E90
        5)E100
        6)E110

Enter desired electrode type (1-6) for welding==>> 2

Maximum stress on the weld= 18.20546 ksi

Minimum shop weld size using the plate length based on the bolt's
design= 5 /16 inches

A smaller weld size may be possible with a greater plate length.
Would you like to try the next smaller size weld? (y or n)==>> n

The plate is thicker than the maximum allowed to maintain ductile behavior

To remedy the situation you may use more bolts, a thicker bolt diameter, or a
different bolt type

        ** DESIGN THE BOLTS **

                BOLTS

            1) A325 bolts
            2) A490 bolts
Which type of bolt do you want to use? Specify by number==>> 1

                CONNECTION TYPE

            1) F-friction type
            2) N-bearing type, threads in shearing plane
            3) I-bearing type, threads not in shearing plane
Which type of connection do you want to use? Specify by number==>> 2

    Checking the Shear Capacity of the Bolts

    Beam Web Connection Bolt Information:
     Number of bolts= 6 A325-N
     Diameter of bolts= .75 inches
     Spacing of bolts (center-center)= 3 inches
     Capacity of bolts= 55.66522 kips versus a reaction= 51.25 kips

The above design of the bolts is based on the smallest diameter bolt
that will work, from .75 in. to 1 in., with a 3 inch center to center spacing

    What would you like to change in the bolt design?
        1) Type of bolt used
        2) Bolt diameter
        3) Spacing
        4) Number of bolts
        5) Nothing

Enter choice by number==>> 5

        ** FIND MOMENT ACTING ON THE SHOP WELD AND THE PLATE **

The computed eccentricity of the loading from the center of the
bolt holes= 8.756619 inches

What is the distance from the center of the bolt holes to the edge of the
column,(in inches)? A distance of 3 inches is the most common==>> 3

    Moment connection is designed for= 602.5267 kip-inches

        ** SELECT A PLATE **

Minimum plate thickness for bending= .3100037 inches
Minimum plate thickness for shear= .1977238 inches

Try an initial plate thickness= 5 /16 inches

Would you like to increase this thickness? (y or n)==>> n

```
     !! DESIGN THE WELDS !!

     Electrode Selection
          1)E60
          2)E70 (most common)
          3)E80
          4)E90
          5)E100
          6)E110

Enter desired electrode type (1-6) for welding==>> 2


 Maximum stress on the weld= 25.48765  ksi

 Minimum shop weld size using the plate length based on the bolt's
 design= 5 /16 inches

 A smaller weld size may be possible with a greater plate length.
 Would you like to try the next smaller size weld? (y or n)==>> y

 The new plate and weld lengths are= 19  inches
 Weld size= 4 /16 inches

 Is this design change satisfactory to you? (if not the initial design
 will be used) (y or n)==>> n


  The plate is thinner than the maximum allowed so continue on with the design


 Would you like to change what you have designed? (y or n)==>> n


 The computer now has to set up the designed connection for later graphic
 output. This takes a while so be patient. While you're waiting you can listen
 to music, when this stops you may continue on to the graphics output

 Would you like to listen to music while you wait? (y or n)==>> n




          !!!!!!!!!!!!!!!!!!!!!!!!!!!!!!!!!!!!!!!!!
          !  Summary of results for a simple    !
          !     shop welded and field bolted    !
          !   beam to column plate connection   !
          !!!!!!!!!!!!!!!!!!!!!!!!!!!!!!!!!!!!!!!!!




 Design end reaction= 51.25 kips

   !! I-Sections !!
 Column section: W12x120   Fy= 36 ksi      Fu== 58  ksi
 Beam section: W24x 68     Fy= 36 ksi      Fu== 58  ksi

   !! T's, Angles and Plates !!
 Plate connecting beam web to column flange:   PL 5 /16   X   4.625
 Length of the plate= 18 inches

   !! Bolts !!
 6  6 /8 inch A325-N bolts for plate to beam web connection

   !! Welds !!
 Shop weld information (plate to column flange)
    Weld size= 5 /16 inches  using an E70-XX  electrode
    Length of weld= 18  inches

 HIT ANY KEY TO CONTINUE ON TO THE GRAPHICS MENU
```

8.4.9 Design of Moment-Resisting Shop-Welded and Field-Bolted Flange Plate Connections

8.4.9.1 Comments. This moment connection type uses flange plates on both the top and bottom flanges of the beam to transfer the horizontal force caused by the applied moment. Both plates are connected to the beam by two rows of bolts. The plate is attached to the column flange by a full penetration groove weld. The shear force caused by the end reaction is transferred to the column by using a single framing plate. The framing plate is connected to the beam by a single row of bolts and to the column by a fillet weld on either side of the plate. A sample interactive session is given in Sec. 8.4.9.2. The design of this type of connection involves the following steps:

1. Determine the allowable bending stress for the beam.
2. Select the bolt type and size.
3. Check the reduced section of the beam for bending.
4. Determine the top and bottom flange plates' thickness and width.
5. Design the bolts for the flange plates.
6. Determine the length of the flange plates and then select them.
7. Design the field bolts connecting the framing plate to the beam.
8. Determine the minimum required framing plate dimensions and then select the plate.
9. Design the shop weld connecting the framing plate to the column.
10. Check if the column needs web reinforcement or stiffeners, and if so, design them.
11. Design the welds for the stiffener plates, if necessary.

The above steps are performed at least once for every design. The user has the option to repeat some of the steps to arrive at a design that he or she is pleased with.

The first step in the design of this connection type is to determine the allowable bending stress F_b for the beam. The user is presented with three possible options for finding F_b for the beam (see Sec. 8.4.9.2). Following the allowable stress determination, the user is prompted to enter the type and size of the bolts used to connect the flange plates to the beam. A short explanation of the common practice for selecting the bolt type is presented to the user.

After the allowable bending stress for the beam and the hole size for the bolts in the beam's flange are determined, a check is performed to make sure that the reduced section modulus S_x of the beam is still adequate for bending. If it is determined that the beam is not strong enough in bending, the user is prompted to select a smaller bolt diameter for the field bolts which will in turn provide a larger section modulus for the beam.

If the beam is satisfactory in bending, the next step is to find the required cross-sectional area for the flange plates. Once this is found, the user is prompted to enter a preliminary thickness for the plate. Based on the preliminary thickness selection, the minimum plate width is found. The minimum width takes into consideration the reduced area of the flange plates due to the two rows of bolt holes. The selection of the plate thickness and determination of the width can be repeated until the user is pleased with the plate design.

Following the establishment of the flange plates' width and thickness, the design of the field bolts—that is, the bolts connecting the flange plates to the beam's flange—may begin. The user must first select the horizontal gage of the two rows of bolts. A check is made to ensure that the gage is sufficiently large to provide clearance for assembly, and small enough to provide adequate end distances for the beam flange and flange plate as specified in AISCS Table 1.16.5.1. The design of the bolts is based on the shear force between the flange plate and the beam flange. Any combination of bolt designation, size, spacing, and number may be tried. It should be noted that any change in the bolt type or bolt diameter will require that steps 3 and 4 described earlier be repeated.

After the completion of the field bolt design, the final dimension for the flange plates is found. The length of the flange plates is found based on the bolt spacing selected previously, minimum end distance from the first bolt to the column, and minimum end distance from the last bolt to the edge of the plate (AISCS 1.16.5). As with other dimensions, the user may increase the minimum value calculated. After the length is found, the top and bottom flange plates are selected. The field bolts connecting the framing plate to the beam's web are designed next.

Following the field bolt design, the framing plate connecting the beam web to the column flange is designed. The thickness for the framing plate is based on providing sufficient area to resist the shear force. The bearing capacity of the framing plate, which is based on the thickness of the plate, is also checked. The width and length of the framing plate are found based on minimum end distances and bolt spacing. The final selection of the framing plate allows the user to increase any of the previously calculated minimum dimensions.

For shop weld—that is, the weld used to connect the framing plate to the column—first minimum and maximum sizes for the weld are found based on AISCS Table 1.17.2.A and AISCS 1.17.3. The length of the weld is established when the framing plate is designed. Therefore, the initial design procedure is to use the minimum weld size permissible with the known weld length to see if the weld is adequate to resist the applied shear force. If it is inadequate, the weld size is increased until it becomes adequate. If the largest weld size permissible is still inadequate, the framing plates' thickness is increased to allow for a greater maximum weld size. The user has the option to increase any selected weld size as long as it is not greater than the previously established maximum.

Each flange plate is connected to the column by means of a full penetration weld. The length of each weld is equal to the width of the flange plates. Backing bars are provided to ensure that a full weld is made in case of any underrun in the beam length.

After the connection has been designed, a check is made to find whether the column needs any web reinforcement or stiffeners based on the provisions of AISCS 1.15.5.2 and 1.15.5.3. If stiffeners are required, their dimensions must conform with the provisions of AISCS 1.15.5.4. When the minimum required stiffener area is very small, it is a common practice to simply forgo the inclusion of the stiffeners. Thus, if the required area for stiffeners is found to be less than 0.25 square inch, the user has the option to exclude the stiffeners from the design. If column web reinforcement is also required, it is brought to the user's attention, but this reinforcement is not designed by STEELCON.

When stiffeners are required, the fillet welds connecting them to the column must be designed. The weld design is divided into the weld for connecting the stiffeners to the column flange and the weld for connecting the stiffeners to the column web. In each case, a minimum weld size is first determined. The user has the option to increase the minimum weld size for each weld. Based on the weld size and the calculated force that the weld must resist, a minimum weld length is established. This minimum length may be increased by the user. An upper bound for the length along the column flange and web is established based on the stiffeners' width and length, respectively, minus 0.75 in. The 0.75-in. dimension is needed because a 0.75-in. cut is made at the corner of the stiffener that frames into the column's web and flange. This allows the stiffener to fit snugly by cutting the corner that would otherwise be obstructed by the column's fillet.

The example presented in this section is similar to the one in the AISC Manual [47]. The interactive session for this example is presented in Sec. 8.4.9.2.

Color plate 1 shows an isometric view of the connection. Color plate 2 shows the front-face view of the connection in medium resolution. Figures 8.56 through 8.58 show the front-face view of the connection in high resolution mode with dimensions and designations for I sections, connecting pieces, and connectors, respectively. Color plate 3 shows part of the front-face view in medium resolution

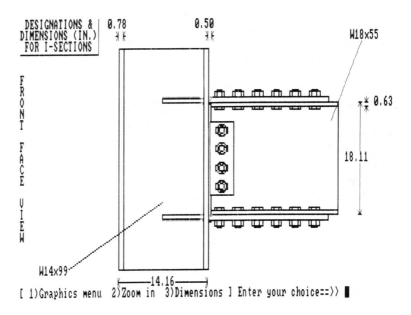

Figure 8.56

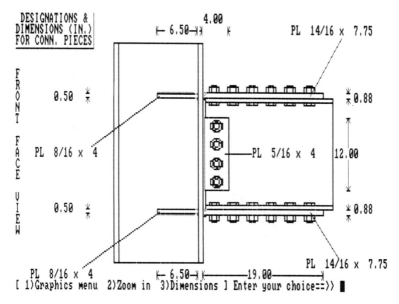

Figure 8.57

after it has been magnified. The magnification clearly shows the use of the backing bar to facilitate the full penetration weld. Figure 8.59 shows the side-face view of the connection in high resolution mode with dimensions and designations for connecting pieces. Color plate 4 shows a magnified side-face view in medium

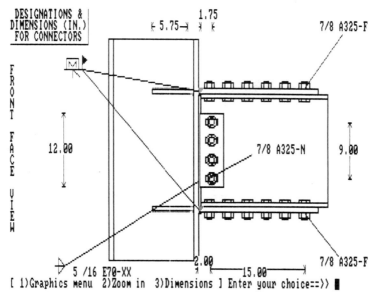

Figure 8.58

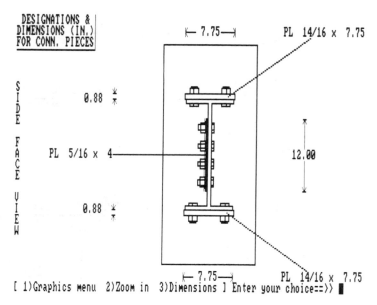

Figure 8.59

resolution. Color plate 5 shows the top-face view of the connection in medium resolution. This view has been magnified in color plate 6. Figure 8.60 shows the top-face view of the connection in high resolution with dimensions and designations for connectors.

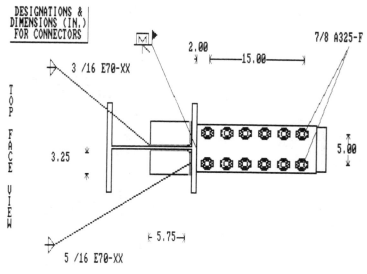

Figure 8.60

8.4.9.2 Interactive session.

```
                          COLUMN INPUT

Enter I-section in the following format: I##x###
(ex. W36x300, S12x40.8, M14x18, HP12x84)==>> W14x99

                 Data for W14x99 steel section

                 Area,  A= 29.10 in.^2
                 Major Moment of Inertia, Ix= 1110.000 in.^4

    Depth,  d= 14.160 in.          Web Thickness, tw=  0.485 in.
    Flange Width, bf= 14.565 in.   Flange Thickness, tf=  0.780 in.

Is this section O.k.? (y or n)==>> y


Setting up the connection and coordinate arrays for the section, please wait

         Possible types of steel

    1) A36 (Fy=36ksi , Fu=58ksi)
    2) A572-grade50 (Fy=50ksi , Fu=65ksi)
    3) A529 (Fy=42ksi , Fu=60ksi)
    4) Different than above types, Fy and Fu specified by user

Enter steel type desired for this section (1-4)==>> 1

                          BEAM INPUT


Enter I-section in the following format: I##x###
(ex. W36x300, S12x40.8, M14x18, HP12x84)==>> W18x55

                 Data for W18x55 steel section

                 Area,  A= 16.20 in.^2
                 Major Moment of Inertia, Ix=  890.000 in.^4

    Depth,  d= 18.110 in.          Web Thickness, tw=  0.390 in.
    Flange Width, bf=  7.530 in.   Flange Thickness, tf=  0.630 in.

Is this section O.k.? (y or n)==>> y


Setting up the connection and coordinate arrays for the section, please wait

         Possible types of steel

    1) A36 (Fy=36ksi , Fu=58ksi)
    2) A572-grade50 (Fy=50ksi , Fu=65ksi)
    3) A529 (Fy=42ksi , Fu=60ksi)
    4) Different than above types, Fy and Fu specified by user

Enter steel type desired for this section (1-4)==>> 1

What is the beam end reaction? (in kips)==>> 40

What is the moment the connection is designed to carry? (in kip-ft)==>> 153


    ** DETERMINE ALLOWABLE BENDING STRESS FOR THE BEAM **

The allowable bending stress for the beam must be found to design the bolts
connecting the flange plates to the beam's flange

Select the appropriate bending stress for the beam
    1) Compact section:  Fb=.66*Fy
    2) Non-compact section:  Fb=.60*Fy
    3) Non-compact section:  Fb<).60*Fy
```

Enter appropriate number(1,2, or 3) for the beam's allowable
bending stress==>> 1

Allowable bending stress (Fb) allowed for the beam=24.0 ksi

** CHECK FOR BEAM ADEQUACY AFTER FLANGE BOLT HOLES ARE CONSIDERED **

A preliminary selection of the size and type of bolts to be used in connecting
the beam's flange to both top and bottom flange plates must be made now.

> NOTE: Friction bolts are usually selected for this type of connection
> because oversized holes are used in the flange plates to facilitate
> any underrun or overrun in the beam. However, if you desire
> bearing type bolts may be used with standard bolt holes 1/16 in.
> larger than the bolt diameter.

BOLTS

1) A325 bolts
2) A490 bolts
Which type of bolt do you want to use? Specify by number==>> 1

CONNECTION TYPE

1) F-friction type
2) N-bearing type, threads in shearing plane
3) X-bearing type, threads not in shearing plane
Which type of connection do you want to use? Specify by number==>> 1

Enter desired preliminary bolt diameter (from .625 inches to 1.5 inches in .125
inch increments)==>> 1.5

The required section modulus for the beam is 76.6 in.^3

The net section modulus for the beam after flange area reduction is
considered (if any) is 75.7 in.^3

The bolt size selected is too large, reselect the bolt diameter

Enter desired preliminary bolt diameter (from .625 inches to 1.5 inches in .125
inch increments)==>> .875

The required section modulus for the beam is 76.6 in.^3

The net section modulus for the beam after flange area reduction is
considered (if any) is 89.0 in.^3

The beam is O.K. for bending

** DESIGN TOP AND BOTTOM FLANGE PLATES **

Possible types of steel for the flange plates
1) A36 (Fy=36 ksi , Fu=58 ksi)
2) A572-grade50 (Fy=50 ksi , Fu=65 ksi)
3) A529 (Fy=42 ksi , Fu=60 ksi)
4) Different than above types, Fy and Fu specified by the user

Enter steel type desired for this section (1-4)==>> 1

Required cross-sectional area for the both flange plates is 4.69 in.^2

You must select a preliminary thickness for the plates, from that value
the required plate width will be determined

Enter desired thickness in 0.0625 inch increments==>> .125

The required width for the plate exceeds the column flange width, specify a
larger plate thickness

Enter desired thickness in 0.0625 inch increments==>> .863

The thickness must be selected in 1/16 inch increments, try again

Enter desired thickness in 0.0625 inch increments==>> 1.125

The required width for the plates based on net area requirements= 6.5 in.

Would you like to select a different plate thickness (this will result in a
new plate width)?==>> y

Enter desired thickness in 0.0625 inch increments==>> .875

The required width for the plates based on net area requirements= 7.75 in.

Would you like to select a different plate thickness (this will result in a
new plate width)?==>> n

Would you like to increase this minimum width? (y or n)==>> n

    ** DESIGN BOLTS CONNECTING THE BEAM TO THE FLANGE PLATES **

Select the horizontal gage for the bolts (in .25 inch increments)==>> 6

The gage selected is too wide for the beam flange, try again

Select the horizontal gage for the bolts (in .25 inch increments)==>> 5

    Flange Plate to Beam Flange Connection Bolt Information
      Number of bolts= 12 A325-F
      Diameter of bolts= .875 inches
      Spacing of bolts (center-center)=2.34 inches
      Capacity of bolts=108.2 kips versus a flange force=101.4 kips

    What would you like to change in the bolt design?
      1) Type of bolt
        Note: will necessitate a redesign of the flange plates
      2) Bolt diameter
        Note: will necessitate a redesign of the flange plates
      3) Spacing
      4) Number of bolts
      5) Nothing

Enter choice by number==>> 3

Enter desired spacing for the bolts in inches==>> 3

    Flange Plate to Beam Flange Connection Bolt Information
      Number of bolts= 12 A325-F
      Diameter of bolts= .875 inches
      Spacing of bolts (center-center)=3.00 inches
      Capacity of bolts=108.2 kips versus a flange force=101.4 kips

    What would you like to change in the bolt design?
      1) Type of bolt
        Note: will necessitate a redesign of the flange plates
      2) Bolt diameter
        Note: will necessitate a redesign of the flange plates
      3) Spacing
      4) Number of bolts
      5) Nothing

Enter choice by number==>> 5

    ** SELECT FLANGE PLATES **

Minimum distance from the column to center of the first holes in the
flange plate= 2 inches

Would you like to increase this distance? (y or n)==>> n

The minimum flange plate length= 19 inches

Would you like to increase this minimum length? (y or n)==>> n

** TOP & BOTTOM FLANGE PLATE SELECTION **

The plate selected is PL 14/16 x 7.75
The length of the plate= 19 inches

** DESIGN FIELD BOLTS CONNECTING THE FRAMING PLATE TO THE BEAM WEB **

BOLTS

1) A325 bolts
2) A490 bolts
Which type of bolt do you want to use? Specify by number==>> 1

CONNECTION TYPE

1) F-friction type
2) N-bearing type, threads in shearing plane
3) X-bearing type, threads not in shearing plane
Which type of connection do you want to use? Specify by number==>> 2

Checking Shear Capacity of the Bolts

Framing Plate to Beam Web Connection Bolt Information:
Number of bolts= 5 A325-N
Diameter of bolts= .75 inches
Spacing of bolts (center-center)=3.00 inches
Capacity of bolts= 46.4 kips versus a reaction= 40 kips

The above design is based on using the smallest diameter bolt that will work
using a 3 inch center to center spacing or 3 times the bolt diameter,
whichever is larger

What would you like to change in the bolt design?
1) Type of bolt
2) Bolt diameter
3) Spacing
4) Number of bolts
5) Nothing

Enter choice by number==>> 2

Enter desired bolt diameter (from .625 inches to 1.5 inches in .125 inch
increments)==>> .875

Framing Plate to Beam Web Connection Bolt Information:
Number of bolts= 4 A325-N
Diameter of bolts= .875 inches
Spacing of bolts (center-center)=2.34 inches
Capacity of bolts= 50.5 kips versus a reaction= 40 kips

What would you like to change in the bolt design?
1) Type of bolt
2) Bolt diameter
3) Spacing
4) Number of bolts
5) Nothing

Enter choice by number==>> 3

Absolute minimum bolt spacing=2.333 inches

Absolute maximum bolt spacing (due to space limitations)=4.212 inches

Enter desired new bolt spacing (center to center) in inches==>> 3

Framing Plate to Beam Web Connection Bolt Information:
Number of bolts= 4 A325-N
Diameter of bolts= .875 inches
Spacing of bolts (center-center)=3.00 inches
Capacity of bolts= 50.5 kips versus a reaction= 40 kips

```
What would you like to change in the bolt design?
    1) Type of bolt
    2) Bolt diameter
    3) Spacing
    4) Number of bolts
    5) Nothing

Enter choice by number==>> 5



       ** CHECK FOR BEAM BEARING CAPACITY **

Allowable bearing capacity in the beam web=115.9 kips  versus a reaction= 40
kips

Meets requirements



       ** CALCULATING DISTANCE BETWEEN WEB END AND CENTER OF BOLT HOLES **

Minimum distance from end of the beam to center of holes in the
web= 1.5 inches

Would you like to increase this distance? (y or n)==>> y

How much do you want to increase the minimum value?
(specify in .25 inch increments)==>> .25

Updated value for the distance from the end of the beam to the
center of the bolt holes= 1.75 inches



       ** DESIGN FRAMING PLATE FOR SHEAR **

 Possible types of steel for the framing plate
    1) A36 (Fy=36 ksi , Fu=58 ksi)
    2) A572-grade50 (Fy=50 ksi , Fu=65 ksi)
    3) A529 (Fy=42 ksi , Fu=60 ksi)
    4) Different than above types, Fy and Fu specified by the user

Enter steel type desired for this section (1-4)==>> 1

 Preliminary thickness for the plate= .3125  inches


 Check Plate Thickness to Make Sure it is Adequate for Bearing

Bearing capacity for framing plate using a thickness of .3125  inches
and an end distance of 1.5  inches is adequate



       ** SELECT THE FRAMING PLATE **

 Minimum Dimensions for Framing Plate
   Thickness= .3125 inches
   Width= 3.375 inches
   Depth= 12 inches



Would you like to increase any of these minimum dimensions? (y or n)==>> y

Enter desired thickness in 0.0625 inch increments==>> .3125

Enter desired width in 0.25 inch increments==>> 4

Enter desired depth in 0.25 inch increments==>> 12

 Framing plate used: PL  5/16 x  4
```

** DESIGN SHOP WELD CONNECTING THE FRAMING PLATE TO THE COLUMN **

Electrode Selection
1)E60
2)E70 (most common)
3)E80
4)E90
5)E100
6)E110

Enter desired electrode type (1-6) for welding==>> 2

Lower and Upper Limits for the Weld Size for the Welds Connecting the
Framing Plate to the Column's Flange
   Minimum weld size= 5 /16 in.
   Maximum weld size= 5 /16 in.

The design of the shop weld requires a weld size of 5 /16 in.
with a length on each side of the plate of 12 inches

Would you like to increase this minimum weld size? (y or n)==>> n

** CHECK IF COLUMN WEB NEEDS REINFORCEMENT OR STIFFENERS **

The column web does not need to be reinforced due to shear

Column web stiffeners required at both top and bottom flanges of the beam

** DESIGN COLUMN WEB STIFFENERS **

Possible types of steel for the stiffener plates
   1) A36 (Fy=36 ksi , Fu=58 ksi)
   2) A572-grade50 (Fy=50 ksi , Fu=65 ksi)
   3) A529 (Fy=42 ksi , Fu=60 ksi)
   4) Different than above types, Fy and Fu specified by the user

Enter steel type desired for this section (1-4)==>> 1

Area of stiffeners required at each flange of the beam=0.67 in.^2

Top Stiffener Minimum Dimensions:
   Stiffener width= 2.5
   Stiffener thickness= .5
   Stiffener length= 6.5
   Area provided by stiffeners= 1.25 in.^2 versus a required
      area of 0.67 in.^2

Would you like to increase any of these minimum dimensions? (y or n)==>> y

Specify width desired for stiffener in 0.25 in. increments==>> 4

Specify the thickness for stiffener in 0.0625 in. increments==>> .5

Specify length desired for stiffener in 0.25 in. increments==>> 6.5

      The plate selected is PL 8/16 x 4
      The length of the plate= 6.5 inches

Bottom Stiffener Minimum Dimensions:
   Stiffener width= 2.5
   Stiffener thickness= .4375
   Stiffener length= 6.5
   Area provided by stiffeners= 1.09 in.^2 versus a required
      area of 0.67 in.^2

Would you like to increase any of these minimum dimensions? (y or n)==>> y

Specify width desired for stiffener in 0.25 in. increments==>> 4

Specify the thickness for stiffener in 0.0625 in. increments==>> .5

Specify length desired for stiffener in 0.25 in. increments==>> 6.5

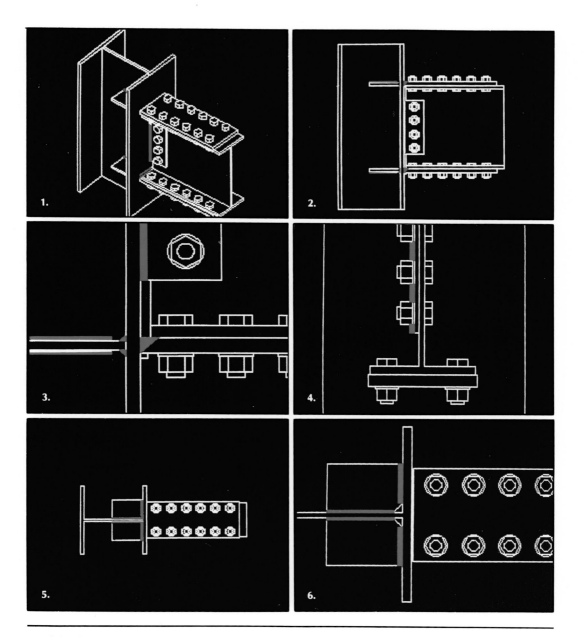

Color Plates 1–6. Color Plate 1. An isometric view of the shop-welded and field-bolted flange plate connection. X rotation = −40; Y rotation = 40. Color Plate 2. The front-face view of the shop-welded and field-bolted flange plate connection in medium resolution. Color Plate 3. The front-face view of color plate 2 after it has been magnified. Color Plate 4. A magnified side-face view of the shop-welded and field-bolted connection in medium resolution. Color Plate 5. The top-face view of the connection in medium resolution. Color Plate 6. A magnified view of the top-face view in medium resolution.

The plate selected is PL 8/16 x 4
The length of the plate= 6.5 inches

STIFFENER WELD REQUIREMENTS

Lower Limit for the Weld Size for the Welds Connecting the Top & Bottom
Stiffeners to the Column Flange is 5 /16 in.

Would you like to use this minimum size weld for the stiffener to
to column connection? (y or n)==>> y

The force the stiffener welds must resist=24.29 kips

 Determine minimum length of the welds along the column flange

Required weld length= 1 inches

Would you like to increase this minimum length? (y or n)==>> y

Enter desired length in 0.25 inch increments==>> 3.25

Lower Limit for the Weld Size for the Welds Connecting the Top & Bottom
Stiffeners to the Column Web is 3 /16 in.

Would you like to use this minimum size weld for the stiffener to
to column connection? (y or n)==>> y

 Determine minimum length of the welds along the column web

Required weld length= 1.5 inches

Would you like to increase this minimum length? (y or n)==>> y

Enter desired length in 0.25 inch increments==>> 5.75

Would you like to change what you have designed? (y or n)==>> n

The computer now has to set up the designed connection for later graphic
output. This takes a while so be patient. While you're waiting you can listen
to music, when this stops you may continue on to the graphics output

Would you like to listen to music while you wait? (y or n)==>> n


```
                    ####################################
                    #        Summary of results for a  #
                    #          moment resisting         #
                    #      shop welded and field bolted #
                    #      beam to column flange plate  #
                    #               connection          #
                    ####################################
```

Design end reaction= 40 kips
Design end moment= 153 kip-ft

I-Sections
Column section: W14x99   Fy= 36 ksi   Fu== 58 ksi
Beam section: W18x55   Fy= 36 ksi   Fu== 58 ksi

T's, Angles, and Plates
Top flange plate is a PL 14/16 x 7.75
Fy= 36 ksi   Fu= 58 ksi
Length of the plate= 19 inches

Bottom flange plate is a PL 14/16 x 7.75
Fy= 36 ksi   Fu= 58 ksi
Length of the plate= 19 inches

Framing plate connecting beam to column is a PL 5/16 x 4
Fy= 36 ksi   Fu= 58 ksi
Length of the plate= 12 inches

Top column web stiffener plates are 2 PL 8/16 x 4
Fy= 36 ksi   Fu= 58 ksi
Length of the plates= 6.5 inches
     Note: These plates are clipped at the corner framing into the column's
           flange and web to accomodate the column's fillet

Bottom column web stiffener plates are 2 PL 8/16 x 4
Fy= 36 ksi   Fu= 58 ksi
Length of the plates= 6.5 inches
     Note: These plates are clipped at the corner framing into the column's
           flange and web to accomodate the column's fillet

Two backing bars will also be provided beneath the top and bottom flange
plates to allow for downhand welding of the full penetration weld
connecting the flange plates to the column flange

** Bolts **
12 7 /8 inch A325-F bolts used in the connecting the beam to
    the top flange plate as well as to the bottom flange plate

Hole size used in the flange plate= 1.0625 in.

4 7 /8 inch A325-N bolts for framing plate to beam web connection

** WELDS **
Shop Welds

Welds connecting the framing plate to the column flange
    Weld size= 5 /16 inches using an E70-XX electrode
    Length of weld= 12 inches

Top stiffener welds
    Stiffener to column flange weld size= 5 /16 in. using an E70 XX electrode
    Length of weld= 3.25 inches

    Stiffener to column web weld size= 3 /16 in. using an E70 XX electrode
    Length of weld= 5.75 inches

Bottom stiffener welds
    Stiffener to column flange weld size= 5 /16 in. using an E70 XX electrode
    Length of weld= 3.25 inches

    Stiffener to column web weld size= 3 /16 in. using an E70 XX electrode
    Length of weld= 5.75 inches

    Full penetration welds connecting flange plates to the column flange
    Length of top weld= 7.75 inches
    Length of bottom weld= 7.75 inches

HIT ANY KEY TO CONTINUE ON TO THE GRAPHICS MENU

8.5 PROBLEMS

8.1 A tension member is connected to the flange of a W shape column through a WT stub
and two vertical rows of 1-in. A325 friction-type bolts, as shown in Fig. 8.61. The
maximum tensile force in the member is 200 K. Find the minimum required number
of bolts. Assume that holes are standard size, that threads are excluded from the shear
planes, and that the tensile force passes through the centroid of bolts.

8.2 A built-up beam is made of four L8 × 6 × 1 angles and two PL40 × $\frac{1}{2}$ bolted together as
shown in Fig. 8.62. Type of steel is A36 with a yield stress of 36 ksi and an ultimate
stress of 58 ksi. Using 1-in. A325 bearing-type bolts, find the spacing of the bolts along
the *length* of the beam for a shear force of $V = 450$ K. Assume that threads are excluded
from the shear planes.

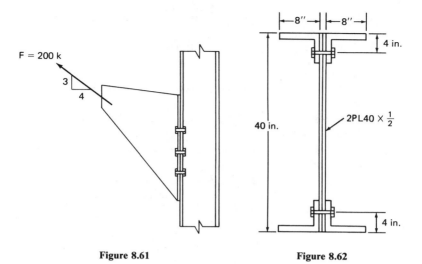

Figure 8.61   **Figure 8.62**

8.3 The web of a W27 × 114 beam is connected to the flange of a W14 × 120 column through two L4 × 4 × $\frac{1}{2}$ and A325 friction-type bolts, as shown in Fig. 8.63. Four 1-in.-diameter bolts are used to connect the angles to the beam web, and ten $\frac{7}{8}$-in.-diameter bolts are used to connect the angles to the column flange. Neglecting the eccentricity, find the shear capacity of the connection. The column, the beam, and the angles are made of A36 steel with a yield stress of 36 ksi.

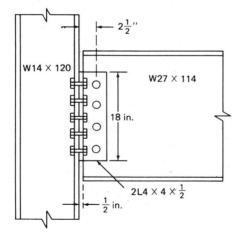

Figure 8.63

8.4 Solve Problem 8.3, using bearing-type connections with threads excluded from the shear plane.

8.5 In Problem 4.2, find the maximum tension capacity of the connection on the basis of the strength of bolts for the following cases:

a. A325 bolts and friction-type connection.

b. A325 bolts and bearing-type connection, assuming that threads are excluded from the shear planes.

    c. A325 bolts and bearing-type connection, assuming that threads are not excluded from the shear plane.

    d. A490 bolts and friction-type connection.

    e. A490 bolts and bearing-type connection, assuming that threads are excluded from the shear planes.

    f. A490 bolts and bearing-type connection, assuming that threads are not excluded from the shear plane.

8.6 Solve Example 2 of this chapter, assuming that the bottom channel is C12 × 25 instead of C12 × 30.

8.7 Solve Example 3 of this chapter, assuming that the force $P = 50$ K makes an angle of 45 degrees with the horizontal (to the right).

8.8 Two 1-in.-thick plates made of A36 steel are connected through fillet weld as shown in Fig. 8.64. Find the maximum force F that can be applied to the connection when $\frac{1}{2}$-in. submerged arc welding and E60 electrode are used.

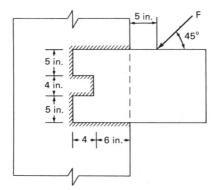

Figure 8.64

8.9 A welded built-up girder is made of a WT12 × 81 and a C10 × 30, as shown in Fig. 8.65.

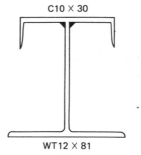

Figure 8.65

The maximum shear force in the girder is 100 K. Design the continuous fillet weld for connecting the two sections. Use E70 electrodes and

    a. shielded metal arc welding (SMAW)

    b. submerged arc welding (SAW)

8.10 Solve Problem 8.9, but use intermittent fillet weld with maximum spacing.

Design
of Plate Girders

9.1 INTRODUCTION

The most common type of plate girder is an I-shaped section built up from two flange plates and one web plate, as shown in Figs. 9.1 and 9.2. The moment-resisting capacities of plate girders lie somewhere between those of deep standard rolled

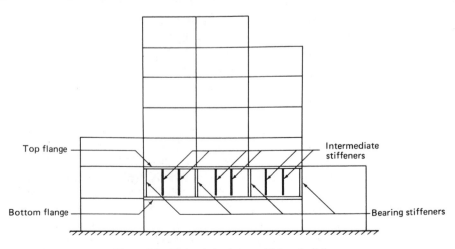

Figure 9.1 Plate girder in a multistory building.

wide-flange sections and those of trusses. Plate girders can be welded (Figs. 9.2 to 9.5), riveted, or bolted (Fig. 9.6). Riveted plate girders are practically obsolete. Very few bolted plated girders are designed nowadays. Therefore, we cover only the design of welded plate girders in this book.

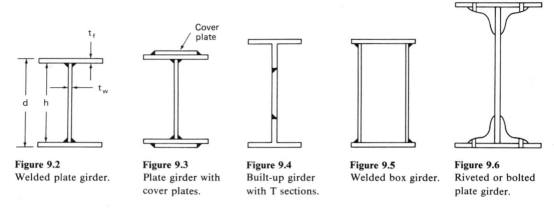

Figure 9.2
Welded plate girder.

Figure 9.3
Plate girder with cover plates.

Figure 9.4
Built-up girder with T sections.

Figure 9.5
Welded box girder.

Figure 9.6
Riveted or bolted plate girder.

Plate girders are used in both buildings and bridges. In buildings, when large column-free spaces are designed to be used as an assembly hall, for example, the plate girder is often the economical solution. In such cases, the designer must choose between a plate girder and a truss. Plate girders, in general, have the following advantages over trusses:

1. Connections are less critical for plate girders than for trusses, particularly statically determinate trusses. In a statically determinate truss, one poor connection is sufficient to cause the possible collapse of the truss.

2. Fabrication cost of plate girders is less than that of trusses.

3. Plate girders can be erected more rapidly and more cheaply than trusses.

4. Depth of a plate girder is less than the height of a comparable truss. Consequently, plate girders need less vertical clearance than trusses. This makes them very attractive for multilevel highway bridges.

5. Plate girders generally vibrate less than trusses under moving loads.

6. Painting of plate girders is easier than painting of trusses. This means less maintenance cost for plate girders.

In contrast, plate girders in general are heavier than trusses, especially for very long spans.

Plate girders basically carry the loads by bending. The bending moment is mostly carried by flange plates. In order to achieve maximum economy, hybrid plate girders are sometimes used. In a hybrid girder, flange plates are made of higher-strength steel than that of the web. Or, in a tee-built-up plate girder, as shown in Fig. 9.4, the two T sections are made of higher-strength steel than the connecting web plate. Design of hybrid plate girders is also covered in this chapter. Allowable bending stress for hybrid girders is limited to $0.60F_y$.

9.2 POSTBUCKLING BEHAVIOR OF THE WEB PLATE

In addition to flange plates and a web plate, a plate girder often consists of intermediate and bearing stiffeners. As mentioned in the previous section, the two flange plates basically carry the bending moment. A web plate is needed to unify the two flange plates and to carry the shear. Thin web plates are susceptible to unstable behavior. Thick web plates make the girder unnecessarily heavy. A relatively thin web plate strengthened by stiffeners often yields the lightest plate girder.

Stiffened plate girders are designed on the basis of the ultimate strength concept. As the magnitude of the load on the girder is increased, the web panels between adjacent vertical stiffeners buckle due to diagonal compression resulting from shear. For a theoretical presentation of the subject the reader should refer to Salmon and Johnson [37]. For the designer of plate girders the detailed knowledge of theoretical development is not essential. He should, however, acquire a feel for the behavior of plate girders under increasing load.

If the plate girder has properly designed stiffeners, the instability of the web plate panels, bounded on all sides by the transverse stiffeners or flanges, will not result in its failure. In fact, after the web panels buckle in shear, the plate girder behaves like the Pratt truss shown in Fig. 9.7(a). It will then be able to carry additional loads. A stiffened plate girder has considerable postbuckling strength. The Pratt truss of Fig. 9.7(a) is subjected to a concentrated load applied at its midspan. In this truss, the vertical members are in compression and the diagonals are in tension. The postbuckling behavior of the plate girder is analogous to the behavior of this truss. As shown in Fig. 9.7(b), after the shear instability of the web plate takes place, a redistribution of stresses occurs; the stiffeners behave like axially

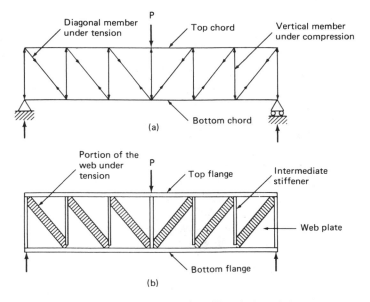

Figure 9.7   Analogy between a truss and a stiffened plate girder.

compressed members, and shaded portions of the web behave like tension diagonals in the truss of Fig. 9.7(a). This truss-like behavior is called *tension-field action* in the literature. The post-buckling strength of the web plate may be three or four times its initial buckling strength. Consequently, designs on the basis of tension-field action yield better economy.

Hybrid girders cannot be designed on the basis of tension-field action, due to the lack of sufficient experimental results.

9.3 PROPORTIONING OF THE WEB PLATE

At the outset, we must initially choose a value for the depth h of the web plate. As a general guideline, experience shows that the ratio of the depth of the web plate h to span length L varies from $\frac{1}{25}$ to $\frac{1}{6}$.

$$\frac{1}{25} \leq \frac{h}{L} \leq \frac{1}{6} \tag{9.1}$$

This ratio, however, is often within the range $\frac{1}{15}$ to $\frac{1}{10}$.

$$\frac{1}{15} \leq \frac{h}{L} \leq \frac{1}{10} \tag{9.2}$$

Deeper girders are generally used when the live loads are heavy (for example, when they need to carry large column loads in high-rise buildings). Very shallow girders with $\frac{1}{25} < h/L < \frac{1}{15}$ are used as continuous plate girders.

In practical design of plate girders, we should design the plate girder with several different values of the web depth-to-span ratios and find the total weight of the plate girder for each case. By drawing the total weight versus the h/L ratio, we can obtain an economical (practical optimum or minimum weight) solution for our design. Of course, repetitive manual design of plate girders is quite cumbersome and time-consuming. However, with the aid of the interactive microcomputer program (to be discussed in Sec. 9.12), the final design can be achieved quickly. Totally automated optimum design of stiffened plate girders is rather complicated and a matter of current research. Due to the highly nonlinear nature of the problem, a final practical design may not be readily obtained without the interaction of the designer. Abuyounes and Adeli [1, 2] presented an algorithm for minimum-weight design of simply supported steel plate girders. This work has recently been extended by Adeli and Chompooming to optimization of continuous prismatic and nonprismatic plate girders [8, 9]. In this book, however, our approach is interactive design, which is presented in Sec. 9.12.

After the h/L ratio has been selected, the depth of the web plate will be known. The next step is to choose the web thickness. The web thickness is chosen based on the following two criteria:

1. The web plate should have sufficient buckling strength to prevent vertical buckling of the compression flange into the web.
2. The web plate should carry all the shearing force.

In calculating the shear strength of the web, it is assumed that the shear stress distribution is uniform throughout the web depth.

During the postbuckling behavior of the web plate, the bending curvature of the plate girder produces compressive forces in the web plate, as shown in Fig. 9.8.

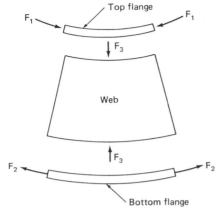

Figure 9.8 Squeezing of the web due to bending of the girder during the tension-field action.

This figure shows a portion of the plate girder located between two neighboring sections. Due to the deflected shape of the girder, the compressive forces F_1 acting on the top compression flange and the tension forces F_2 acting on the bottom tension flange create compressive forces F_3 on the web plate. This plate should have sufficient vertical buckling strength to resist the compressive forces F_3. To satisfy this requirement, according to AISCS Sec. 1.10.2, the web depth-thickness ratio should not be greater than α_1, which is a decreasing function of *the yield stress of the compression flange F_y.*

$$\frac{h}{t_w} \leq \alpha_1 = \frac{14{,}000}{[F_y(F_y + 16.5)]^{1/2}} \tag{9.3}$$

The variation of α_1 with F_y is shown in Fig. 9.9. This equation is derived from a stability analysis of the web plate, taking into account the effect of residual stresses but without including the transverse stiffeners. For closely spaced stiffeners—that is, when spacing of the transverse stiffeners a is not greater than 1.5 times the girder depth d—the limiting ratio α_1 is increased to α_2.

$$\frac{h}{t_w} \leq \alpha_2 = \frac{2000}{\sqrt{F_y}} \qquad \text{when } a \leq 1.5d \tag{9.4}$$

The variable α_2 is also shown in Fig. 9.9. Note that the difference between α_2 and α_1 increases with the yield stress. For high-strength steel, α_2 is much larger than α_1.

To satisfy the second criterion, we should have

$$t_w \geq \frac{V}{hF_v} \tag{9.5}$$

where V is the shear force and F_v is the allowable shear stress given in AISCS Sec. 1.10.5.2.

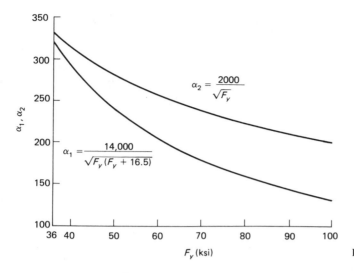

$$\alpha_2 = \frac{2000}{\sqrt{F_y}}$$

$$\alpha_1 = \frac{14,000}{\sqrt{F_y\,(F_y + 16.5)}}$$

F_y (ksi)   **Figure 9.9**

$$F_v = \frac{F_y C_v}{2.89} \leq 0.40 F_y \tag{9.6}$$

where

$$C_v = \begin{cases} \dfrac{45,000k}{F_y(h/t_w)^2} & \text{when } C_v \leq 0.8 \\[2mm] \dfrac{190\sqrt{k}}{h/t_w\sqrt{F_y}} & \text{when } C_v > 0.8 \end{cases} \tag{9.7}$$

$$k = \begin{cases} 4.00 + \dfrac{5.34}{(a/h)^2} & \text{when } a/h < 1.0 \quad (k > 9.34) \\[2mm] 5.34 + \dfrac{4.00}{(a/h)^2} & \text{when } a/h > 1.0 \end{cases} \tag{9.8}$$

Note that C_v is the ratio of shear stress at buckling to the shear yield stress [37]. For hybrid girders, F_y in Eqs. (9.6) and (9.7) is the yield stress of the web steel.

We can increase the allowable shear stress by relying on the postbuckling behavior and tension-field action of the web plate, provided that the following conditions are met:

1. The plate girder is not a hybrid one.
2. Intermediate stiffeners are provided.
3. $C_v \leq 1.0$ $\qquad\qquad\qquad\qquad\qquad\qquad\qquad\qquad\qquad\qquad$ (9.9)
4. $a/h \leq [260/(h/t_w)]^2$ $\qquad\qquad\qquad\qquad\qquad\qquad\qquad$ (9.10)
5. $a/h \leq 3.0$ $\qquad\qquad\qquad\qquad\qquad\qquad\qquad\qquad\qquad\qquad$ (9.11)

The last two conditions are somewhat arbitrarily chosen limits on the panel aspect ratio a/h to facilitate handling during fabrication and erection. When the effect of tension-field action is taken into account, the allowable shear stress is given by

$$F_v = \frac{F_y}{2.89}\left[C_v + \frac{1 - C_v}{1.15(1 + a^2/h^2)^{1/2}}\right] \leqslant 0.40F_y \qquad (9.12)$$

Note that the second term within the brackets is the tension-field contribution.

One may select the web thickness based on the first criterion [Eqs. (9.3) and (9.4)] and then check for the second criterion [Eq. (9.5)]. In this case the maximum computed shear stress $(f_v)_{max}$ must be less than the allowable shear stress F_v.

$$(f_v)_{max} = \frac{V_{max}}{ht_w} \leqslant F_v \qquad (9.13)$$

After preliminary proportioning of the web plate, we may check if intermediate stiffeners are needed. According to AISCS Sec. 1.10.5.3, intermediate stiffeners are not required if

$$\frac{h}{t_w} < 260 \qquad (9.14)$$

and the maximum shear stress in the web is less than the allowable shear stress given by Eq. (9.6). Equation (9.6) can be specialized for the case of no stiffeners. For very large a/h, Eq. (9.8) yields $k = 5.34$. Substituting this value of k into Eq. (9.7) and the resulting values into Eq. (9.6), we finally find the following equation for the allowable shear stress when intermediate stiffeners are not needed:

$$F_v = \begin{cases} \dfrac{83{,}150}{(h/t_w)^2} & \text{when } \dfrac{h}{t_w} \geqslant \dfrac{548}{\sqrt{F_y}} \\[3mm] \dfrac{152\sqrt{F_y}}{h/t_w} & \text{when } \dfrac{380}{\sqrt{F_y}} \leqslant \dfrac{h}{t_w} \leqslant \dfrac{548}{\sqrt{F_y}} \\[3mm] 0.40F_y & \text{when } \dfrac{h}{t_w} \leqslant \dfrac{380}{\sqrt{F_y}} \end{cases} \qquad (9.15)$$

It should be noted that plate girders with intermediate stiffeners are generally lighter than plate girders without intermediate stiffeners.

To prevent the undesirable consequences of corrosion, an absolute minimum web thickness is usually specified in practice. A minimum thickness of $\frac{3}{8}$ in. is recommended for bridge plate girders. For plate girders used in buildings which are not exposed to the harsh corrosive environment, a smaller absolute minimum web thickness of $\frac{1}{4}$ in. is suggested.

9.4 PROPORTIONING OF THE FLANGES

9.4.1 Preliminary Calculation of Flange Area

The aim is to select a flange plate of area sufficient to carry the maximum bending moment M_{max}.

$$\text{Required } S = \frac{M_{\text{max}}}{F_b} \qquad (9.16)$$

In this equation, S is the elastic section modulus with respect to the major axis and F_b is the allowable bending stress given in AISCS Sec. 1.5.1.4 and discussed in Chapter 5. We can find an approximate relation for the section modulus. The moment of inertia of the section with respect to the major axis is

$$I = \tfrac{1}{12}t_w h^3 + 2b_f t_f (h/2 + t_f/2)^2 + \tfrac{2}{12}b_f t_f^3 \qquad (9.17)$$

$$I \simeq \tfrac{1}{12}t_w h^3 + 2A_f(h/2)^2$$

where b_f is the width of the flange plate, t_f is the thickness of the flange plate, and $A_f = b_f t_f$ = area of the flange. The elastic section modulus is then approximately equal to

$$S = \frac{I}{d/2} \simeq \frac{I}{h/2} = \frac{t_w h^2}{6} + A_f h \qquad (9.18)$$

By equating Eqs. (9.16) and (9.18) and solving for A_f, we find an equation for the preliminary estimate of the area of the flange.

$$A_f = b_f t_f \simeq \frac{M_{\text{max}}}{F_b h} - \frac{t_w h}{6} = \frac{M_{\text{max}}}{F_b h} - \frac{A_w}{6} \qquad (9.19)$$

In this equation, $A_w = t_w h$ is the web area.

9.4.2 Preliminary Selection of the Flange Plate

In order to avoid local flange buckling, the width-thickness ratio of the flange plate is limited by AISCS Sec. 1.9.1.2 (Table 5.1 in the text).

$$\frac{b_f}{t_f} < \frac{190}{\sqrt{F_y}} \qquad (9.20)$$

To find the minimum flange thickness required, we set

$$b_f = \frac{190}{\sqrt{F_y}} t_f \qquad (9.21)$$

Substituting Eq. (9.21) into Eq. (9.19) and solving for t_f, we obtain

$$t_f = \left[\frac{\sqrt{F_y}}{190} \left(\frac{M_{\text{max}}}{F_b h} - \frac{t_w h}{6} \right) \right]^{1/2} = \left[\frac{\sqrt{F_y}}{190} A_f \right]^{1/2} \qquad (9.22)$$

This equation roughly gives the minimum flange thickness required to prevent the local buckling of the flange plate. We may round this thickness to a commercially available size, for example, as a fraction of $\frac{1}{16}$ in., and use it as the trial design thickness of the flange plate. However, in many cases, this design would result in very thin and wide flange plates. Therefore, the designer may wish to choose a flange thickness larger than that obtained by Eq. (9.22). After selecting the thickness of the flange plate, we find the required flange width b_f from Eq. (9.19).

At this point we can calculate the exact value of the moment of inertia of the section from Eq. (9.17) and check if the available section modulus is at least equal to M_{max}/F_b.

9.4.3 Reduction of the Allowable Bending Stress

During the postbuckling behavior in tension-field action, the lateral displacement of the web plate on the compression side reduces its bending capacity and consequently additional stresses are transmitted to the compression flange. To take into account this loss of bending capacity, AISCS Sec. 1.10.6 requires that the allowable bending stress in the compression flange be reduced to F_b' when the web depth-thickness ratio exceeds $760/\sqrt{F_b}$.

$$F_b' \leq F_b \left[1.0 - 0.0005 \frac{A_w}{A_f} \left(\frac{h}{t_w} - \frac{760}{\sqrt{F_b}} \right) \right] \qquad \text{when } \frac{h}{t_w} > \frac{760}{\sqrt{F_b}} \qquad (9.23)$$

In this equation, A_w is the web area. When $h/t_w < 760/\sqrt{F_b}$, no reduction of the allowable bending stress is necessary. Note that when the compression flange does not have sufficient lateral support, the allowable bending stress F_b must be reduced according to AISCS Sec. 1.5.1.4.5, as discussed in Sec. 5.5, to take into account the possibility of lateral torsional buckling.

In the case of hybrid girders, the allowable bending stress also should not be greater than the value given by the following equation:

$$F_b' \leq F_b \left[\frac{12 + \dfrac{A_w}{A_f} \left(\dfrac{3 F_{yw}}{F_{yf}} - \dfrac{F_{yw}^3}{F_{yf}^3} \right)}{2 \left(6 + \dfrac{A_w}{A_f} \right)} \right] \qquad (9.24)$$

where F_{yw} is the web yield stress, F_{yf} is the flange yield stress, and F_b is the allowable bending stress after the lateral-torsional buckling has been considered, when it is assumed that the entire member is made of the grade of steel used in the flanges. This equation is intended to account for the effect on the strength of a hybrid girder with a web of low yield strength. Equation (9.24) is applicable only when the area and grade of steel in both flanges are the same. Otherwise, a more complicated analysis is required.

If reduction of the allowable bending stress is necessary, we should check if the computed bending stress is less than the reduced allowable bending stress.

$$f_b = \frac{M_{max}}{S} < F_b' \qquad (9.25)$$

9.5 INTERMEDIATE STIFFENERS

Intermediate stiffeners are provided to stiffen the web plate against buckling and to resist compressive forces transmitted from the web during tension-field action. They are designed based on the following requirements:

1. *When the design of the plate girder is based on tension-field action*, the gross area of each intermediate stiffener or the total area of a pair of stiffeners, when they are used in pairs, should be at least equal to (AISCS Sec. 1.10.5.4)

$$A_{st} = \frac{1}{2} Dht_w(1 - C_v)\left[\frac{a}{h} - \frac{a^2/h^2}{(1 + a^2/h^2)^{1/2}}\right]\left(\frac{F_{yw}}{F_{ys}}\right)\left(\frac{f_v}{F_v}\right) \qquad (9.26)$$

where f_v is the greatest computed shear stress in the panel under consideration, F_{ys} is the yield stress of the stiffener, and

$$D = \begin{cases} 2.4 & \text{for single plate stiffeners} \\ 1.0 & \text{for stiffeners used in pairs} \\ 1.8 & \text{for single angle stiffeners} \end{cases}$$

During tension-field action the intermediate stiffeners behave as short struts. The required area by Eq. (9.26) ensures sufficient compression capacity of the stiffeners. Due to eccentric transfer of load with respect to the web, single-sided stiffeners are subject to considerable bending moment in addition to axial load and consequently are substantially less efficient than double-sided stiffeners. This consideration is reflected in Eq. (9.26) by the variable D.

2. The moment of inertia of a single intermediate stiffener or a pair of intermediate stiffeners (I_{st}) with respect to an axis in the plane of the web and perpendicular to the plane of the stiffener(s) (axis X–X in Fig. 9.10) should be at least equal to

$$I_{st} = (h/50)^4 \qquad (9.27)$$

where h, the depth of the web plate, is in inches. This requirement is intended to provide adequate lateral support for the web plate and prevent it from deflecting out of its plane when web buckling takes place.

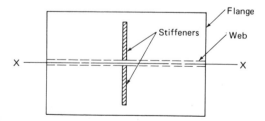

Figure 9.10   Plan of portion of a plate girder with a pair of stiffeners.

3. Each stiffener should be checked for the buckling requirement of AISCS Sec. 1.9.1.2. Stiffeners are free on one edge and consequently considered as unstiffened elements. Denoting the width and thickness of the stiffener by b_s and t_s, respectively, we should have

$$\frac{b_s}{t_s} \leq \frac{95}{\sqrt{F_y}} \qquad (9.28)$$

4. When intermediate stiffeners are required, their spacing should be such that
   a. The computed shear stress in the web does not exceed the allowable shear stress given by Eq. (9.6) or Eq. (9.12), whichever is applicable.

b. $a/h \leq [260/(h/t_w)]^2$ (9.10)

c. $a/h \leq 3.0$ (9.11)

d. $f_b \leq F_b = \left(0.825 - 0.375\frac{f_v}{F_v}\right)F_y \leq 0.60F_y$ (9.29)

The last requirement should be met only when the design of the web plate is based on tension-field action. In this case, due to large shear stresses in the web, the maximum tensile stress which acts at an angle to the girder axis could be considerably larger than the maximum tensile stress parallel to the girder axis. In lieu of a lengthy analysis for finding the maximum tensile stress based on the combined shear and tension stresses, AISCS Sec. 1.10.7 requires that Eq. (9.29) be satisfied, in which f_b is the maximum bending tensile stress due to moment in the plane of the girder web. The commentary in AISCS (Sec. 1.10.7) recommends that the interaction equation (9.29) need not be checked in the following two cases:

1. $f_v \leq 0.6F_v$ and $f_b \leq F_b$
2. $f_v \leq F_v$ and $f_b \leq 0.75F_b$

The two end panels adjacent to the supports are designed without the advantage of tension-field action. They are expected to act as anchor panels for the neighboring panels with tension-field action. For these two panels, the computed shear stress should not exceed the allowable shear stress given by Eq. (9.6).

9.6 WEB CRIPPLING

Web crippling in beams was discussed in Sec. 5.7. The same requirements must also be satisfied for plate girders.

In addition, to guard against instability of thin web plates, the amount of load that can be carried by the girder over a panel is limited by AISCS Sec. 1.10.10.2. In this regard, the portion of the web plate in one panel between adjacent stiffeners is considered as a short strut and the bearing compressive stresses in the web plate (f_{cw} in ksi) due to concentrated and distributed loads are checked.

If the flange is restrained against rotation about its longitudinal axis (for example, if it is in contact with a rigid slab), we should have

$$f_{cw} \leq F_a = \frac{10,000}{(h/t_w)^2}\left[5.5 + \frac{4}{(a/h)^2}\right]$$ (9.30)

If the flange is not restrained against rotation, we should have

$$f_{cw} \leq \frac{10,000}{(h/t_w)^2}\left[2 + \frac{4}{(a/h)^2}\right]$$ (9.31)

If the intensity of distributed load over a panel is w, the bearing compressive stress in the web is

$$f_{cw} = \frac{w}{t_w}$$ (9.32)

If, in addition to a distributed load, concentrated loads are also acting on the flange over the panel, they are treated as distributed loads by dividing the magnitude of the concentrated load by the smaller of the panel length or girder depth. Equations (9.30) and (9.31) are derived on the basis of the elastic buckling analysis of the stiffened web plate subjected to edge loading.

If Eq. (9.30) or Eq. (9.31) is not satisfied, we may choose one of the following solutions:

1. Add more stiffeners to decrease their spacing, a.
2. Increase the thickness of the web.

9.7 BEARING STIFFENERS

According to AISCS Sec. 1.10.5.1, bearing stiffeners should always be provided in pairs at the ends of plate girders and, if required, at points of application of concentrated loads. These bearing stiffeners should extend roughly to the edges of the flange plates, and their length should be close to the depth of the web plate in order to have close bearing with the flange plates. They are designed as columns with a cross-sectional area which includes a centrally located strip of the web.

For end bearing stiffeners, the width of the central strip of the web is taken as 12 times the thickness of the web (Fig. 9.11). Therefore, the effective area for checking the axial compressive stresses is

$$A_{\text{eff}} = 2A_{bs} + 12t_w^2 \tag{9.33}$$

where A_{bs} is the cross-sectional area of each bearing stiffener.

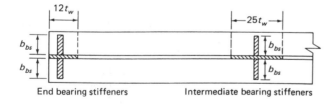

Figure 9.11   Equivalent cross-sectional areas for column design of bearing stiffeners.

For an interior bearing stiffener the width of the central strip of the web is taken as 25 times the thickness of the web (Fig. 9.11). Thus, the effective area becomes

$$A_{\text{eff}} = 2A_{bs} + 25t_w^2 \tag{9.34}$$

If the bearing stiffener is subjected to a concentrated load (or reaction) of magnitude P, the compressive stress in the bearing stiffener, f_{cb}, shall not exceed the allowable axial compressive stress F_a.

$$f_{cb} = \frac{P}{A_{\text{eff}}} \leq F_a \tag{9.35}$$

Evaluation of the allowable axial stress requires determination of the slenderness ratio KL/r. Because the bearing stiffeners are connected to the web, the effective length factor K may be taken as low as 0.75.

Since the allowable axial compressive stress F_a depends on the radius of gyration, r, the bearing stiffeners must be designed by the trial-and-error procedure.

Buckling of the web will conceivably occur about a horizontal axis parallel to the plane of the web. So it is customarily assumed that the web and stiffeners together will possibly buckle about the same axis; otherwise each stiffener will buckle about its own axis, which is perpendicular to the previously mentioned axis. Buckling of each stiffener, however, is checked by its width-thickness ratio. As a result, the radius of gyration r is calculated about a horizontal axis parallel to the plane of the web.

9.8 DESIGN OF WELDED CONNECTIONS

9.8.1 Connection of Flange to Web

Flange and web plates are connected to each other by fillet welds. Figure 9.12 shows a disassembled portion of the plate girder between two neighboring sections. Flange-to-web fillet welds are designed to transmit horizontal shear due to the variation of the bending moment over the girder and the direct pressure due to applied distributed load.

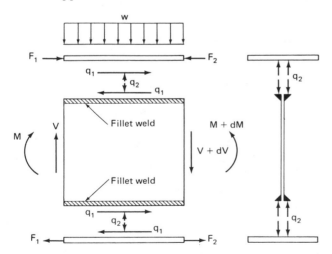

Figure 9.12   Connection of flange to web.

From elementary beam theory (Sec. 5.2), the horizontal (longitudinal) shear per unit length of the fillet weld is

$$q_1 = \frac{VQ}{I} = \frac{VA_f(h + t_f)}{2I} \tag{9.36}$$

where Q is the first moment of the flange area about the neutral axis and V is the shear force at the section under consideration.

The direct pressure due to applied load creates vertical shear q_2 per unit length of the fillet weld, which is in practice assumed to be equal to the intensity of the distributed load w. The resultant design force per unit length of the weld is

$$q = (q_1^2 + q_2^2)^{1/2} \tag{9.37}$$

If we denote the size of the fillet weld by w_w and the allowable shear stress of the weld electrode by F_v, noting that there are two lines of fillet weld on each side of the web plate, the allowable strength of the fillet weld will be

$$q_a = (0.707)(2w_w)F_v \qquad \text{(for SMAW)} \tag{9.38}$$

Substituting for values of q_1, from Eq. (9.36), and $q_2 = w$ into Eq. (9.37) and equating the resulting equation to Eq. (9.38), we obtain the following equation for the size of the continuous fillet weld:

$$w_w = \frac{\left[\dfrac{V^2 A_f^2}{4I^2}(h + t_f)^2 + w^2\right]^{1/2}}{1.414 F_v} \qquad \text{(for SMAW)} \tag{9.39}$$

Instead of continuous flange-to-web weld, intermittent fillet welds are sometimes used, as shown in Fig. 9.13. If we denote the length of each portion of the

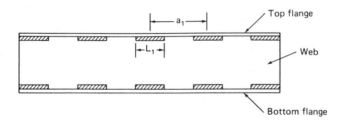

Figure 9.13   Intermittent flange-to-web fillet weld.

fillet weld by L_1 and the spacing of the intermittent weld by a_1, the following relation holds between these two variables:

$$L_1 q_a = a_1 q \tag{9.40}$$

Substituting for q and q_a from Eqs. (9.37) and (9.38), respectively, we obtain

$$\frac{L_1 w_w}{a_1} = \frac{\left[\dfrac{V^2 A_f^2}{4I^2}(h + t_f)^2 + w^2\right]^{1/2}}{1.414 F_v} \qquad \text{(for SMAW)} \tag{9.41}$$

By choosing two of the three parameters a_1, L_1, and w_w, the designer can find the third parameter from Eq. (9.41).

9.8.2 Connection of Intermediate Stiffeners to the Web

The magnitude of the shear transfer between the web and stiffeners is in general very small. As a result, a minimum amount of welding is used. When the tension-field action is the design basis [Eq. (9.12)], however, a conservative formula is provided by the AISCS for the amount of shear to be transferred between the web and stiffeners. According to AISCS Sec. 1.10.5.4, the connection of the intermediate stiffeners to the web plate should be designed for a total shear transfer, in Kips per linear inch of single stiffener or pair of stiffeners, at least equal to

$$f_{vs} = h\left(\frac{F_y}{340}\right)^{3/2}\frac{f_v}{F_v} \tag{9.42}$$

where F_y is the yield stress of the web steel in ksi, and f_v and F_v are the maximum computed shear stress and the allowable shear stress in the adjacent panels, respectively. Furthermore, welds in stiffeners which are required to transmit a concentrated load or reaction should be designed for the larger of the corresponding load (or reaction) and the shear given by Eq. (9.42).

If intermediate stiffeners are used in pairs, noting that there are four lines of fillet weld at each stiffener-web connection, we find that the required continuous weld size is

$$w_w = \frac{f_{vs}/4}{0.707F_v} = \frac{f_{vs}}{2.828F_v} \qquad \text{(for SMAW)} \tag{9.43}$$

When single stiffeners are used, either alternated on the sides of the web plate (Fig. 9.14) or placed on one side of the web (Fig. 9.15) possibly for a better look,

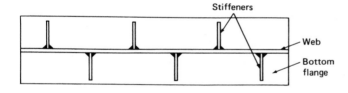

Figure 9.14 Horizontal section of a plate girder with alternated stiffeners.

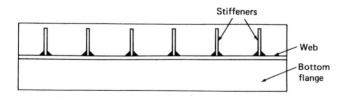

Figure 9.15 Horizontal section of a plate girder with stiffeners placed on one side.

two lines of fillet welds are available at each stiffener-web connection and the required weld size will be

$$w_w = \frac{f_{vs}/2}{0.707F_v} = \frac{f_{vs}}{1.414F_v} \qquad \text{(for SMAW)} \tag{9.44}$$

We may also use an intermittent weld for the stiffener-web connections, as shown in Fig. 9.16. Denoting the weld length by L_1 and the spacing by a_1, we find that the required size of the fillet weld for the case of double stiffeners is

$$w_w = \frac{a_1 f_{vs}}{2.828 L_1 F_v} \qquad \text{(for SMAW)} \tag{9.45}$$

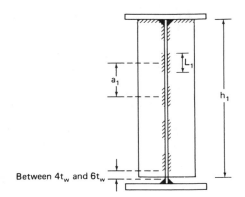

Figure 9.16 Connection of stiffeners to web by intermittent welds.

Similarly, for the case of single stiffeners, we obtain

$$w_w = \frac{a_1 f_{vs}}{1.414 L_1 F_v} \qquad \text{(for SMAW)} \tag{9.46}$$

The clear distance between welds should not be greater than 16 times the web thickness or greater than 10 inches (AISCS Sec. 1.10.5.4):

$$(a_1 - L_1) \le 16 t_w \text{ and } 10 \text{ in.} \tag{9.47}$$

Welding of the stiffeners to the compression flange keeps them normal to the web and consequently makes them more stable. Moreover, such welding causes the stiffeners to resist any uplift tendency due to torsion and thus provides restraint against torsional buckling of the compression flange.

Welding of the stiffeners to the tension flange is not necessary (Fig. 9.16). In fact, such welding increases the chance of fatigue or brittle failure. Intermediate stiffeners not transmitting a concentrated load or reaction can be stopped short of the tension flange (AISCS Sec. 1.10.5.4) (Fig. 9.16). The distance between the point of termination of the stiffener-to-web weld and the near toe of the web-to-flange weld should not be smaller than four times the web thickness or greater than six times the web thickness.

9.9 DESIGN OF A SIMPLE HOMOGENEOUS PLATE GIRDER BASED ON AISCS

9.9.1 Problem Description

Design of the doubly symmetric steel plate girder shown in Fig. 9.17 is covered in this section. The girder has a span of $L = 150$ ft. The loading on the girder consists of a uniform load of $w = 4$ K/ft and a concentrated load of $P = 500$ Kips applied at a distance of 100 ft from the left support. Use A36 steel for the flange and web plates as well as the double stiffeners. For welds, use E70 electrodes with an allowable shear stress of 21 ksi. Lateral support is provided at supports, at the point of application of concentrated load, and at point D located at a distance of 50 ft

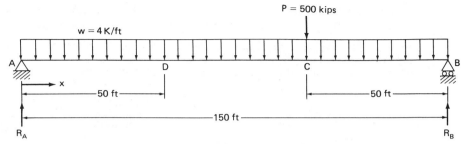

Figure 9.17

from the left support A. Since the compression flange carries a uniform load, assume that it is restrained against rotation.

9.9.2 Shear and Bending Moment Diagrams

Reactions at supports A and B (Fig. 9.17) are

$$R_A = \frac{(4)(150)(75) + (500)(50)}{150} = 466.67 \text{ Kips}$$

$$R_B = (4)(150) + 500 - 466.67 = 633.33 \text{ Kips}$$

The shear diagram is shown in Fig. 9.18.

Figure 9.18   Shear diagram.

Bending moment over the girder:

For $0 < x < 100$ ft:

$$M(x) = R_A x - wx^2/2$$

$$\frac{dM(x)}{dx} = R_A - wx_1 = 0 \Rightarrow x_1 = \frac{R_a}{w} = \frac{466.67}{4} = 116.67 \text{ ft} > 100 \text{ ft}$$

$\therefore$ There is no maximum between A and C.

For $100 \text{ ft} < x < 150 \text{ ft}$:

$$M(x) = R_A x - wx^2/2 - P(x - 100)$$

$$\frac{dM(x)}{dx} = R_A - wx_1 - P = 0 \Rightarrow x_1 = \frac{R_A - P}{w} < 0$$

$\therefore$ There is no maximum between C and B. Therefore, the maximum bending moment over the girder is at point C under the concentrated load.

$$M_{\max} = M_c = 466.67(100) - 4(100)(50) = 26667 \text{ K-ft}$$

The bending moment diagram is shown in Fig. 9.19.

Figure 9.19   Bending moment diagram.

9.9.3 Selection of the Web Plate

Referring to Eq. (9.2), we choose the following depth for the girder web:

$$h = \frac{L}{12} = \frac{(150)(12)}{12} = 150 \text{ in.}$$

From Eq. (9.3):

$$\frac{h}{t_w} \le \frac{14,000}{[36(36 + 16.5)]^{1/2}} = 322.0 \Rightarrow t_w \ge \frac{150}{322} = 0.47 \text{ in.}$$

For closely spaced stiffeners, that is, when $a \le 1.5d$, from Eq. (9.4):

$$\frac{h}{t_w} \le \frac{2000}{\sqrt{36}} = 333.3 \Rightarrow t_w \ge \frac{150}{333.3} = 0.45 \text{ in.}$$

It is seen that for A36 steel with a yield stress of 36 ksi, Eqs. (9.3) and (9.4) yield values close to each other. In other words, the minimum thickness of the web plate cannot be substantially reduced by closely spacing stiffeners. Try $t_w = 0.5$ in. or PL 150 in. × 0.5 in. for the web plate.

$$\frac{h}{t_w} = 300; \qquad A_w = 75 \text{ in.}^2$$

Tentatively assuming the allowable bending stress to be $F_b = 0.60F_y = 22$ ksi, check if reduction of the allowable bending stress is required.

$$\frac{h}{t_w} = 300 > \frac{760}{\sqrt{F_b}} = 162$$

$\therefore$ Allowable bending stress must be reduced according to Eq. (9.23).

9.9.4 Selection of Flange Plates

Because the allowable bending stress in the flange plates must be reduced, we may decrease $F_b = 0.60F_y = 22$ ksi somewhat, say, to $F_b = 20$ ksi. The required flange area, A_f, can be computed approximately from Eq. (9.19).

$$A_f = b_f t_f = \frac{(26,667)(12)}{(20)(150)} - \frac{75}{6} = 94.17 \text{ in.}^2$$

The minimum flange thickness in order to prevent local flange buckling is obtained from Eq. (9.22).

$$t_f = \left[\frac{\sqrt{36}}{190} (94.17) \right]^{1/2} = 1.72 \text{ in.}$$

Try PL 38 in. × 2.5 in. for each flange; $A_f = 95$ in.^{2}. The girder section is shown in Fig. 9.20.

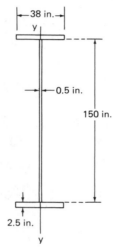

Figure 9.20

From Eq. (9.17):

$$I = \tfrac{1}{12}(0.5)(150)^3 + 2(38)(2.5)(75 + 1.25)^2 + \tfrac{2}{12}(38)(2.5)^3 = 1,245,396 \text{ in.}^4$$

$$S = \frac{I}{h/2 + t_f} = \frac{1,245,396}{75 + 2.5} = 16,069.6 \text{ in.}^3$$

Find the allowable bending stress F_b (Sec. 5.5).

$r_T =$ radius of gyration of the flange plus one-sixth of the web area about the y-axis (Fig. 9.20)

$$r_T = \left[\frac{6t_f b_f^3 + h t_w^3}{12(6t_f b_f + h t_w)} \right]^{1/2}$$

$$r_T \simeq \left[\frac{6t_f b_f^3}{12(6t_f b_f + h t_w)} \right]^{1/2} = \frac{b_f}{(12 + 2A_w/A_f)^{1/2}} \qquad (9.48)$$

$$r_T = \frac{38}{(12 + 150/95)^{1/2}} = 10.31 \text{ in.}$$

For portions AD and CB of the girder the ratio of the smaller to larger end moments,

M_1/M_2, is zero. This ratio for portion DC is equal to

$$M_1/M_2 = M_D/M_C = -18{,}333.5/26{,}667 = -0.687$$

Portion DC is the critical portion, so we find the allowable bending stress for this portion.

$$C_b = 1.75 + 1.05(M_1/M_2) + 0.3(M_1/M_2)^2$$

$$= 1.75 + 1.05(-0.687) + 0.3(-0.687)^2 = 1.17 < 2.30$$

$$\left(\frac{510{,}000 C_b}{F_y}\right)^{1/2} = \left[\frac{(510{,}000)(1.17)}{36}\right]^{1/2} = 128.74$$

$$\left(\frac{102{,}000 C_b}{F_y}\right)^{1/2} = \left[\frac{(102{,}000)(1.17)}{36}\right]^{1/2} = 57.58$$

$$L_u = \text{unbraced length} = \overline{AD} = \overline{DC} = \overline{CB} = 50 \text{ ft}$$

$$\left(\frac{102{,}000 C_b}{F_y}\right)^{1/2} < \frac{L_u}{r_T} = \frac{(50)(12)}{10.31} = 58.2 < \left(\frac{510{,}000 C_b}{F_y}\right)^{1/2}$$

$$F_b = \left[\frac{2}{3} - \frac{F_y(L_u/r_T)^2}{1{,}530{,}000 C_b}\right]F_y = \left[\frac{2}{3} - \frac{36(58.2)^2}{1{,}530{,}000(1.17)}\right](36) = 21.55 \text{ ksi}$$

Calculate the reduced allowable bending stress F_b' from Eq. (9.23).

$$F_b' = \left[1.0 - 0.0005\left(\frac{75}{95}\right)\left(\frac{150}{0.5} - \frac{760}{\sqrt{21.55}}\right)\right]21.55 = 20.39 \text{ ksi}$$

The maximum bending stress in the girder is

$$f_b = \frac{M_{\max}}{S} = \frac{(26{,}667)(12)}{16{,}069.6} = 19.91 \text{ ksi} < 20.39 \text{ ksi} \qquad \underline{\underline{\text{O.K.}}}$$

9.9.5 Intermediate Stiffeners

1. Check if intermediate stiffeners are required [Eq. (9.14)].

$$\frac{h}{t_w} = \frac{150}{0.5} = 300 > 260$$

∴ Stiffeners are required.

2. Find the location of the first stiffener from each end.
   a. *At the left end*

$$f_v = \frac{R_A}{A_w} = \frac{466.67}{75} = 6.22 \text{ ksi}$$

Substitute for $F_v = f_v = 6.22$ ksi in Eq. (9.6) and solve for C_v.

$$C_v = \frac{2.89 F_v}{F_y} = \frac{(2.89)(6.22)}{36} = 0.5 < 0.8$$

From Eq. (9.7):

$$k = \frac{F_y C_v (h/t_w)^2}{45,000} = \frac{(36)(0.5)(300)^2}{45,000} = 36 > 9.34$$

From Eq. (9.8):

$$\frac{a}{h} = \left(\frac{5.34}{k-4}\right)^{1/2} = \left(\frac{5.34}{36-4}\right)^{1/2} = 0.41$$

$$a_{max} = 0.41(150) = 61.3 \text{ in.}$$

Tentatively, place the first intermediate stiffener at a distance of 50 in. from the left end A.

b. *At the right end*

$$f_v = \frac{R_B}{A_w} = \frac{633.33}{75} = 8.44 \text{ ksi}$$

$$C_v = \frac{2.89 F_v}{F_y} = \frac{(2.89)(8.44)}{36} = 0.678 < 0.8$$

$$k = \frac{F_y C_v (h/t_w)^2}{45,000} = \frac{(36)(0.678)(300)^2}{45,000} = 48.81 > 9.34$$

$$\frac{a}{h} = \left(\frac{5.34}{k-4}\right)^{1/2} = \left(\frac{5.34}{48.81-4}\right)^{1/2} = 0.345 \text{ in.}$$

$$a_{max} = 0.345(150) = 51.8 \text{ in.}$$

Tentatively, place the first intermediate stiffener at a distance of 40 in. from the right end B.

3. Find the spacing of the remaining stiffeners.
    From Eq. (9.10):

$$a \leq [260/(h/t_w)]^2 h = (260/300)^2(150) = 112.6 \text{ in.}$$

$$a_{max} = 112.6 \text{ in.}$$

Equation (9.10) controls over Eq. (9.11), so the latter need not be checked.

We choose to use uniform spacing between point E (at the location of the first stiffener away from the left end) and point C (at the location of the concentrated load) and also between point C and point F (at the location of the first stiffener away from the right end) (Fig. 9.22).

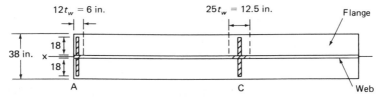

Figure 9.21   Equivalent areas for bearing stiffeners at A and C.

Flanges: 2 PL 38 in. $\times$ 2$\frac{1}{2}$ in.   Intermediate stiffeners between E and C: 2 PL6.5 in. $\times$ $\frac{7}{16}$ in. $\times$ 12 ft 4 in.

Web: PL150 in. $\times$ $\frac{1}{2}$ in.   Intermediate stiffeners between C and F: 2 PL7 in. $\times$ $\frac{9}{16}$ in. $\times$ 12 ft 4 in.

Bearing stiffeners at A, B, and C: 2 PL18 in. $\times$ 1$\frac{3}{16}$ in. $\times$ 12 ft 6 in.

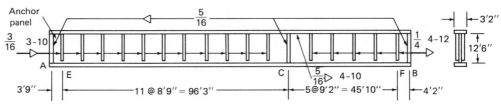

Figure 9.22   Final design.

a. *Spacing of the stiffeners between points E and C (Fig. 9.22)*

$$V_{\max} = \text{maximum shear} = R_A - \tfrac{50}{12}(w) = 466.67 - \tfrac{50}{12}(4) = 450.0 \text{ Kips}$$

$$f_v = \frac{V_{\max}}{A_w} = \frac{450}{75} = 6.00 \text{ ksi}$$

Try $a = 105$ in. and change the spacing of the first stiffener away from end A from 50 in. to 45 in. (Fig. 9.22).

$$V_{\max} = R_A - \tfrac{45}{12}(w) = 466.67 - \tfrac{45}{12}(4) = 451.67 \text{ Kips}$$

$$f_v = \frac{V_{\max}}{A_w} = \frac{451.67}{75} = 6.02 \text{ ksi}$$

$$\frac{a}{h} = \frac{105}{150} = 0.7 < 1$$

$$k = 4.0 + \frac{5.34}{(a/h)^2} = 4.0 + \frac{5.34}{(0.7)^2} = 14.90$$

$$C_v = \frac{45,000k}{F_y(h/t_w)^2} = \frac{(45,000)(14.90)}{(36)(300)^2} = 0.207 < 0.8$$

Because all five conditions mentioned in Sec. 9.3 are satisfied, we can take advantage of the tension-field action and use Eq. (9.12) for calculating the allowable shear stress.

$$F_v = \frac{36}{2.89}\left[0.207 + \frac{1 - 0.207}{1.15(1 + 0.7^2)^{1/2}}\right]$$

$$= 2.58 + 7.04 = 9.62 \text{ ksi} > f_v = 6.02 \text{ ksi} \qquad \underline{\text{O.K.}}$$

Note that the contribution of the tension-field action is substantial in this example.

b. *Spacing of the stiffeners between points C and F (Fig. 9.22)*

$$V_{\max} = R_B - \tfrac{40}{12}(w) = 633.33 - \tfrac{40}{12}(4) = 620.0 \text{ Kips}$$

$$f_v = \frac{V_{\max}}{A_w} = \frac{620}{75} = 8.27 \text{ ksi}$$

Try $a = 110$ in. and change the spacing of the first stiffener away from end B from 40 in. to 50 in. (Fig. 9.22).

$$V_{\max} = R_B - \tfrac{50}{12}(w) = 633.33 - \tfrac{50}{12}(4) = 616.66 \text{ Kips}$$

$$f_v = \frac{V_{\max}}{A_w} = \frac{616.66}{75} = 8.22 \text{ ksi}$$

$$\frac{a}{h} = \frac{110}{150} = 0.733 < 1$$

$$k = 4.0 + \frac{5.34}{(a/h)^2} = 4.0 + \frac{5.34}{(0.733)^2} = 13.93$$

$$C_v = \frac{45,000k}{F_y(h/t_w)^2} = \frac{(45,000)(13.93)}{(36)(300)^2} = 0.193 < 0.8$$

From Eq. (9.12):

$$F_v = \frac{36}{2.89}\left[0.193 + \frac{1 - 0.193}{1.15(1 + 0.733^2)^{1/2}}\right]$$

$$= 2.40 + 7.05 = 9.45 \text{ ksi} > f_v = 8.22 \text{ ksi} \qquad \underline{\text{O.K.}}$$

Spacing of the stiffeners over the girder span is shown in Fig. 9.22.

4. Check combined shear and bending in the web.

The critical section for this check is either at point C, where the bending moment has the largest value and the shear force is considerable, or somewhere close to but to the right of this point, where the bending moment is slightly smaller than the maximum value but the shear force is larger than that at point C (Figs. 9.18 and 9.19).

Let us first check the combined shear and bending at point C under the concentrated load [Eq. (9.29)].

$$f_v = \frac{V_C}{A_w} = \frac{433.33}{75} = 5.78 \text{ ksi}$$

The allowable bending tensile stress in the web [Eq. (9.29)]:

$$F_b = \left(0.825 - 0.375\,\frac{f_v}{F_v}\right)F_y = \left(0.825 - 0.375\,\frac{5.78}{9.45}\right)(36) = 21.44 \text{ ksi}$$

The maximum bending tensile stress in the web (at the junction of web and flange):

$$f_b = \frac{M_C y}{I} = \frac{(26,667)(12)(75)}{1,245,396} = 19.27 \text{ ksi} < F_b = 21.44 \text{ ksi} \qquad \underline{\text{O.K.}}$$

Note that if this condition is not satisfied, the spacing of the intermediate stiffeners is normally reduced. Also, because $f_b/F_b = 0.90$ is considerably less than one, we do not need to check the other locations to the right of point C. Whenever f_b/F_b is close to one, such a check may be necessary.

5. Select the intermediate stiffeners.

Spacing of the stiffeners between E and C is $a = 105$ in. and between C and F is $a = 110$ in. (Fig. 9.22). The cross-sectional area of a pair of stiffeners is found by Eq. (9.26), which is a function of aspect ratio a/h.

a. *For region EC.* From Eq. (9.26):

$$A_{st} = \frac{1}{2}(150)(0.5)(1 - 0.207)\left[0.7 - \frac{0.7^2}{(1 + 0.7^2)^{1/2}}\right]\left(\frac{6.02}{9.62}\right)$$

$$= 5.56 \text{ in.}^2$$

$$b_s t_s = 5.56/2 = 2.78 \text{ in.}^2 \tag{9.49}$$

From Eq. (9.28):

$$\frac{b_s}{t_s} \leqslant \frac{95}{\sqrt{36}} = 15.8$$

Substituting for $b_s = 15.8 t_s$ in Eq. (9.49) will yield the minimum thickness required for the intermediate stiffeners.

$$t_s = (2.78/15.8)^{1/2} = 0.419 \text{ in.} = 6.7/16 \text{ in.}$$

Try 2PL $6\frac{1}{2}$ in. $\times \frac{7}{16}$ in. for the intermediate stiffeners. Check the moment of inertia requirement (Eq. 9.27).

$$\left(\frac{h}{50}\right)^4 = \left(\frac{150}{50}\right)^4 = 81 \text{ in.}^4$$

$$I_{st} = \frac{1}{12}(\tfrac{7}{16})(6.5 \times 2 + 0.5)^3 = 89.7 \text{ in.}^4 > 81 \text{ in.}^4 \qquad \underline{\text{O.K.}}$$

Use 2PL $6\frac{1}{2}$ in. $\times \frac{7}{16}$ in. for intermediate stiffeners from E to C (excluding C).
    Intermediate stiffeners need not be extended to the tension flange (Fig. 9.22) and their length h_1 can be four times the web thickness shorter than the depth of the web.

$$h_1 = h - 4t_w = 150 - 4(0.5) = 148 \text{ in.} = 12 \text{ ft 4 in.}$$

b. *For region CF.* From Eq. (9.26):

$$A_{st} = \frac{1}{2}(150)(0.5)(1 - 0.193)\left[0.733 - \frac{0.733^2}{(1 + 0.733^2)^{1/2}}\right]\left(\frac{8.22}{9.45}\right)$$

$$= 7.89 \text{ in.}^2$$

$$b_s t_s = 7.89/2 = 3.94 \text{ in.}^2$$

$$b_s = 15.8 t_s$$

Minimum $t_s = (3.94/15.8)^{1/2} = 0.5$ in.

Use 2PL 7 in. $\times \frac{9}{16}$ in. for the intermediate stiffeners from C to F (excluding C). Moment of inertia check is not needed, because these plates are larger than those used in region EC.
    Length of stiffeners:

$$h_1 = h - 4t_w = 12 \text{ ft 4 in.}$$

9.9.6 Web Crippling

The compressive load transmitted to the web is 4 K/ft plus the weight of the top flange. We temporarily neglect the flange weight. The compressive stress in the web is

$$f_{cw} = \frac{w}{t_w} = \frac{4/12}{0.5} = 0.667 \text{ ksi}$$

The allowable compressive stress at the location where a/h is critical (where a/h is the largest) is [Eq. (9.30)]

$$F_a = \frac{10,000}{(h/t_w)^2}\left[5.5 + \frac{4}{(a/h)^2}\right] = \frac{10,000}{300^2}\left[5.5 + \frac{4}{(110/150)^2}\right]$$

$$= 1.44 \text{ ksi} > f_{cw} = 0.667 \text{ ksi} \qquad \underline{\text{O.K.}}$$

The allowable stress is considerably larger than the actual stress; therefore, neglect of flange weight is justified.

Note that if $F_a < f_{cw}$, we have to either add more stiffeners to decrease a/h or increase the web thickness, t_w.

9.9.7 Bearing Stiffeners

Bearing stiffeners should extend roughly to the edges of the flange plates. Noting that the width of the flange plates is 38 in. and the thickness of the web is 0.5 in., we choose a width of $b_{bs} = 18$ in. for the three pairs of bearing stiffeners (Fig. 9.22).

1. Bearing stiffeners at the left support
   a. *Check buckling (width-thickness ratio).*

$$\frac{b_{bs}}{t_{bs}} \leq \frac{95}{\sqrt{F_y}} = \frac{95}{\sqrt{36}} = 15.8$$

$$t_{bs} \geq \frac{18}{15.8} = 1.14 \text{ in.} = \frac{18.2}{16} \text{ in.}$$

Try 2PL 18 in. $\times$ $1\frac{3}{16}$ in. for the left support. $A_{bs} = 21.375 \text{ in.}^2$
   b. *Check axial compressive stress due to reaction $R_A = 466.67$ Kips.* From Eq. (9.33):

$$A_{\text{eff}} = 2A_{bs} + 12t_w^2 = 2(21.375) + 12(0.5)^2 = 45.75 \text{ in.}^2$$

Moment of inertia of the equivalent area about the x-axis (Fig. 9.21):

$$I \simeq \tfrac{1}{12}(\tfrac{19}{16})(36.5)^3 = 4812.06 \text{ in.}^4$$

Radius of gyration about the x-axis:

$$r = \sqrt{I/A_{\text{eff}}} = \sqrt{4812.06/45.75} = 10.26 \text{ in.}$$

$$\frac{KL}{r} = \frac{0.75(150)}{10.26} = 10.96$$

$$C_c = \left(\frac{2\pi^2 E}{F_y}\right)^{1/2} = \left[\frac{2\pi^2(29,000)}{36}\right]^{1/2} = 126.1 > \frac{KL}{r}$$

$$\text{F.S.} = \text{factor of safety} = \frac{5}{3} + \frac{3(KL/r)}{8C_c} - \frac{(KL/r)^3}{8C_c^3}$$

$$= \frac{5}{3} + \frac{3}{8}\left(\frac{10.96}{126.1}\right) - \frac{1}{8}\left(\frac{10.96}{126.1}\right)^3 = 1.699$$

Allowable axial stress:

$$F_a = \left[1 - \frac{(KL/r)^2}{2C_c^2}\right]F_y/\text{F.S.} = \left[1 - \frac{1}{2}\left(\frac{10.96}{126.1}\right)^2\right]\left(\frac{36}{1.699}\right) = 21.11 \text{ ksi}$$

Actual axial stress:

$$f_a = \frac{R_A}{A_{\text{eff}}} = \frac{466.67}{45.75} = 10.20 \text{ ksi} < 21.11 \text{ ksi} \qquad \underline{\text{O.K.}}$$

Use 2PL 18 in. $\times$ $1\frac{3}{16}$ in. $\times$ 12 ft 6 in. for the bearing stiffeners at the left support.

Sometimes the height of the bearing stiffeners is chosen slightly less, say $\frac{1}{4}$ in., than the depth of the web plate. The bearing stiffener, however, should be in contact with the flange receiving the concentrated load.

2. Bearing stiffeners at the right support

Reaction at the right support:

$$R_B = 633.33 \text{ Kips}$$

Try 2PL 18 in. $\times$ $1\frac{3}{16}$ in., the same as for the left support.

$$f_a = \frac{R_B}{A_{\text{eff}}} = \frac{633.33}{45.75} = 13.84 \text{ ksi} < F_a = 21.11 \text{ ksi} \qquad \underline{\text{O.K.}}$$

Use 2PL 18 in. $\times$ $1\frac{3}{16}$ in. $\times$ 12 ft 6 in. for bearing stiffeners at the right support.

3. Bearing stiffeners at the concentrated load

Because the axial load for these stiffeners, $P = 500$ Kips, is less than R_B, and the effective area (Fig. 9.21) is larger than that of the end bearing stiffeners, we can use the same 2 PL 18 in. $\times$ $1\frac{3}{16}$ in. $\times$ 12 ft 6 in. for bearing stiffeners at the location of the concentrated load.

9.9.8 Web-to-Flange Fillet Weld

We design the web-to-flange connection based on the maximum shear over the girder length.

$$V_{\max} = R_B = 633.33 \text{ Kips}$$

We use intermittent SMAW welds. The relation between the width of the fillet weld w_w, the length of weld segment L_1, and the spacing a_1 is given by Eq. (9.41).

$$\frac{L_1 w_w}{a_1} = \frac{\left[\dfrac{(633.33)^2(95)^2}{4(1,245,396)^2}(150 + 2.5)^2 + \left(\dfrac{4}{12}\right)^2\right]^{1/2}}{1.414(21)} = 0.1246$$

Minimum weld size for a 2.5-in. flange plate is $\frac{5}{16}$ in. (Sec. 8.3.5 and AISCS Table 1.17.2A).

Try $w_w = 5/16$ in. Substituting this value into the previous equation, we obtain

$$a_1 = 2.51 L_1 \qquad (9.50)$$

The minimum length of a segment of intermittent weld is the larger of four times the weld size ($4w_w = 1.25$ in.) and 1.5 in. (AISCS Sec. 1.17.5).

$$L_{1\min} = 1.5 \text{ in.}$$

The maximum longitudinal spacing of the intermittent weld is the smaller of 24 times the thickness of the thinner plate and 12 in. (AISCS 1.18.3.1).

$$\text{Thickness of the thinner plate} = 0.5 \text{ in.}$$

$$a_{\max} = 12 \text{ in.}$$

Try $\frac{5}{16}$-in. weld, 4 in. long. From Eq. (9.50), we obtain

$$a_1 = 10.04 \text{ in.}$$

Use $\frac{5}{16}$-in. weld, 4 in. long, 10-in. spacing.

9.9.9 Stiffener-to-Web Fillet Weld

1. For segment EC (Fig. 9.22)

$$f_v = 6.02 \text{ ksi}; \qquad F_v = 9.62 \text{ ksi}$$

From Eq. (9.42):

$$f_{vs} = h\left(\frac{F_y}{340}\right)^{3/2}\left(\frac{f_v}{F_v}\right) = (150)\left(\frac{36}{340}\right)^{3/2}\left(\frac{6.02}{9.62}\right) = 3.23 \text{ ksi}$$

We use intermittent welds. From Eq. (9.45):

$$\frac{L_1 w_w}{a_1} = \frac{f_{vs}}{2.828 F_v} = \frac{3.23}{2.828(21)} = 0.05439 \qquad (9.51)$$

The minimum weld size for a $\frac{1}{2}$-in.-thick plate is $\frac{3}{16}$ in. The maximum weld spacing is

$$a_{\max} = 24 t_s \quad \text{or} \quad 12 \text{ in.} = 24(\tfrac{7}{16}) = 10.5 \text{ in.}$$

Try a $\frac{3}{16}$-in. fillet weld with a spacing of 10 in. From Eq. (9.51) we obtain $L_1 = 2.90$ in.

Minimum length of the weld segment $= 4w_w$ or   1.5 in. $= 1.5$ in.

Use $\frac{3}{16}$-in. weld, 3.0 in. long, 10-in. spacing.

2. For segment CF (Fig. 9.22)

$$f_v = 8.22 \text{ ksi}; \qquad F_v = 9.45 \text{ ksi}$$

$$f_{vs} = 150\left(\frac{36}{340}\right)^{3/2}\left(\frac{8.22}{9.45}\right) = 4.5 \text{ ksi}$$

$$\frac{L_1 w_w}{a_1} = \frac{4.50}{2.828(21)} = 0.0757 \tag{9.52}$$

The minimum weld size for a $\frac{9}{16}$-in.-thick plate is $\frac{1}{4}$ in.

$$a_{\max} = 12 \text{ in.}$$

$$L_{1\min} = 1.5 \text{ in.}$$

Try a $\frac{1}{4}$-in. intermittent fillet weld with a spacing of $a_1 = 12$ in. From Eq. (9.52) we obtain $L_1 = 3.63$ in.

Use $\frac{1}{4}$-in. weld, 4 in. long, 12-in. spacing.

3. Bearing stiffeners

We use 2PL 18 in. $\times 1\frac{3}{16}$ in. for each pair of bearing stiffeners.
Minimum weld size: $\frac{5}{16}$ in.
Try $w_w = \frac{5}{16}$ in. Use continuous welds on both sides of each stiffener plate.

$$\text{Shear strength of } \tfrac{5}{16}\text{-in. fillet weld} = 0.707(\tfrac{5}{16})(21)$$

$$= 4.64 \text{ K/in.}$$

Total strength of four lines of weld $= 4(150)(4.64) = 2784$ Kips

$$> V_{\max} = 633.33 \text{ Kips} \qquad\qquad \underline{\text{O.K.}}$$

and

$$> hf_{vs} = (150)(4.5) = 675 \text{ Kips} \qquad\qquad \underline{\text{O.K.}}$$

Use $\frac{5}{16}$-in. weld continuously on both sides of all bearing stiffeners.

9.9.10 Girder Weight

$$\gamma = \text{specific gravity of steel} = 0.490 \text{ Kips/ft}^3$$

$$\text{Weight of the flange plates} = 2A_f L\gamma = 2(\tfrac{95}{144})(150)(0.49) = 96.98 \text{ Kips}$$

$$\text{Weight of the web plate} = ht_w L\gamma = \left(\frac{150}{12}\right)\left(\frac{0.5}{12}\right)(150)(0.49) = 38.28 \text{ Kips}$$

$$\text{Weight of the flange and web plates} = 135.26 \text{ Kips}$$

$$\text{Weight of the stiffeners} = \left[3(2)(18)(\tfrac{19}{16})(150) + 11(2)(6.5)(\tfrac{7}{16})(148.0) \right.$$
$$\left. + 5(2)(7)\left(\frac{9}{16}\right)(148.0) \right]\left(\frac{0.49}{12^3}\right) = 9.73 \text{ Kips}$$

$$\text{Total weight of the plate girder} = 135.26 + 9.73 = 145.0 \text{ Kips}$$

Note that the weight of the stiffeners is 6.7 percent of the total weight of the plate girder.

9.10 DESIGN OF A HYBRID GIRDER BASED ON AISCS

9.10.1 Problem Description

Design of the same plate girder described in Sec. 9.9.1 is desired except that

1. The yield stress of steel used in the flange plates is 50 ksi (the yield stress of the web plate and stiffeners is the same, 36 ksi).
2. The compression flange is laterally supported throughout its length.
3. Single intermediate stiffeners are used.

The shear and bending moment diagrams are the same as before (Figs. 9.18 and 9.19).

9.10.2 Selection of the Web Plate

Depth of the girder web: $h = L/12 = 150$ in.
   From Eq. (9.3):

$$\frac{h}{t_w} \leqslant \frac{14{,}000}{[50(50 + 16.5)]^{1/2}} = 242.8 \Rightarrow t_w \geqslant \frac{150}{242.8} = 0.62 \text{ in.}$$

For closely spaced stiffeners, i.e., when $a \leqslant 1.5d$, from Eq. (9.4):

$$\frac{h}{t_w} \leqslant \frac{2000}{\sqrt{50}} = 282.8 \Rightarrow t_w \geqslant \frac{150}{282.8} = 0.53 \text{ in.}$$

Try $t_w = \tfrac{9}{16}$ in. or PL 150 in. $\times \tfrac{9}{16}$ in. for the web plate. Note that spacing of the intermediate stiffeners a should not be greater than $1.5d$. For the selected web plate, we have

$$\frac{h}{t_w} = \frac{800}{3} = 266.67; \qquad A_w = (150)(\tfrac{9}{16}) = 84.375 \text{ in.}^2$$

$$\frac{760}{\sqrt{F_b}} = \frac{760}{\sqrt{0.6(50)}} = 138.76 < \frac{h}{t_w} = 266.67$$

The allowable bending stress must be reduced according to Eq. (9.23).

9.10.3 Selection of Flange Plates

First, we find an approximate relation for the area of one flange plate for hybrid girders, taking into account the reduction of the allowable bending stress according to Eq. (9.24). This equation at the limit can be written as

$$F_b' = \frac{6A_f + \alpha A_w}{6A_f + A_w} F_b \tag{9.53}$$

where

$$\alpha = 0.5(F_{yw}/F_{yf})[3 - (F_{yw}/F_{yf})^2] \leqslant 1.0 \tag{9.54}$$

From Eq. (9.18):

$$S = \frac{M_{max}}{F_b'} = \frac{h}{6}(6A_f + A_w) \tag{9.55}$$

Combining Eqs. (9.53) and (9.55) and solving for A_f, we obtain

$$A_f = \frac{M_{max}}{F_b h} - \frac{\alpha A_w}{6} \tag{9.56}$$

In this example:

$$\alpha = 0.5(36/50)[3 - (36/50)^2] = 0.893$$

$$A_f = \frac{(26,667)(12)}{(0.60)(50)(150)} - \frac{0.893(84.375)}{6} = 58.55 \text{ in.}^2$$

Minimum flange thickness from Eq. (9.22):

$$t_f = \left[\frac{\sqrt{50}}{190}(58.55) \right]^{1/2} = 1.48 \text{ in.}$$

Try PL 39 in. × $1\frac{1}{2}$ in. for each flange plate; $A_f = 58.50$ in.^{2}
    Check Eq. (9.20):

$$\frac{b_f}{t_f} = \frac{39}{1.5} = 26 < \frac{190}{\sqrt{F_y}} = 26.9 \qquad \text{O.K.}$$

Properties of the section:

$$I = 829,578 \text{ in.}^4$$

$$S = \frac{I}{h/2 + t_f} = \frac{829,578}{75 + 1.5} = 10,844.2 \text{ in.}^3$$

Allowable bending stress $F_b = 0.60F_{yf} = 0.60(50) = 30$ ksi

From Eq. (9.23):

$$F_b' = (30)[1.0 - (0.0005)\frac{84.375}{58.50}(266.67 - 138.76)] = 27.23 \text{ ksi}$$

From Eq. (9.53):

$$F'_b = \frac{6(58.50) + 0.893(84.375)}{6(58.50) + 84.375}(30) = 29.38 \text{ ksi}$$

Therefore, $F'_b = 27.23$ ksi governs.

$$f_b = \frac{M_{\max}}{S} = \frac{(26,667)(12)}{10,844.2} = 29.51 \text{ ksi} > 27.23 \text{ ksi} \qquad \underline{\text{N.G.}}$$

Try PL 40 in. $\times 1\frac{5}{8}$ in. for each flange.

$$A_f = 65 \text{ in.}^2; \qquad I = 905,411 \text{ in.}^4; \qquad S = 11,816.1 \text{ in.}^3$$

Check Eq. (9.20):

$$\frac{b_f}{t_f} = \frac{40}{\frac{13}{8}} = 24.62 < \frac{190}{\sqrt{F_y}} = 26.9 \qquad \underline{\text{O.K.}}$$

From Eq. (9.23):

$$F'_b = 27.51 \text{ ksi}$$

$$f_b = \frac{M_{\max}}{S} = \frac{(26,667)(12)}{11,816.1} = 27.08 \text{ ksi} < 27.51 \text{ ksi} \qquad \underline{\text{O.K.}}$$

9.10.4 Intermediate Stiffeners

Intermediate stiffeners are required, and their spacing should not be greater than

$$a_{\max} = 1.5d = 1.5(h + 2t_f) = 1.5(150 + 3.25) = 230 \text{ in.}$$

Also, from Eq. (9.10):

$$a_{\max} = [260/(h/t_w)]^2 h = (260/266.67)^2(150) = 142.60 \text{ in.} \quad \text{(governs)}$$

For hybrid girders, the design cannot be made on the basis of tension-field action, and we must use Eq. (9.6) for the allowable shear stress. We use uniform spacing between A and C and between C and B (Fig. 9.17).

1. Spacing of the stiffeners between A and C.

$$f_v = \frac{V_{\max}}{A_w} = \frac{R_A}{A_w} = \frac{466.67}{84.375} = 5.53 \text{ ksi} < 0.40F_y = 14.40 \text{ ksi}$$

From Eq. (9.6):

$$C_v = 2.89F_v/F_y = 2.89(5.53)/36 = 0.444 < 0.8$$

From Eq. (9.7):

$$k = F_y C_v (h/t_w)^2/45,000 = 36(0.444)(266.67)^2/45,000$$

$$= 25.26 > 9.34$$

From Eq. (9.8):

$$\frac{a}{h} = \left(\frac{5.34}{k-4}\right)^{1/2} = \left(\frac{5.34}{25.26-4}\right)^{1/2} = 0.501$$

$$a_{max} = 0.501(150) = 75.2 \text{ in.}$$

Use $a = 75$ in. between A and C (Fig. 9.23).

Flanges: 2PL 40 in. $\times$ $1\frac{5}{8}$ in.

Web: PL 150 in. $\times$ $\frac{9}{16}$ in.

Intermediate stiffeners: PL 7.5 in. $\times$ $\frac{9}{16}$ in. $\times$ 12 ft $3\frac{3}{4}$ in. (on one side)

Bearing stiffeners at A, B, and C: 2PL 19 in. $\times$ $1\frac{1}{4}$ in. $\times$ 12 ft 6 in.

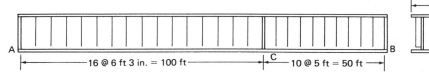

Figure 9.23 Final design of the hybrid plate girder.

2. Spacing of the stiffeners between C and B.

$$f_v = \frac{V_{max}}{A_w} = \frac{R_B}{A_w} = \frac{633.33}{84.375} = 7.51 \text{ ksi} < 0.40F_y = 14.40 \text{ ksi}$$

$$C_v = 2.89F_v/F_y = 2.89(7.51)/36 = 0.603 < 0.8$$

$$k = F_yC_v(h/t_w)^2/45,000 = 36(0.603)(266.67)^2/45,000 = 34.30 > 9.34$$

$$\frac{a}{h} = \left(\frac{5.34}{k-4}\right)^{1/2} = \left(\frac{5.34}{34.30-4}\right)^{1/2} = 0.42$$

$$a_{max} = 0.42(150) = 63 \text{ in.}$$

Use $a = 60$ in. between C and B (Fig. 9.23).

Note that because the design of hybrid girders is not based on tension-field action, no stress reduction due to the interaction of simultaneous bending and shear stresses is necessary. In other words, the combined shear and bending check [Eq. (9.29)] is not required for hybrid girders.

3. Selection of intermediate stiffeners.

Selection of the size of intermediate stiffeners in hybrid girders is based upon Eqs. (9.27) and (9.28). We can find an approximate formula for the minimum required thickness of the stiffeners.

First, let us consider the case of single stiffeners. The moment of inertia of a single stiffener with respect to an axis in the plane of the web and perpendicular to the plane of the stiffeners is approximately equal to

$$I_{st} \simeq \tfrac{1}{3}t_sb_s^3 \tag{9.57}$$

Substituting for b_s from the limiting case of Eq. (9.28) into Eq. (9.57) yields

$$I_{st} = \frac{285,792}{(F_y)^{3/2}} t_s^4 \tag{9.58}$$

Finally, equating Eqs. (9.27) and (9.58) and solving for t_s, we obtain the following approximate formula for the minimum required thickness of the intermediate stiffeners:

$$t_s = 0.000865 h (F_y)^{0.375} \qquad (9.59)$$

In the case of double stiffeners, we must mutliply the right-hand side of Eqs. (9.57) and (9.58) by 2, and the resulting equation for the minimum thickness is

$$t_s = 0.000727 h (F_y)^{0.375} \qquad (9.60)$$

If we assume single stiffeners, the minimum stiffener thickness for our problem is

$$t_s = 0.000865(150)(36)^{0.375} = 0.50 \text{ in.}$$

Try $t_s = \frac{9}{16}$ in. From Eq. (9.28):

$$b_s \leqslant \frac{95 t_s}{\sqrt{F_y}} = \frac{95(9/16)}{\sqrt{36}} = 8.90 \text{ in.}$$

Try PL 7.5 in. $\times \frac{9}{16}$ in. for stiffeners on one side.

$$\text{Furnished } I_{st} = \frac{1}{3} t_s \left(b_s + \frac{t_w}{2} \right)^3 = \frac{1}{3} \left(\frac{9}{16} \right) \left(7.5 + \frac{9}{32} \right)^3 = 88.3 \text{ in.}^4$$

$$\text{Required } I_{st} = \left(\frac{h}{50} \right)^4 = \left(\frac{150}{50} \right)^4 = 81 \text{ in.}^4 < 88.3 \text{ in.}^4 \qquad \underline{\text{O.K.}}$$

Length of intermediate stiffeners:

$$h_1 = h - 4 t_w = 150 - 4(9/16) = 147.75 \text{ in.} = 12 \text{ ft } 3\tfrac{3}{4} \text{ in.}$$

Use PL 7.5 in. $\times \frac{9}{16}$ in. $\times$ 12 ft $3\tfrac{3}{4}$ in. stiffeners on one side.

9.10.5 Web Crippling

The compressive stress in the web:

$$f_{cw} = \frac{w}{t_w} = \frac{\frac{4}{12}}{\frac{9}{16}} = 0.593 \text{ ksi}$$

The allowable compressive stress at the location where a/h is critical (that is, a/h has the largest value) from Eq. (9.30) is

$$F_a = \frac{10,000}{(h/t_w)^2} \left[5.5 + \frac{4}{(a/h)^2} \right] = \frac{10,000}{(266.67)^2} \left[5.5 + \frac{4}{(75/150)^2} \right]$$

$$= 3.02 \text{ ksi} > f_{cw} = 0.593 \text{ ksi} \qquad \underline{\text{O.K.}}$$

9.10.6 Bearing Stiffeners

The procedure for design of bearing stiffeners is the same as for the nonhybrid girder example of Sec. 9.9.7 and therefore will not be repeated here. The answer:

Use 2PL 19 in. $\times$ $1\frac{1}{4}$ in. $\times$ 12 ft 6 in. for bearing stiffeners at each support and at the location of the concentrated load.

9.10.7 Connections

The procedure for design of web-to-flange fillet welds is the same as for nonhybrid plate girders. This portion of the design is left as an exercise for the reader.

For stiffener-to-web fillet welds, however, Eq. (9.42) will not apply and only the minimum amount of welding, as given in AISCS Sec. 1.17.2 and covered in Sec. 8.3, needs to be used.

9.10.8 Girder Weight

$$\text{Weight of the flange plate} = 2A_f L\gamma = 2(65)(150)(0.49)/144 = 66.35 \text{ Kips}$$

$$\text{Weight of the web plate} = ht_w L\gamma = (150)(\tfrac{9}{16})(150)(0.49)/144$$

$$= 43.07 \text{ Kips}$$

$$\text{Weight of the flange and web plate} = 66.35 + 43.07 = 109.42 \text{ Kips}$$

$$\text{Weight of the stiffeners} = [2(3)(\tfrac{5}{4})(19)(150) + (24)(\tfrac{9}{16})(7.5)$$

$$\times (147.75)](0.49)/(12^3) = 10.30 \text{ Kips}$$

$$\text{Total weight of the plate girder} = 109.42 + 10.30 = 119.7 \text{ Kips}$$

Note that this hybrid plate girder is 17.4 percent lighter than the corresponding homogeneous plate girder designed in Sec. 9.9. However, this does not necessarily mean a more economical design.

9.11 ADDITIONAL EXAMPLES OF DESIGN OF PLATE GIRDERS ACCORDING TO AISCS

Example 1

A simply supported hybrid steel plate girder consists of two WT18 $\times$ 115 sections with $F_y = 50$ ksi and a 100 in. $\times$ 0.75 in. plate with $F_y = 36$ ksi as shown in Fig. 9.24. The length of the span is 150 ft. What is the maximum uniformly distributed load that can be supported by this girder? The compression flange is embedded in a concrete floor deck. Consider bending only.

Solution   The centroidal axis of the top WT section is specified by the axis $\bar{x}$ in Fig. 9.24. Properties of a WT 18 $\times$ 115 section are

$$t_f = 1.26 \text{ in.} \qquad t_w = 0.760 \text{ in.}$$

$$A = 33.8 \text{ in.}^2 \qquad I_{\bar{x}} = 934 \text{ in.}^4$$

Moment of inertia of the built-up section:

$$I_x = 2[934 + 33.8(50 + 17.95 - 4.01)^2] + \tfrac{1}{12}(0.75)(100)^3$$

$$= 340,739 \text{ in.}^4$$

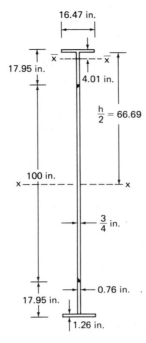

16.47 in.

17.95 in.

4.01 in.

$\frac{h}{2} = 66.69$

100 in.

$\frac{3}{4}$ in.

0.76 in.

17.95 in.

1.26 in.   **Figure 9.24**

Section modulus of the built-up section:

$$S_x = \frac{I_x}{50 + 17.95} = \frac{340{,}739}{67.95} = 5014.5 \text{ in.}^3$$

Allowable bending stress:

$$F_b = 0.60 F_{yf} = 0.60(50) = 30 \text{ ksi}$$

$$\frac{h}{t_w} = \frac{2(66.69)}{0.75} = 177.8 > \frac{760}{\sqrt{F_b}} = \frac{760}{\sqrt{30}} = 138.8$$

∴ Reduction of allowable bending stress in the compression flange is necessary (see Sec. 9.4.3).

$$A_f = (16.47)(1.26) = 20.75 \text{ in.}^2$$

$$A_w = (100)(\tfrac{3}{4}) + 2(33.8 - 20.75) = 101.1 \text{ in.}^2$$

$$F_b' = F_b \left[1.0 - 0.0005 \frac{A_w}{A_f} \left(\frac{h}{t_w} - \frac{760}{\sqrt{F_b}} \right) \right]$$

$$= 30 \left[1.0 - 0.0005 \left(\frac{101.1}{20.75} \right) \left(\frac{133.38}{0.75} - \frac{760}{\sqrt{30}} \right) \right] = 27.14 \text{ ksi}$$

$$\alpha = F_{yw}/F_{yf} = \tfrac{36}{50} = 0.72$$

$$F_b' = F_b \left[\frac{12 + (A_w/A_f)(3\alpha - \alpha^3)}{12 + 2(A_w/A_f)} \right] = 30 \left[\frac{12 + (101.1/20.75)(3 \times 0.72 - 0.72^3)}{12 + 2(101.1)/(20.75)} \right]$$

$$= 28.57 \text{ ksi}$$

Governing allowable bending stress: $F'_b = 27.14$ ksi

w_o = weight of the girder per unit length

$$= 2(0.115) + (100)(0.75)(0.490)/144 = 0.23 + 0.26 = 0.49 \text{ K/ft}$$

Maximum moment acting on the girder is (Fig. 9.25)

$$M_{\max} = \frac{(w_{\max} + w_o)L^2}{8} = F'_b S_x$$

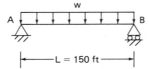

Figure 9.25

Therefore, the maximum uniformly distributed load that can be supported by this girder is

$$w_{\max} = \frac{8F'_b S_x}{L^2} - w_o = \frac{8(27.14)(5014.5)}{(150)^2(12)} - 0.49 = 3.54 \text{ Kips/ft}$$

Example 2

Find the spacing of the intermediate stiffeners in the previous example, assuming that the beam must carry a uniformly distributed load of 3.54 K/ft in addition to its own weight. Choose the minimum number of uniformly spaced stiffeners using A36 steel.

Solution   The design load intensity (Fig. 9.25):

$$w = 3.54 + 0.49 = 4.03 \text{ Kips/ft}$$

Since uniformly spaced stiffeners are provided, the spacing of stiffeners must be based on the maximum shear along the beam.

$$V_{\max} = \frac{wL}{2} = \frac{(4.03)(150)}{2} = 302.25 \text{ Kips}$$

The maximum shear stress in the girder is

$$f_v = \frac{V_{\max}}{A_w} = \frac{302.25}{101.1} = 2.99 \text{ ksi} < 0.40 \, F_y = 14.4 \text{ ksi}$$

We set this maximum shear stress to the allowable shear stress given by Eq. (9.6) and solve for C_v.

$$F_v = \frac{F_y C_v}{2.89} = f_v$$

$$C_v = \frac{2.89 f_v}{F_y} = \frac{(2.89)(2.99)}{36} = 0.240 < 0.80$$

Calculate k from Eq. (9.7).

$$k = \frac{C_v F_y (h/t_w)^2}{45,000} = \frac{0.24(36)(177.8)^2}{45,000} = 6.07 < 9.34$$

Now, we can calculate maximum spacing from Eq. (9.8).

$$\frac{a}{h} = \left(\frac{4}{k - 5.34}\right)^{1/2} = \left(\frac{4}{6.07 - 5.34}\right)^{1/2} = 2.34$$

$$a_{\max} = 2.34h = 2.34(2)(66.69) = 312.1 \text{ in.} = 26.01 \text{ ft}$$

Spacing of the stiffeners is also limited by Eqs. (9.10) and (9.11).

$$a_{\max} = 3h = 3(133.38) = 400.14 \text{ in.}$$

and

$$a_{\max} = \left(\frac{260}{h/t_w}\right)^2 h = \left(\frac{260}{177.8}\right)^2 (133.38) = 285.22 \text{ in.} = 23.77 \text{ ft}$$

Therefore, the governing limit is $a_{\max} = 23.77$ ft. Use

$$a = \frac{(150)(12)}{7} = 257.1 \text{ in.}$$

Use six intermediate stiffeners and two bearing stiffeners at the supports.

Example 3

Determine the allowable shear capacity (kips) of the plate girder interior panel shown in Fig. 9.26 if the bending tensile stress in the web is 27.5 ksi. What percentage of the shear capacity of the panel comes from the beam action prior to the buckling and what percentage is the share of the tension-field action after the shear buckling of the web? Yield stress of the steel is 50 ksi.

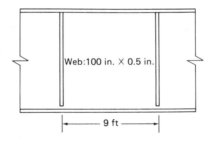

Web:100 in. × 0.5 in.

9 ft

Figure 9.26

Solution

$$a = 9 \text{ ft} = 108 \text{ in.} \qquad h = 100 \text{ in.} \qquad t_w = 0.5 \text{ in.}$$

$$\frac{a}{h} = 1.08 > 1$$

$$\frac{h}{t_w} = 200 < \frac{2000}{\sqrt{F_y}} = 283$$

Find k from Eq. (9.8):

$$k = 5.34 + \frac{4}{(a/h)^2} = 5.34 + \frac{4}{(1.08)^2} = 8.77$$

Find C_v from Eq. (9.7):

$$C_v = \frac{45,000k}{F_y(h/t_w)^2} = \frac{45,000(8.77)}{50(200)^2} = 0.197 < 0.8$$

Find the allowable shear stress from Eq. (9.12). (All the five conditions given in Sec. 9.3 for relying on the postbuckling behavior and tension-field action of the web plate are met.)

$$F_v = \frac{F_y}{2.89}\left[C_v + \frac{1 - C_v}{1.15\sqrt{1 + (a/h)^2}}\right] = 0.068F_y + 0.164F_y = 0.232F_y$$

$$= 11.61 \text{ ksi} < 0.40F_y$$

Allowable bending stress:

$$F_b = \left(0.825 - 0.375\frac{f_v}{F_v}\right)F_y = 27.5 \text{ ksi} = 0.55F_y$$

Solve for f_v:

$$f_v = 0.733F_v = 0.733(11.61) = 8.51 \text{ ksi}$$

Shear capacity of the plate girder = $f_v A_w = (8.51)(100)(0.5) = 425.7$ Kips

Percentage of the shear capacity for the beam action

$$= \left(\frac{C_v}{2.89}F_y A_w\right)\Big/(f_v A_w) = 0.068F_y/f_v = 0.068(50)/8.51 = 0.4 = 40 \text{ percent}$$

Percentage of the shear capacity from the tension-field action = $100 - 40 = 60$ percent

9.12 INTERACTIVE MICROCOMPUTER-AIDED DESIGN OF WELDED PLATE GIRDERS ACCORDING TO AISCS

9.12.1 Program Structure and Menus

The microcomputer program presented in this section interactively designs a simply supported welded steel plate girder according to AISCS. It consists of a main unit and 24 modules. A complete list of variables is given at the outset of the program listing. Numerous comment statements are provided within the program that explain the logic and flow of the program.

The program is menu-driven, from which the user can select an action by entering its number. At the beginning, an input menu will be displayed as shown in Fig. 9.27. The input data may be provided through the keyboard or read from a disk by specifying the name of the data file. The input data may be displayed on screen or modified. In the modify action, a modify-input-data menu will be displayed as shown in Fig. 9.28, and the user may select from this menu to change one or several parts of the data. At the same time, the old data will also be displayed. The input data may be saved on a disk. In reading or saving the input data the user may choose to use disk drive A or disk drive B. Immediately after this choice, the directory of the corresponding diskette will be displayed on the screen to help the

```
*** INPUT DATA MENU ***
~~~~~~~~~~~~~~~~~~~~~~~~

1.  INPUT DATA FROM KEYBOARD

2.  READ DATA FROM A DISK

3.  DISPLAY INPUT DATA

4.  MODIFY INPUT MENU

5.  SAVE INPUT DATA

6.  PROCEED TO DESIGN

7.  QUIT

ENTER THE NUMBER PLEASE  ==>> ▮
```

```
*** MODIFY DATA MENU ***
~~~~~~~~~~~~~~~~~~~~~~~~~

1.  PROBLEM HEADING & DESCRIPTION

2.  CONTROL DATA

3.  LOAD DATA

4.  LATERAL SUPPORT

5.  INTERMEDIATE STIFFENERS

6.  YIELD/ULTIMATE STRESSES

7.  WIDTH/THICKNESS INCREMENT

ENTER THE NUMBER PLEASE ==> ▮
```

PRESS 'ENTER' TO GO BACK TO INPUT MENU !

Figure 9.27   **Figure 9.28**

user in choosing the name of the file to be provided for the input data. Once the input data are ready, the user may start the process of interactive design of the plate girder.

The program takes full advantage of the interactive structure of the BASIC language for fulfilling the unavoidable trial and error process in plate girder design. After the design of each portion or element of the girder, say, flange plate, the user/designer will be asked for possible alterations according to his or her judgment. Then the program will modify the design accordingly and inquire again about further alterations. This feature will give the user/designer full control of the program without the burden of numerous repetitive calculations and design checks.

Several string variables are used for specifying different flags, as defined at the outset of the program. For the sake of convenience to the user, the program is written in such a manner that input for string variables can be typed in as either lowercase or uppercase characters.

To increase the reliability, the program has a number of built-in error messages. For example, if there is an error in input data, the program will often inform the user by explaining the possible source of error and through sound effects.

The program is capable of designing hybrid girders made of a web plate of low-strength steel and flange plates of high-strength steel. It designs a doubly symmetric section. For nonhybrid girders, it takes into account the postbuckling behavior and tension-field action of the web. Provisions for lateral torsional buckling when the compression flange is not laterally supported are also incorporated in the program.

The loading on the girder may consist of a uniformly distributed load plus a specified number of concentrated loads. The concentrated loads are numbered from left to right. A portion of the girder between two neighboring concentrated loads or between a reaction and a concentrated load is called a "segment." Segments are numbered from left to right. The spacing of stiffeners is assumed to be uniform in each segment (except at the two ends of nonhybrid plate girders when designed on the basis of tension-field action).

Single or double plates may be used as transverse stiffeners. For flange-to-web connections, either continuous or intermittent fillet weld may be specified. For intermediate stiffener-to-web connections, intermittent welds are used. For bearing stiffeners, continuous fillet welds are used as required by the AISC specification.

The program rounds off the thickness and width of the flange plates and stiffeners and the thickness of the web plate to practical sizes according to the increment specified by the user.

```
        *** MAIN MENU ***
        NNNNNNNNNNNNNNNNNNNN

    1.  DISPLAY SUMMARY OF RESULTS

    2.  DISPLAY MENU

    3.  REDESIGN

    4.  INPUT DATA MENU

    5.  QUIT

    ENTER THE NUMBER PLEASE  ==) █
```

Figure 9.29

```
        * DISPLAY  MENU *
        NNNNNNNNNNNNNNNNN

    1.  DISPLAY THE LOADINGS

    2.  DISPLAY THE SHEAR DIAGRAM

    3.  DISPLAY THE MOMENT DIAGRAM

    4.  DISPLAY THE SECTIONS

    5.  DISPLAY THE ELEVATION

    6.  MAIN MENU

    7.  QUIT

    ENTER THE NUMBER PLEASE  ==)) █
```

Figure 9.30

When the design of the plate girder is completed, a main menu will be displayed as shown in Fig. 9.29. From this menu, the user may call the display menu shown in Fig. 9.30. From the display menu, the user may select to display/plot the loading on the girder, the shear force diagram, the bending moment diagram, the plate

```
        * SECTION  MENU *
        NNNNNNNNNNNNNNNNN

    1.  SECTION AT LEFT SUPPORT

    2.  SECTION AT RIGHT SUPPORT

    3.  SECTION AT CONCENTRATED LOAD

    4.  SECTION WITHIN SEGMENT

    5.  DISPLAY MENU

    ENTER THE NUMBER PLEASE  ==)) █
```

Figure 9.31

PLATE GIRDER LOADINGS

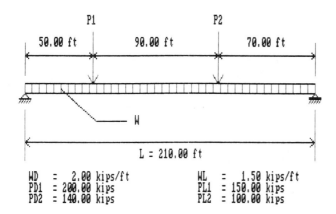

WD =   2.00 kips/ft   WL =   1.50 kips/ft
PD1 = 200.00 kips   PL1 = 150.00 kips
PD2 = 140.00 kips   PL2 = 100.00 kips

*** PRESS ANY KEY TO GO BACK TO DISPLAY MENU ***   **Figure 9.32**

girder elevation, different sections of the girder, or to go back to the main menu. There is also a section menu for displaying the sections of the girder, as shown in Fig. 9.31. Sample plots for the plate girder loading, shear force diagram, bending moment diagram, elevation, and sections are presented in Figs. 9.32 through 9.37. These plots are for the plate girder designed in the sample interactive session of Sec. 9.12.2.

In the graphic module, a value acting as a divider is provided, and the height and the width of a pixel are calculated by dividing the height of the screen by the number of pixels in one column and the width of the screen by the number of pixel

SHEAR FORCE DIAGRAM

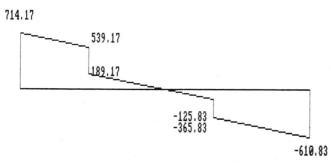

NOTE : VALUES OF SHEAR FORCES ARE IN kips.

* PRESS ANY KEY TO CONTINUE *   **Figure 9.33**

BENDING MOMENT DIAGRAM

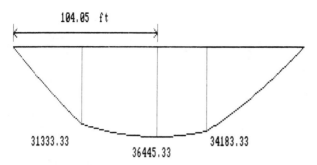

104.05 ft

31333.33   34183.33

36445.33

NOTE : VALUES OF BENDING MOMENTS ARE IN kips.ft

* PRESS ANY KEY TO CONTINUE *   **Figure 9.34**

in one row. These two pixel sizes together with the divider are used in the graphic modules to find the equivalent lengths of the horizontal and vertical lines to be displayed. This feature enables the designer to use the graphic modules on various display monitors with different screen resolutions by simply changing the dimensions of the pixel.

For displaying/plotting the girder elevation or sections, the user can zoom into one portion of the display by providing the coordinates of the center point of the object to be displayed on the screen and a scaling factor. Default values for the coordinates of the center point of the object are the coordinates of the center of the screen. These features are particularly useful when the user/designer wishes to enlarge a portion of the plate girder elevation or section. For example, elevation

ELEVATION OF THE PLATE GIRDER

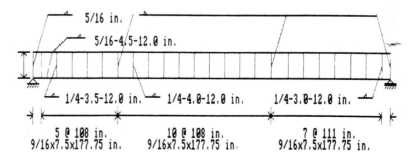

5/16 in.

5/16-4/5-12.0 in.

1/4-3.5-12.0 in.   1/4-4.0-12.0 in.   1/4-3.0-12.0 in.

5 @ 108 in.   10 @ 108 in.   7 @ 111 in.
9/16x7.5x177.75 in.   9/16x7.5x177.75 in.   9/16x7.5x177.75 in.

Figure 9.35 Elevation of the plate girder.

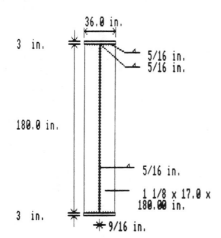

PLATE GIRDER SECTION AT RIGHT SUPPORT

Figure 9.36 Plate girder section at right support.

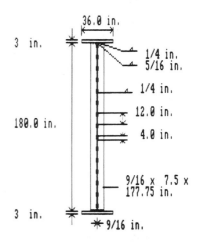

PLATE GIRDER SECTION IN SEGMENT # 2

Figure 9.37 Plate girder section in segment 2.

of the plate girder shown in Fig. 9.35 has been enlarged in Fig. 9.38. Also, an enlarged portion of the plate girder section is shown in Fig. 9.39. Intermediate stiffeners are not normally welded to the tension flange to prevent the possible undesirable consequences of moving loads and fatigue. This detail is well presented in Figs. 9.38 and 9.39. Note that all the necessary details are presented on the plots so that they are even suitable for manufacturing the plate girder.

A sample interactive session is presented in Sec. 9.12.2.

ELEVATION OF THE PLATE GIRDER

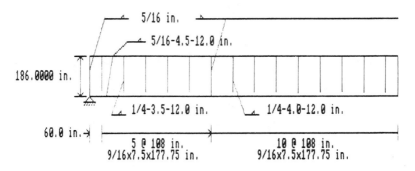

Figure 9.38 Elevation of the plate girder enlarged.

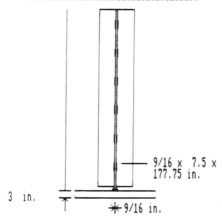

PLATE GIRDER SECTION IN SEGMENT # 1

9/16 x 7.5 x
177.75 in.

3 in.

9/16 in.

Figure 9.39   Plate girder section enlarged.

9.12.2 Sample Interactive Session

***** PLATE GIRDER DESIGN *****

(((Based on AISC Specification 1980)))

This is a program written in Advanced BASIC that will design a simply supported hybrid or non-hybrid steel plate girder interactively, based on 1980 AISC specification.
Sign convention : positive moment produces tension at the bottom fiber, positive vertical shear acts clockwise on the surface.
A portion of a plate girder between two lateral supports is called a region and a portion of a plate girder between two concentrated loads or between a reaction and a concentrated load is called a segment.
The segments and regions are numbered from left to right.
Single stiffeners or stiffeners furnished in pair are used.
Continuous or intermittent weld may be used for flange-to-web and intermediate stiffeners-to-web connections.

* PRESS ANY KEY TO CONTINUE *

*** INPUT DATA ***

* PROBLEM HEADING & DESCRIPTION *

ENTER PROBLEM HEADING ==> NON-HYBRID GIRDER BASED ON AISC SPECIFICATION

ENTER PROBLEM DESCRIPTION ==> Fyw=Fyf=Fys=36 ksi

* CONTROL DATA *

ENTER NUMBER OF DESIGN ITERATIONS ==> 2

IS PLATE GIRDER HYBRID OR NON-HYBRID (H/NH) ? nh

IS THE PLATE GIRDER USE IN BUILDING OR BRIDGE (BD/BG) ? bd

ENTER LENGTH OF THE SPAN (ft) ==> 210

IS THE PLATE GIRDER FRAMED AT BOTH ENDS (Y/N) ? n

IS THE PLATE GIRDER USE FOR FLOOR OR ROOF (F/R) ? f

IS THE DEAD WEIGHT OF THE GIRDER AN ESTIMATED PART
OF THE DESIGN LOAD (Y/N) ? y

IS FLANGE-TO-WEB CONNECTION WELDED CONTINUOUSLY (Y/N) ? n

LOAD DATA

ENTER INTENSITY OF DISTRIBUTED DEAD-LOAD (kips/ft) ==> 2

ENTER INTENSITY OF DISTRIBUTED LIVE-LOAD (kips/ft) ==> 1.5

ENTER NUMBER OF CONCENTRATED LOADS ==> 2

ENTER DEAD-CONCENTRATED LOAD NO. 1 FROM LEFT SUPPORT (kips) ==> 200

ENTER LIVE-CONCENTRATED LOAD NO. 1 FROM LEFT SUPPORT (kips) ==> 150

ENTER DISTANCE OF LOAD NO. 1 FROM LEFT SUPPORT (ft) ==> 50

ENTER DEAD-CONCENTRATED LOAD NO. 2 FROM LEFT SUPPORT (kips) ==> 140

ENTER LIVE-CONCENTRATED LOAD NO. 2 FROM LEFT SUPPORT (kips) ==> 100

ENTER DISTANCE OF LOAD NO. 2 FROM LEFT SUPPORT (ft) ==> 140

LATERAL SUPPORT

DOES THE PLATE GIRDER HAVE TOTAL LATERAL SUPPORT (Y/N) ? y

INTERMEDIATE STIFFENERS

ARE INTERMEDIATE STIFFENERS PROVIDED (Y/N) ? y

SINGLE OR DOUBLE PLATE STIFFENERS (S/D) ? d

YIELD/ULTIMATE STRESSES

ENTER YIELD STRESS OF FLANGE AND WEB PLATES (kips/in/in) ==> 36

ENTER YIELD STRESS OF STIFFENER PLATES (kips/in/in) ==> 36

ENTER TENSILE STRENGTH OF ELECTRODE (kips/in/in) ==> 70

WIDTH/THICKNESS INCREMENT

ENTER INCREMENT OF WIDTH FOR FLANGE-PLATE WIDTH (in.)
(e.g. 0.5 or 1 in.) ==> 1

ENTER INCREMENT OF THICKNESS FOR WEB AND FLANGE PLATES (in.)
(e.g. 1/16 = 0.0625 in.) ==> .0625

ENTER INCREMENT OF THICKNESS FOR STIFFENER PLATES (in.)
(e.g. 1/16 = 0.0625 in.) ==> .0625

••• CALCULATION OF SHEAR FORCES •••

LEFT SUPPORT REACTION : +714.17 kips

RIGHT SUPPORT REACTION : -610.83 kips

SHEAR FORCE TO THE LEFT OF POINT LOAD NO. 1 : +539.17 kips

SHEAR FORCE TO THE RIGHT OF POINT LOAD NO. 1 : +189.17 kips

SHEAR FORCE TO THE LEFT OF POINT LOAD NO. 2 : -125.83 kips

SHEAR FORCE TO THE RIGHT OF POINT LOAD NO. 2 : -365.83 kips

• PRESS ANY KEY TO CONTINUE •

••• CALCULATION OF BENDING MOMENT •••

BENDING MOMENT AT LOAD NO. 1   : +31333.33 kips.ft

BENDING MOMENT AT LOAD NO. 2   : +34183.33 kips.ft

MAXIMUM BENDING MOMENT   : +36445.33 kips.ft

LOCATION OF MAXIMUM BENDING MOMENT : 104.05 ft FROM LEFT SUPPORT

• PRESS ANY KEY TO CONTINUE •

••• DESIGN OF WEB PLATE •••

THE COMMON WEB DEPTH IS IN THE RANGE :

1/32 - 1/8 OF THE SPAN LENGTH

(78.75 - 315.00 in. for L = 210 ft.)

ENTER DEPTH OF THE WEB PLATE (in.) ==> 180

THICKNESS OF THE WEB : .5625 in.

THIS DESIGN IS BASED ON THE MINIMUM WEB THICKNESS REQUIREMENT

DO YOU WANT TO INCREASE THE WEB THICKNESS
BY ONE INCREMENT (Y/N) ? n

••• DESIGN OF FLANGE PLATE •••

WIDTH OF THE FLANGE   : 55 in.

THICKNESS OF THE FLANGE : 1.75 in.

AREA OF THE FLANGE   : 96.25 sq.in

DO YOU WANT ANY ALTERATION IN FLANGE DESIGN (Y/N) ? y

CHOOSE EITHER FLANGE WIDTH OR FLANGE THICKNESS (W/T) ==> t

ENTER FLANGE THICKNESS (in.) ==> 3

```
WIDTH OF THE FLANGE      :  32  in.

THICKNESS OF THE FLANGE  :  3  in.

AREA OF THE FLANGE       :  96  sq.in

DO YOU WANT ANOTHER ALTERATION IN FLANGE DESIGN (Y/N) ? n


ALLOWABLE BENDING STRESS OF FLANGE IS EXCEEDED.

WIDTH OF THE FLANGE      :  32  in.

THICKNESS OF THE FLANGE  :  3  in.

     CALCULATED STRESS :  21.62317  kips/in/in
     ALLOWABLE STRESS  :  19.81766  kips/in/in

DO YOU WANT TO INCREASE THE FLANGE WIDTH, FLANGE THICKNESS OR
WEB THICKNESS BY ONE INCREMENT (W/TF/TW) ? w

WIDTH OF THE FLANGE      :  33  in.

THICKNESS OF THE FLANGE  :  3  in.

AREA OF THE FLANGE       :  99  sq.in

DO YOU WANT ANOTHER ALTERATION IN FLANGE DESIGN (Y/N) ? n

ALLOWABLE BENDING STRESS OF FLANGE IS EXCEEDED.

WIDTH OF THE FLANGE      :  33  in.

THICKNESS OF THE FLANGE  :  3  in.

     CALCULATED STRESS :  21.06068  kips/in/in
     ALLOWABLE STRESS  :  19.87167  kips/in/in

DO YOU WANT TO INCREASE THE FLANGE WIDTH, FLANGE THICKNESS OR
WEB THICKNESS BY ONE INCREMENT (W/TF/TW) ? w

WIDTH OF THE FLANGE      :  34  in.

THICKNESS OF THE FLANGE  :  3  in.

AREA OF THE FLANGE       :  102  sq.in

DO YOU WANT ANOTHER ALTERATION IN FLANGE DESIGN (Y/N) ? n

ALLOWABLE BENDING STRESS OF FLANGE IS EXCEEDED.

WIDTH OF THE FLANGE      :  34  in.

THICKNESS OF THE FLANGE  :  3  in.

     CALCULATED STRESS :  20.52671  kips/in/in
     ALLOWABLE STRESS  :  19.92251  kips/in/in

DO YOU WANT TO INCREASE THE FLANGE WIDTH, FLANGE THICKNESS OR
WEB THICKNESS BY ONE INCREMENT (W/TF/TW) ? w

WIDTH OF THE FLANGE      :  35  in.

THICKNESS OF THE FLANGE  :  3  in.

AREA OF THE FLANGE       :  105  sq.in

DO YOU WANT ANOTHER ALTERATION IN FLANGE DESIGN (Y/N) ? n

ALLOWABLE BENDING STRESS OF FLANGE IS EXCEEDED.

WIDTH OF THE FLANGE      :  35  in.

THICKNESS OF THE FLANGE  :  3  in.
```

```
     CALCULATED STRESS :   20.01914   kips/in/in
     ALLOWABLE STRESS  :   19.97043   kips/in/in
```

DO YOU WANT TO INCREASE THE FLANGE WIDTH, FLANGE THICKNESS OR
WEB THICKNESS BY ONE INCREMENT (W/TF/TW) ? w

WIDTH OF THE FLANGE   : 36 in.

THICKNESS OF THE FLANGE : 3 in.

AREA OF THE FLANGE   : 108 sq.in

DO YOU WANT ANOTHER ALTERATION IN FLANGE DESIGN (Y/N) ? n

```
            TENTATIVE SIZES OF PLATE GIRDER SECTION  :

            HEIGHT OF THE WEB PLATE      :  180  in.
            WEB THICKNESS                :  .5625  in.
            WIDTH OF THE FLANGE PLATES   :  36  in.
            FLANGE THICKNESS             :  3  in.
            DEPTH OF THE GIRDER          :  186  in.
```

‡‡‡ SPACING OF INTERMEDIATE STIFFENER ‡‡‡

STIFFENER SPACING IN SEGMENT 1 : 108 in.

LOCATION OF FIRST INTERMEDIATE STIFFENER FROM LEFT SUPPORT : 60 in.

NUMBER OF SPACING   : 5

DO YOU WANT TO INCREASE THE WEB THICKNESS OR
REDUCE THE STIFFENER SPACING BY ONE INCREMENT (Y/N) ? n

STIFFENER SPACING IN SEGMENT 2 : 108 in.

NUMBER OF SPACING   : 10

DO YOU WANT TO INCREASE THE WEB THICKNESS OR
REDUCE THE STIFFENER SPACING BY ONE INCREMENT (Y/N) ? n

STIFFENER SPACING IN SEGMENT 3 : 111 in.

LOCATION OF FIRST INTERMEDIATE STIFFENER FROM RIGHT SUPPORT : 63 in.

NUMBER OF SPACING   : 7

DO YOU WANT TO INCREASE THE WEB THICKNESS OR
REDUCE THE STIFFENER SPACING BY ONE INCREMENT (Y/N) ? n

‡‡‡ DESIGN OF INTERMEDIATE STIFFENER ‡‡‡

‡‡‡ STIFFENER IN SEGMENT 1 ‡‡‡

SIZE OF INTERMEDIATE STIFFENERS IN SEGMENT 1 :

WIDTH OF STIFFENERS   : 8 in.

THICKNESS OF STIFFENERS : .5625 in.

AREA OF ONE PLATE   : 4.5 sq.in

DO YOU WANT TO ALTER THE STIFFENER WIDTH (Y/N) ? y

ENTER STIFFENER WIDTH (in.) ==> 7.5

SIZE OF INTERMEDIATE STIFFENERS IN SEGMENT 1 :

WIDTH OF STIFFENERS   : 7.5 in.

THICKNESS OF STIFFENERS : .5625 in.

AREA OF ONE PLATE   : 4.21875 sq.in

DO YOU STILL WANT TO ALTER THE STIFFENER WIDTH (Y/N) ? n

SIZE OF STIFFENERS IN SEGMENT 1 : .5625 x 7.5 x 177.75 in.

‡‡‡ STIFFENER IN SEGMENT 2 ‡‡‡

SIZE OF INTERMEDIATE STIFFENERS IN SEGMENT 2 :

WIDTH OF STIFFENERS   : 7.5 in.

THICKNESS OF STIFFENERS : .5625 in.

AREA OF ONE PLATE   : 4.21875 sq.in

DO YOU WANT TO ALTER THE STIFFENER WIDTH (Y/N) ? n

SIZE OF STIFFENERS IN SEGMENT 2 : .5625 x 7.5 x 177.75 in.

‡‡‡ STIFFENER IN SEGMENT 3 ‡‡‡

SIZE OF INTERMEDIATE STIFFENERS IN SEGMENT 3 :

WIDTH OF STIFFENERS   : 8 in.

THICKNESS OF STIFFENERS : .5625 in.

AREA OF ONE PLATE   : 4.5 sq.in

DO YOU WANT TO ALTER THE STIFFENER WIDTH (Y/N) ? y

ENTER STIFFENER WIDTH (in.) ==> 7.5

SIZE OF INTERMEDIATE STIFFENERS IN SEGMENT 3 :

WIDTH OF STIFFENERS   : 7.5 in.

THICKNESS OF STIFFENERS : .5625 in.

AREA OF ONE PLATE   : 4.21875 sq.in

DO YOU STILL WANT TO ALTER THE STIFFENER WIDTH (Y/N) ? n

SIZE OF STIFFENERS IN SEGMENT 3 : .5625 x 7.5 x 177.75 in.

‡‡‡ DESIGN OF BEARING STIFFENER ‡‡‡

BEARING STIFFENERS MUST EXTEND APPROXIMATELY TO THE EDGES
OF THE FLANGE PLATES AND SHOULD BE PROVIDED IN PAIR !

SIZE OF BEARING STIFFENER AT LEFT SUPPORT :

  17 x 1.125 x 180 in.

SIZE OF BEARING STIFFENER AT RIGHT SUPPORT :

  17 x 1.125 x 180 in.

SIZE OF BEARING STIFFENER AT LOAD NO. 1   :

  17 x 1.125 x 180 in.

SIZE OF BEARING STIFFENER AT LOAD NO. 2   :

  17 x 1.125 x 180 in.

DO YOU WANT TO ALTER THE WIDTH OF BEARING STIFFENERS (Y/N) ? n

‡‡‡ DESIGN OF WELDS CONNECTION ‡‡‡

‡‡‡ FLANGE-TO-WEB CONNECTION ‡‡‡

WELD WITH MINIMUM SIZE : 5/16 in., IS SUFFICIENT.

DO YOU WANT TO INCREASE THE WELD SIZE BY ONE INCREMENT (Y/N) ? n

MINIMUM LENGTH OF FILLET WELD : 1.5 in.

MAXIMUM SPACING BETWEEN WELDS : 12 in.

CHOOSE EITHER LENGTH OF FILLET-WELD SEGMENTS OR SPACING (L/A) ==> a

ENTER SPACING BETWEEN FILLET WELDS (in.) ==> 12

FLANGE-TO-WEB CONNECTION :

```
    WELD SIZE       :  5/16 in.
    WELD LENGTH     :  2.5 in.
    WELD SPACING    :  12 in.
```

DO YOU WANT TO REDESIGN (Y/N) ? n

‡‡‡ INTERMEDIATE STIFFENER-TO-WEB CONNECTION ‡‡‡

DESIGN OF WELD SIZE IN SEGMENT 1 :

WELD WITH MINIMUM SIZE : 1/4 in., IS SUFFICIENT.

DO YOU WANT TO INCREASE THE WELD SIZE BY ONE INCREMENT (Y/N) ? n

MINIMUM LENGTH OF FILLET WELD : 1.5 in.

MAXIMUM SPACING BETWEEN WELDS : 12 in.

CHOOSE EITHER LENGTH OF FILLET-WELD SEGMENTS OR SPACING (L/A) ==> a

ENTER SPACING BETWEEN FILLET WELDS (in.) ==> 12

STIFFENERS-TO-WEB CONNECTION IN SEGMENT : 1

```
    WELD SIZE       :  1/4 in.
    WELD LENGTH     :  2 in.
    WELD SPACING    :  12 in.
```

DO YOU WANT TO REDESIGN (Y/N) ? n

DESIGN OF WELD SIZE IN SEGMENT 2 :

WELD WITH MINIMUM SIZE : 1/4 in., IS SUFFICIENT.

DO YOU WANT TO INCREASE THE WELD SIZE BY ONE INCREMENT (Y/N) ? n

MINIMUM LENGTH OF FILLET WELD : 1.5 in.

MAXIMUM SPACING BETWEEN WELDS : 12 in.

CHOOSE EITHER LENGTH OF FILLET-WELD SEGMENTS OR SPACING (L/A) ==> a

ENTER SPACING BETWEEN FILLET WELDS (in.) ==> 12

STIFFENERS-TO-WEB CONNECTION IN SEGMENT : 2

```
    WELD SIZE       :  1/4 in.
    WELD LENGTH     :  2 in.
    WELD SPACING    :  12 in.
```

DO YOU WANT TO REDESIGN (Y/N) ? n

DESIGN OF WELD SIZE IN SEGMENT 3 :

WELD WITH MINIMUM SIZE : 1/4 in., IS SUFFICIENT.

DO YOU WANT TO INCREASE THE WELD SIZE BY ONE INCREMENT (Y/N) ? n

MINIMUM LENGTH OF FILLET WELD : 1.5 in.

MAXIMUM SPACING BETWEEN WELDS : 12 in.

CHOOSE EITHER LENGTH OF FILLET-WELD SEGMENTS OR SPACING (L/A) ==> a

ENTER SPACING BETWEEN FILLET WELDS (in.)   ==> 12

STIFFENERS-TO-WEB CONNECTION IN SEGMENT : 3

    WELD SIZE   : 1/4 in.
    WELD LENGTH   : 1.5 in.
    WELD SPACING   : 12 in.

DO YOU WANT TO REDESIGN (Y/N) ? n

‡‡‡ BEARING STIFFENERS TO WEB CONNECTION ‡‡‡

WELD WITH MINIMUM SIZE : 5/16 in., IS SUFFICIENT.

DO YOU WANT TO INCREASE THE WELD SIZE BY ONE INCREMENT (Y/N) ? n

BEARING STIFFENERS TO WEB CONNECTION :

WELD SIZE : 5/16 in. WELD CONTINOUSLY !

‡‡‡ CALCULATION OF TOTAL WEIGHT ‡‡‡

WEIGHT OF FLANGE AND WEB PLATES   : 226.70 kips

TOTAL WEIGHT OF THE PLATE GIRDER   : 243.44 kips

PRESS ANY KEY TO CONTINUE

‡‡‡ SUMMARY OF FINAL DESIGN OF PLATE GIRDER ‡‡‡

DESIGN NUMBER 1
NON-HYBRID GIRDER BASED ON AISC-SPECIFICATION

DEPTH OF WEB PLATE   : 180 in.

THICKNESS OF WEB PLATE   : .5625 in.

WIDTH OF FLANGE PLATES   : 36 in.

THICKNESS OF FLANGE PLATES : 3 in.

DEPTH OF PLATE GIRDER   : 186 in.

PRESS ANY KEY TO CONTINUE

LOCATION OF FIRST STIFFENER FROM LEFT SUPPORT   : 60 in.

SPACING OF INTERMEDIATE STIFFENERS IN SEGMENT 1 : 108 in.

SIZE OF INTERMEDIATE STIFFENERS IN SEGMENT 1 :

```
        WIDTH OF STIFFENERS    : 7.5   in.
        THICKNESS OF STIFFENERS : .5625 in.
        LENGTH OF STIFFENERS   : 177.75 in.

NUMBER OF INTERMEDIATE STIFFENERS IN SEGMENT  1  :  5

                PRESS ANY KEY TO CONTINUE


SPACING OF INTERMEDIATE STIFFENERS IN SEGMENT  2  :  108  in.

SIZE OF INTERMEDIATE STIFFENERS IN SEGMENT  2  :

        WIDTH OF STIFFENERS    : 7.5   in.
        THICKNESS OF STIFFENERS : .5625 in.
        LENGTH OF STIFFENERS   : 177.75 in.

NUMBER OF INTERMEDIATE STIFFENERS IN SEGMENT  2  :  9

                PRESS ANY KEY TO CONTINUE


LOCATION OF FIRST STIFFENER FROM RIGHT SUPPORT  :  63  in.

SPACING OF INTERMEDIATE STIFFENERS IN SEGMENT  3  :  111  in.

SIZE OF INTERMEDIATE STIFFENERS IN SEGMENT  3  :

        WIDTH OF STIFFENERS    : 7.5   in.
        THICKNESS OF STIFFENERS : .5625 in.
        LENGTH OF STIFFENERS   : 177.75 in.

NUMBER OF INTERMEDIATE STIFFENERS IN SEGMENT  3  :  7

                PRESS ANY KEY TO CONTINUE


SIZE OF BEARING STIFFENERS AT LEFT SUPPORT :

        WIDTH OF STIFFENER    : 17    in.
        THICKNESS OF STIFFENER : 1.125 in.
        LENGTH OF STIFFENER   : 180   in.

SIZE OF BEARING STIFFENER AT RIGHT SUPPORT :

        WIDTH OF STIFFENER    : 17    in.
        THICKNESS OF STIFFENER : 1.125 in.
        LENGTH OF STIFFENER   : 180   in.

SIZE OF BEARING STIFFENER AT LOAD NO. 1  :

        WIDTH OF STIFFENER    : 17    in.
        THICKNESS OF STIFFENER : 1.125 in.
        LENGTH OF STIFFENER   : 180   in.

                PRESS ANY KEY TO CONTINUE


SIZE OF BEARING STIFFENER AT LOAD NO. 2  :

        WIDTH OF STIFFENER    : 17    in.
        THICKNESS OF STIFFENER : 1.125 in.
        LENGTH OF STIFFENER   : 180   in.

FLANGE-TO-WEB CONNECTION :

        WELD SIZE    : 5/16 in.
        WELD LENGTH  : 2.5  in.
        WELD SPACING : 12   in.

INTERMEDIATE STIFFENERS TO WEB CONNECTION IN SEGMENT :  1

        WELD SIZE    : 1/4  in.
        WELD LENGTH  : 2    in.
        WELD SPACING : 12   in.
```

```
INTERMEDIATE STIFFENERS TO WEB CONNECTION IN SEGMENT :  2

        WELD SIZE    :  1/4  in.
        WELD LENGTH  :  2  in.
        WELD SPACING :  12  in.

INTERMEDIATE STIFFENERS TO WEB CONNECTION IN SEGMENT :  3

        WELD SIZE    :  1/4  in.
        WELD LENGTH  :  1.5  in.
        WELD SPACING :  12  in.

BEARING STIFFENERS TO WEB CONNECTION :

        WELD SIZE    :  5/16  in.


WEIGHT OF FLANGE AND WEB PLATES : 226.70 kips

TOTAL WEIGHT OF PLATE GIRDER    : 243.44 kips
```

9.13 LOAD AND RESISTANCE FACTOR DESIGN OF PLATE GIRDERS

9.13.1 Introduction

In this section we cover the LRFD of plate girders with tension-field action according to the Proposed LRFD Specification for Structural Steel Buildings [48]. Economical design of stiffened plate girders is normally achieved by taking advantage of the tension-field action.

The LRFD of plate girders can be based on tension-field action if

$$\frac{h}{t_w} > \frac{970}{\sqrt{F_{yf}}} \tag{9.61}$$

If this requirement is not satisfied, the plate girder must be designed as a beam. The LRFD of beams is covered in Chapter 5 and thus will not be repeated here.

9.13.2 Web Buckling Strength

The following requirement must be satisfied in order to maintain sufficient web buckling strength:

$$\frac{h_c}{t_w} \leq \begin{cases} \dfrac{14{,}000}{[F_{yf}(F_{yf} + 16.5)]^{1/2}} & \text{for } a > 1.5h_c \\[2mm] \dfrac{2000}{\sqrt{F_{yf}}} & \text{for } a \leq 1.5h_c \end{cases} \tag{9.62}$$

Note that these equations are similar to Eqs. (9.3) and (9.4), except that F_{yf} is the yield stress of the flange steel in ksi and

h_c = twice the distance from the neutral axis of the girder cross section to the inside face of the compression flange minus the fillet or corner radius

For doubly symmetric sections, h_c can be conservatively assumed equal to the depth of the web plate, h.

9.13.3 Flexural Design

The limit states to be considered in the design are tension-flange yield and compression flange buckling. If we denote the nominal flexural strength of the girder by M_n and the resistance factor by ϕ, the design flexure strength will be ϕM_n. At present, the recommended resistance factor is $\phi = 0.90$. The nominal flexural strength should be taken as the smaller value obtained from the limit states of the tension-flange yield and compression flange buckling.

1. **Limit state of tension-flange yield.** The nominal flexural strength based on the limit state of tension-flange yield is given by

$$M_n = R_{PG}R_eS_{xt}F_{yt} \tag{9.63}$$

The coefficient R_{PG} takes care of the reduction of the allowable bending stress resulting from the lateral displacement of the web plate on its compression side during the tension-field action and is determined by

$$R_{PG} = 1 - 0.0005\, a_r\left(\frac{h_c}{t_w} - \frac{970}{\sqrt{F_{cr}}}\right) \leq 1.0 \tag{9.64}$$

where F_{cr} is the critical compression flange stress to be discussed later in this section. Note that Eq. (9.64) corresponds to Eq. (9.23) of AISCS covered in Sec. 9.4.3.

Coefficient R_e in Eq. (9.63) is the hybrid girder factor. It is equal to 1 for nonhybrid girders and is given by

$$R_e = 1.0 - 0.1(1.3 + a_r)(0.81 - m) \leq 1.0 \tag{9.65}$$

for hybrid girders where a_r is the ratio of the web area to the compression flange area and m is the ratio of the web yield stress to the flange yield stress ($m = F_{yw}/F_{yf}$).

Finally, S_{xt} is the section modulus of the girder corresponding to the tension flange in in.^{3} and F_{yt} is the yield stress of the tension flange in ksi.

2. **Limit states of buckling.** The nominal flexural strength based on the limit states of buckling is given by

$$M_n = R_{PG}R_eS_{xc}F_{cr} \tag{9.66}$$

where S_{xc} is the section modulus of the girder corresponding to the compression flange in in.^{3}

In general, there exist three limit states of buckling. They are lateral-torsional buckling (LTB), flange local buckling (FLB), and web local buckling

(WLB). Because the design is based on the postbuckling behavior of the web plate, the limit state of WLB does not apply. However, we must evaluate the critical stress F_{cr} corresponding to the limit states of LTB and FLB and use the lower value in Eq. (9.66).

3. **Evaluation of critical stress F_{cr}.**   The critical stress F_{cr} is given as a function of slenderness parameters λ, λ_p, and λ_r, and a plate girder coefficient C_{PG}. λ_p and λ_r are the limiting slenderness parameters for compact and noncompact elements. If the width-thickness ratio of a compression element is greater than λ_p, then the section is noncompact. If the width-thickness ratio of a compression element in a noncompact section is larger than λ_r, then the element is identified as a slender compression element.

$$F_{cr} = \begin{cases} F_{yf} & \text{for } \lambda \leq \lambda_p \\ C_b F_{yf}\left[1 - \frac{1}{2}\left(\frac{\lambda - \lambda_p}{\lambda_r - \lambda_p}\right)\right] \leq F_{yf} & \text{for } \lambda_p < \lambda \leq \lambda_r \quad (9.67) \\ \dfrac{C_{PG}}{\lambda^2} & \text{for } \lambda > \lambda_r \end{cases}$$

The quantities λ, λ_p, λ_r, and C_{PG} are specified for two limit states of LTB and FLB separately.

a. For the limit state of LTB:

$$\lambda = \frac{L_u}{r_T} \tag{9.68}$$

$$\lambda_p = \frac{300}{\sqrt{F_{yf}}} \tag{9.69}$$

$$\lambda_r = \frac{756}{\sqrt{F_{yf}}} \tag{9.70}$$

$$C_{PG} = 286{,}000 C_b \tag{9.71}$$

In these equations, C_b is the same as that defined by Eq. (5.17), but r_T is defined as the radius of gyration of the compression flange plus one-sixth of the web. The latter definition will be identical to the earlier definition for doubly symmetric sections.

b. For the limit state of FLB:

$$\lambda = \frac{b_f}{2t_f} \tag{9.72}$$

$$\lambda_p = \frac{65}{\sqrt{F_{yf}}} \tag{9.73}$$

$$\lambda_r = \frac{150}{\sqrt{F_{yf}}} \tag{9.74}$$

$$C_{PG} = 11{,}200 \tag{9.75}$$

$$C_b = 1.0$$

Note that the governing slenderness parameters will be the ones that render the lower value of the critical stress F_{cr}.

9.13.4 Preliminary Proportioning of the Flange Plate

Equations (9.63) and (9.66) cannot be used directly for the design of the flange plate. We need to develop an approximate formula for the area of the flange to be used for preliminary proportioning of the flange plates. We will derive such an equation for the more common case of doubly symmetric girders and based on the limit state of tension-flange yield.

Equation (9.64) can be written as

$$R_{PG} = 1 - \frac{K_1}{A_f} \tag{9.76}$$

where

$$K_1 = 0.0005 A_w \left(\frac{h_c}{t_w} - \frac{970}{\sqrt{F_{cr}}} \right) \simeq 0.0005 A_w \left(\frac{h}{t_w} - \frac{970}{\sqrt{F_{yf}}} \right) \tag{9.77}$$

For the case of doubly symmetric girders, we can write from Eq. (9.63)

$$M_u = \phi M_n = \phi(1 - K_1/A_f) R_e S F_{yf}$$

Substituting for the section modulus from Eq. (9.18) into this equation and solving for A_f, we finally find the following approximate formula for the flange area of a doubly symmetric girder:

$$A_f = \frac{1}{2}\left(K_1 - \frac{A_w}{6} + K_2 \right) + \frac{1}{2}\left[\left(K_1 + \frac{A_w}{6} \right)^2 + K_2^2 + 2K_2\left(K_1 - \frac{A_w}{6} \right) \right]^{1/2} \tag{9.78}$$

where

$$K_2 = \frac{M_u}{\phi h F_{yf} R_e} \tag{9.79}$$

For the case of nonhybrid girders, $R_e = 1$ and Eq. (9.78) is an explicit equation for the flange area A_f. For hybrid girders, however, R_e is a function of the flange area as follows:

$$R_e = 1.0 - 0.1(1.3 + A_w/A_f)(0.81 - F_{yw}/F_{yf}) \leq 1.0 \tag{9.80}$$

and consequently Eq. (9.78) is not an explicit equation and must be solved iteratively. For the first iteration, a value must be assumed for the ratio A_w/A_f and substituted into Eq. (9.80). The resulting R_e is then replaced in Eq. (9.78). This iterative scheme converges very quickly. In the microcomputer program presented in Sec. 9.15, for the very first iteration, the ratio A_w/A_f is assumed to be one, and the following equation is used

$$R_e = 0.81 + 0.23 F_{yw}/F_{yf} \leq 1.0$$

Now, let us find bounds on b_f and t_f for the case when the critical stress F_{cr} is equal to the yield stress F_{yf}. For the limit state of LTB, we must have [Eqs. (9.67), (9.68), and (9.69)]:

$$\frac{L_u}{r_T} \le \frac{300}{\sqrt{F_{yf}}} \tag{9.81}$$

Substituting for r_T from Eq. (9.48) into this equation, we obtain the following bound on b_f in order to have $F_{cr} = F_{yf}$:

$$b_f \ge \frac{L_u}{300} \sqrt{2F_{yf}(6 + A_w/A_f)} \tag{9.82}$$

For the limit state of FLB, we must have [Eqs. (9.67), (9.72), and (9.73)]:

$$\frac{b_f}{2t_f} \le \frac{65}{\sqrt{F_{yf}}}$$

Substituting for $b_f = A_f/t_f$ into this equation and solving for t_f, we find the following bound on t_f in order to have $F_{cr} = F_{yf}$:

$$t_f \ge (A_f/130)^{1/2}(F_{yf})^{1/4} \tag{9.83}$$

Note that for a given required A_f, the requirements of Eqs. (9.82) and (9.83) often cannot be met simultaneously, and a compromise must be made. For the economical design of plate girders, the values of critical stress F_{cr} obtained from the limit states of LTB and FLB should be close to each other. In Sec. 9.15, an iterative scheme is used for achieving this goal in the microcomputer program.

9.13.5 Shear Design

If the nominal shear strength is denoted by V_n, the design shear strength is $\phi_v V_n$, where the resistance factor ϕ_v is 0.9. The nominal shear strength is given by

$$V_n = \begin{cases} 0.6A_w F_{yw} & \text{for } \dfrac{h_c}{t_w} \le \dfrac{187\sqrt{k}}{\sqrt{F_{yw}}} \\[4mm] 0.6A_w F_{yw}\left[C_v + \dfrac{1 - C_v}{1.15(1 + a^2/h_c^2)^{1/2}} \right] & \text{for } \dfrac{h_c}{t_w} > \dfrac{187\sqrt{k}}{\sqrt{F_{yw}}} \end{cases} \tag{9.84}$$

where C_v is the ratio of critical web stress according to linear buckling theory to the shear yield stress of web material. It is given by

$$C_v = \begin{cases} \dfrac{187\sqrt{k}}{(h_c/t_w)\sqrt{F_{yw}}} & \text{for } \dfrac{187\sqrt{k}}{\sqrt{F_{yw}}} \le \dfrac{h_c}{t_w} \le \dfrac{234\sqrt{k}}{\sqrt{F_{yw}}} & \text{(or } 1.0 \ge C_v > 0.8) \\[4mm] \dfrac{44,000k}{(h_c/t_w)^2 F_{yw}} & \text{for } \dfrac{h_c}{t_w} > \dfrac{234\sqrt{k}}{\sqrt{F_{yw}}} & \text{(or } C_v \le 0.8) \end{cases} \tag{9.85}$$

$$k = 5.0 + \frac{5.0}{(a/h_c)^2} \tag{9.86}$$

The web plate buckling coefficient k shall be taken as 5 whenever

$$\frac{a}{h_c} > 3 \quad \text{or} \quad [260/(h_c/t_w)]^2 \tag{9.87}$$

Equation (9.85) shall not be used for hybrid girders or for end-panels in nonhybrid girders. In these cases, as well as when relation (9.87) holds, the tension-field action is not permitted, and the following equation shall apply:

$$V_n = 0.6 A_w F_{yw} C_v \tag{9.88}$$

9.13.6 Transverse Stiffeners

Transverse stiffeners are not required in the following two cases:

1.
$$\frac{h_c}{t_w} \leq \frac{418}{\sqrt{F_{yw}}} \tag{9.89}$$

2.
$$V_u \leq 0.6 \phi_v A_w F_{yw} C_v \tag{9.90}$$

In Eq. (9.90), V_u is the required shear based on the factored loads, ϕ_v is 0.90, and C_v must be evaluated for $k = 5$.

When transverse stiffeners are required, their design should be based on the following requirements:

1. The gross area of each transverse stiffener or a pair of stiffeners should be at least equal to

$$A_{st} = \left[0.15 h_c t_w D(1 - C_v) \frac{V_u}{\phi_v V_n} - 18 t_w^2 \right] \left(\frac{F_{yw}}{F_{ys}} \right) \tag{9.91}$$

The coefficient D is the same as that defined in Sec. 9.5. V_u is the required factored shear at the location of the stiffener.

2. The moment of inertia of a single stiffener or a pair of stiffeners with respect to an axis in the plane of the web and perpendicular to the plane of the stiffeners should be at least equal to

$$I_{st} = a t_w^3 j \tag{9.92}$$

where

$$j = \left[\frac{2.5}{(a/h_c)^2} - 2 \right] \geq 0.5 \tag{9.93}$$

9.13.7 Combined Bending and Shear

For plate girders with transverse stiffeners, depending on the ratio of the required shear V_u and required moment M_u, an interaction check may be necessary as specified by the following equation (see also the last paragraph of Sec. 9.5):

$$\frac{M_u}{M_n} + 0.625 \frac{V_u}{V_n} \leq 1.375 \phi \quad \text{when } 0.6 \frac{V_n}{M_n} \leq \frac{V_u}{M_u} \leq 1.33 \frac{V_n}{M_n} \tag{9.94}$$

In this equation, M_u and V_u may not be greater than $\phi M_n = 0.9 M_n$ and $\phi V_n = 0.9 V_n$, respectively.

9.14 DESIGN OF A SIMPLE PLATE GIRDER BASED ON LRFDS

9.14.1 Problem Description

We design the same example of Sec. 9.9 on the basis of LRFD, but only for one load combination. Assume that the distributed and concentrated loads are resolved into dead and live loads as follows:

Intensity of distributed dead load: $w_D = 3$ K/ft
Intensity of distributed live load: $w_L = 1$ K/ft
Concentrated dead load: $P_D = 400$ Kips
Concentrated live load: $P_L = 100$ Kips

Using the load factors given in Sec. 2.7, Eq. (2.20), we find the design factored distributed and concentrated loads are as follows:

$$w = 1.2w_D + 1.6w_L = 1.2(3) + 1.6(1) = 5.2 \text{ K/ft}$$

$$P = 1.2P_D + 1.6P_L = 1.2(400) + 1.6(100) = 640 \text{ Kips}$$

9.14.2 Shear and Bending Moment Diagrams

Shear and bending moment diagrams are shown in Figs. 9.40 and 9.41, respectively. Maximum design bending moment and shear are

$$M_{\max} = M_u = 34{,}333.3 \text{ K-ft} = 412{,}000 \text{ K-in.}$$

$$V_{\max} = V_u = 816.67 \text{ Kips}$$

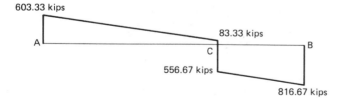

Figure 9.40   Shear diagram.

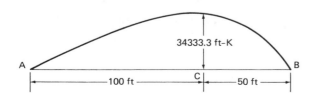

Figure 9.41   Bending moment diagram.

9.14.3 Selection of the Web Plate

Select the depth of the girder web:

$$h = \frac{L}{12} = \frac{(150)(12)}{12} = 150 \text{ in.}$$

From Eq. (9.62)

$$\frac{h_c}{t_w} \leqslant \begin{cases} 322.0 & \text{for } a > 1.5h_c \\ 333.3 & \text{for } a \leqslant 1.5h_c \end{cases}$$

The relation between h_c and h:

$$h_c = h - 2w_w$$

The size or width of the fillet weld w_w is not known at this stage. So we can approximately use $h_c \simeq h$.

$$t_w \geqslant \frac{h}{322} = \frac{150}{322} = 0.47 \text{ in.}$$

Try $t_w = 0.50$ in. or PL 150 in. $\times$ 0.5 in. for the web plate.

$$\frac{h}{t_w} = 300; \qquad A_w = 75 \text{ in.}^2$$

9.14.4 Selection of the Flange Plates

1. Preliminary selection of the flange plates.

    We find an approximate flange area from Eq. (9.78). First, from Eq. (9.77):

$$K_1 = 0.0005 A_w \left(\frac{h}{t_w} - \frac{970}{\sqrt{F_y}} \right) = 0.0005(75)\left(300 - \frac{970}{\sqrt{36}} \right) = 5.1875$$

and noting that

$$K_2 = \frac{M_u}{\phi h F_y} = \frac{412{,}000}{0.9(150)(36)} = 84.77$$

we find the approximate flange area from Eq. (9.78) as follows:

$$A_f = \tfrac{1}{2}(5.1875 - \tfrac{75}{6} + 84.77)$$
$$+ \tfrac{1}{2}[(\tfrac{75}{6} + 5.1875)^2 + (84.77)^2 - 2(84.77)(\tfrac{75}{6} - 5.1875)]^{1/2} = 78.29 \text{ in.}^2 \quad (9.95)$$

$$L_u = \text{unbraced length} = 50 \text{ ft} = 600 \text{ in.}$$

For the critical stress F_{cr} to be equal to the yield stress, from Eq. (9.82):

$$b_f \geqslant \tfrac{600}{300}\sqrt{2(36)(6 + 75/78.29)} = 44.8 \text{ in.} \quad (9.96)$$

and from Eq. (9.83):

$$t_f \geqslant (78.29/130)^{1/2}(36)^{1/4} = 1.90 \text{ in.} \quad (9.97)$$

To begin with, based on Eq. (9.97), try $t_f = 2$ in. Then, based on Eq. (9.95), try $b_f = 40$ in.; $A_f = 80$ in.^{2}

$$I_x = \tfrac{1}{12}(0.5)(150)^3 + \tfrac{2}{12}(40)(2)^3 + 2(80)(75+1)^2 = 1{,}064{,}838 \text{ in.}^4$$

$$S_x = \frac{I_x}{77} = \frac{1{,}064{,}838}{77} = 13{,}829.1 \text{ in.}^3$$

$$r_T = \frac{b_f}{(12 + 2A_w/A_f)^{1/2}} = \frac{40}{[12 + 2(75)/(80)]^{1/2}} = 10.74 \text{ in.}$$

For doubly symmetric nonhybrid girders, the nominal flexural strength on the basis of the limit state of tension-flange yield is always greater than or equal to the flexural strength on the basis of the limit state of buckling. Therefore, only the latter needs to be checked.

2. Check for the limit state of buckling.
   a. *Lateral torsional buckling*

$$\lambda = \frac{L_u}{r_T} = \frac{600}{10.74} = 55.87$$

$$\lambda_p = \frac{300}{\sqrt{F_{yf}}} = \frac{300}{\sqrt{36}} = 50.0$$

From the LTB point of view, the middle portion of the girder (portion DC in Fig. 9.17) is the critical portion. Bending moment at point D is equal to

$$M_D = 603.33(50) - 5.2(50)(25) = 23{,}666.5 \text{ ft-K}$$

$$M_1/M_2 = M_D/M_C = -23{,}666.5/34{,}333.3 = -0.689$$

$$C_b = 1.75 + 1.05(M_1/M_2) + 0.3(M_1/M_2)^2$$

$$= 1.75 + 1.05(-0.689) + 0.3(-0.689)^2 = 1.17 < 2.3$$

$$\lambda_r = \frac{756}{\sqrt{F_{yf}}} = \frac{756}{\sqrt{36}} = 126.0$$

$$F_{cr} = C_b F_{yf}\left[1 - \frac{1}{2}\left(\frac{\lambda - \lambda_p}{\lambda_r - \lambda_p}\right)\right] = (1.17)(36)\left[1 - \frac{1}{2}\left(\frac{55.87 - 50.00}{126.0 - 50.00}\right)\right]$$

$$= 40.49 \text{ ksi} > F_{yf} = 36 \text{ ksi}$$

$$\therefore \; F_{cr} = 36 \text{ ksi.}$$

   b. *Flange local buckling*

$$\lambda = \frac{b_f}{2t_f} = \frac{40}{2(2)} = 10$$

$$\lambda_p = \frac{65}{\sqrt{F_{yf}}} = \frac{65}{\sqrt{36}} = 10.83 > \lambda = 10$$

$$F_{cr} = F_{yf} = 36 \text{ ksi}$$

Therefore, the governing critical stress is $F_{cr} = 36$ ksi.

$$R_{PG} = 1 - 0.0005 \frac{A_w}{A_f}\left(\frac{h_c}{t_w} - \frac{970}{\sqrt{F_{cr}}}\right) = 1 - 0.0005\left(\frac{75}{80}\right)\left(300 - \frac{970}{\sqrt{36}}\right)$$

$$= 0.935$$

$$\phi M_n = \phi R_{PG} S_X F_{cr} = 0.9(0.935)(13,829.1)(36.00)$$

$$= 418,939 \text{ K-in.} > M_u = 412,000 \text{ K-in.} \qquad \underline{\text{O.K.}}$$

If the section were not O.K., we would have to increase the area of the flange. We choose the second trial flange plate on the basis of the following estimate of the new required flange area:

$$\text{New } A_f = (\text{old } A_f)(M_u / \phi M_n) \qquad (9.98)$$

Note that, in general, we can use Eq. (9.98) again to possibly reduce the area of the flange. In this example, however, ϕM_n is only 1.7 percent greater than M_u, and consequently we stop the iteration at this point.

9.14.5 Intermediate Stiffeners

1. Check if intermediate stiffeners are required.
   a. *Check Eq. (9.89):*

$$\frac{h_c}{t_w} \simeq \frac{h}{t_w} = 300 > \frac{418}{\sqrt{F_{yw}}} = \frac{418}{\sqrt{36}} = 69.67$$

   b. *Check Eq. (9.90) (with k = 5):*

$$\frac{h_c}{t_w} \simeq \frac{h}{t_w} = 300 > \frac{234\sqrt{k}}{\sqrt{F_{yw}}} = \frac{234\sqrt{5}}{\sqrt{36}} = 87.2$$

$$C_v = \frac{44,000k}{(h_c/t_w)^2 F_{yw}} = \frac{(44,000)(5)}{(300)^2(36)} = 0.0679$$

$$0.6\phi_v A_w F_{yw} C_v = 0.6(0.9)(75)(36)(0.0679) = 99 \text{ Kips}$$

This value is smaller than the maximum shear in segments AC and CB of the plate girder (Fig. 9.40).

∴ Intermediate transverse stiffeners are required.

2. Find the location of the first stiffener away from each end.
   a. *At the left end:*

$$V_u = 603.33 \text{ Kips} = \phi_v V_n = 0.9 V_n$$

From Eq. (9.88):

$$C_v = V_u / [0.6\phi_v A_w F_{yw}] = 603.33 / [0.6(0.9)(75)(36)] = 0.4138 < 0.8$$

From Eq. (9.85):

$$k = (h_c/t_w)^2 C_v F_{yw} / 44,000 = (300)^2(0.4138)(36)/44,000 = 30.47$$

From Eq. (9.86):

$$a/h_c = [5/(k-5)]^{1/2} = (5/25.47)^{1/2} = 0.443$$

$$a_{max} \simeq 0.443h = 0.443(150) = 66.4 \text{ in.}$$

Tentatively, place the first intermediate stiffener at a distance 45 in. from the left end A.

b. *At the right end*:

$$a_{max} = 55.7 \text{ in.}$$

Tentatively, place the first intermediate stiffener at a distance 50 in. from the right end B.

3. Find the spacing of the remaining stiffeners.

As in the example of Sec. 9.9, we use uniform spacing between point E (at the location of the first stiffener from the left end) and point C (the location of the concentrated load) and also between point C and point F (at the location of the first stiffener from the right end), similar to Fig. 9.22. Whenever the aspect ratio a/h_c is greater than 3 or $[260/(h_c/t_w)]^2$, the coefficient k shall be taken as 5. We limit the aspect ratio a/h_c to these values.

$$\frac{a}{h_c} \leq [260/(h_c/t_w)]^2 = [260/300]^2 = 0.75 < 3$$

$$a_{max} \simeq 0.75h = 0.75(150) = 112.5 \text{ in.}$$

a. *Spacing of the stiffeners between points E and C (Fig. 9.22)*:
Try $a = 105$ in.; $a/h_c \simeq 0.7$.
From Eq. (9.86):

$$k = 5 + \frac{5}{(a/h_c)^2} = 5 + \frac{5}{(0.7)^2} = 15.20$$

$$\frac{234\sqrt{k}}{\sqrt{F_{yw}}} = \frac{234\sqrt{15.20}}{\sqrt{36}} = 152.1 < \frac{h_c}{t_w} \simeq 300$$

From Eq. (9.85):

$$C_v = \frac{44,000k}{(h_c/t_w)^2 F_{yw}} = \frac{44,000(15.20)}{(300)^2(36)} = 0.206$$

From Eq. (9.84):

$$V_n = 0.6A_w F_{yw} \left[C_v + \frac{1 - C_v}{1.15(1 + a^2/h_c^2)^{1/2}} \right]$$

$$= 0.6(75)(36) \left[0.206 + \frac{1 - 0.206}{1.15(1 + 0.7^2)^{1/2}} \right] = 1250 \text{ Kips}$$

$$\phi_v V_n = 0.9(1250.0) = 1125 \text{ Kips}$$

$$V_u = R_A - \tfrac{45}{12}w = 603.33 - \tfrac{45}{12}(5.2) = 583.83 \text{ Kips} < \phi_v V_n \qquad \underline{\text{O.K.}}$$

b. *Spacing of the stiffeners between C and F* (*Fig. 9.22*).
Similar to case a, we find $a = 110$ in. to be satisfactory.

$$C_v = 0.194; \qquad V_n = 1230.0 \text{ Kips}; \qquad V_u = 795 \text{ Kips}$$

4. Check shear and bending interaction.

We perform the interaction check at point C under the concentrated load where the bending moment is the largest (actually, a very small distance to the right of point C).

$$M_u = 412,000 \text{ K-in.}; \qquad V_u = 556.67 \text{ Kips}$$

$$M_n = 465,487 \text{ K-in.}; \qquad V_n = 1230 \text{ Kips}$$

$$0.6 \frac{V_n}{M_n} = 0.00159 > \frac{V_u}{M_u} = 0.00135$$

According to Eq. (9.94), an interaction check is not necessary.

Note that the spacing of the stiffeners in this example is the same as that for the example of Sec. 9.9 designed according to AISCS and shown in Fig. 9.22.

5. Select intermediate stiffeners.

a. *Region EC*:

$$a = 105 \text{ in.}; \qquad a/h = 0.7; \qquad C_v = 0.206$$

From Eq. (9.91):

$$A_{st} = \left[0.15(1.0)(150)(0.5)(1 - 0.206) \frac{583.83}{1125} - 18(0.5)^2 \right] \left(\frac{36}{36} \right)$$

$$= 0.14 \text{ in.}^2$$

$$b_s t_s = 0.07 \text{ in.}^2 \tag{9.99}$$

From Eq. (9.93):

$$j = \frac{2.5}{(a/h_c)^2} - 2 = \frac{2.5}{(0.7)^2} - 2 = 3.10 > 0.5$$

From Eq. (9.92):

$$I_{st} = \tfrac{1}{12} t_s (2b_s + t_w)^3 \geqslant a t_w^3 j = (105)(0.5)^3(3.10)$$

$$= 40.69 \text{ in.}^4 \tag{9.100}$$

$$\frac{b_s}{t_s} \leqslant \frac{95}{\sqrt{36}} = 15.8 \tag{9.101}$$

The area requirement of Eq. (9.99) is minimal. Therefore, b_s and t_s should be selected based on Eqs. (9.100) and (9.101). The minimum thickness to satisfy Eqs. (9.100) and (9.101) is $t_s = \tfrac{3}{8}$ in. From Eq. (9.101):

$$b_s \leqslant 5.92 \text{ in.}$$

Try 2PL 5.5 in. $\times \tfrac{3}{8}$ in.

From Eq. (9.100):

$$I_{st} = \tfrac{1}{12}(0.375)(11 + 0.5)^3 = 47.53 \text{ in.}^4 > 40.69 \text{ in.}^4 \quad \underline{\text{O.K.}}$$

Use 2PL 5.5 in. $\times \frac{3}{8}$ in. for intermediate stiffeners from E to C.

b. *Region CF*:

$$a = 110 \text{ in.;} \qquad a/h = 0.733$$

The answer is the same as for region EC.
Use 2PL 5.5 in $\times \frac{3}{8}$ in. for intermediate stiffeners from C to F.

9.14.6 Bearing Stiffeners

Noting that the width of the flange plates is 40 in. and the thickness of the web is 0.5 in., we choose a width of $b_{bs} = 19$ in. for all bearing stiffeners.

1. Bearing stiffeners at supports.

    We design the support bearing stiffeners based on the larger shear at support B, $R_B = 816.67$ Kips. To satisfy the buckling requirement we must have

$$t_{bs} \geqslant \frac{\sqrt{F_y}}{95} b_{bs} = \frac{\sqrt{36}}{95}(19) = 1.20$$

Try 2PL $19 \times 1\frac{1}{4}$ in.; $\qquad A_{bs} = 23.75$ in.^{2}

Check the axial compressive stress due to reaction $R_B = 816.67$ Kips according to Sec. 6.7. The effective area for checking the axial compressive stresses is found from Eq. (9.33):

$$A_{\text{eff}} = 2A_{bs} + 12t_w^2 = 2(23.75) + 12(0.5)^2 = 50.50 \text{ in.}^2$$

$$I = \tfrac{1}{12}(\tfrac{5}{4})(19 + 19 + 0.5)^3 = 5944.4 \text{ in.}^4$$

$$r = \sqrt{I/A_{\text{eff}}} = \sqrt{5944.4/50.50} = 10.85 \text{ in.}$$

$$\frac{KL}{r} = \frac{(0.75)(150)}{10.85} = 10.37$$

$$\lambda_c = \frac{KL}{r}\left(\frac{F_y}{\pi^2 E}\right)^{1/2} = 10.37\left(\frac{36}{29{,}000\pi^2}\right)^{1/2} = 0.116 < 1.5$$

$$F_{cr} = F_y \exp\left(-0.419\lambda_c^2\right) = (36)\exp\left(-0.419 \times 0.116^2\right)$$

$$= 35.80 \text{ ksi}$$

$$P_n = A_{\text{eff}}F_{cr} = (50.50)(35.80) = 1807.9 \text{ Kips}$$

$$\phi_c P_n = 0.85(1807.9) = 1536.7 \text{ Kips} > R_B = 816.17 \text{ Kips} \qquad \underline{\text{O.K.}}$$

Use 2PL 19 in. $\times 1\frac{1}{4}$ in. $\times 12$ ft 6 in. for the bearing stiffeners at the left and right supports.

2. Bearing stiffener at concentrated load.

   Buckling requirement governs. Therefore,

   Use 2PL 19 in. $\times$ $1\frac{1}{4}$ in. $\times$ 12 ft 6 in. at the location of concentrated load.

   Note. In all the computations so far, we assumed $h_c \simeq h$. After designing the web-to-flange fillet weld, we can more accurately obtain $h_c = h - 2w_w$, where w_w is the width or size of the fillet weld, and check the design equations once more. However, such a check is rarely necessary because in most cases h_c is greater than $0.99h$. Design of welded connections is left as an exercise for the reader.

9.14.7 Girder Weight

γ = specific gravity of steel = 0.490 Kips/ft^3

Weight of the flange plates = $2A_f L\gamma = 2(\frac{80}{144})(150)(0.49) = 81.7$ Kips

Weight of the web plate = $A_w L\gamma = (\frac{75}{144})(150)(0.49) = 38.3$ Kips

Weight of the flange and web plates = $81.7 + 38.3 = 120.0$ Kips

Weight of the stiffeners = $[(3)(2)(19)(\frac{5}{4})(150) + (16)(2)(5.5)(\frac{3}{8})(148)]\left(\frac{0.49}{12^3}\right)$

$= 8.8$ Kips

Total weight of the plate girder = $120.0 + 8.8 = 128.8$ Kips

Comparing the weight of this design based on LRFD specification with the corresponding one based on AISCS (Sec. 9.9.10), we observe that the former is about 11 percent lighter than the latter.

9.15 INTERACTIVE MICROCOMPUTER-AIDED DESIGN OF WELDED PLATE GIRDERS ACCORDING TO LRFDS

9.15.1 Program Structure and Menus

The program structure and capabilities of this program are similar to those presented in Sec. 9.12 for interactive design of plate girders according to AISCS. The program designs a simply supported welded plate girder according to LRFDS. It consists of a main unit and 24 modules. Two load combinations are considered: $1.4D_n$ and $1.2D_n + 1.6L_n$, where D_n and L_n are the dead and live loads acting on the girder, respectively. The loading on the girder may consist of a uniformly distributed loading and any given number of concentrated loads.

A sample interactive session is presented in Sec. 9.15.2. The program is menu-driven. In addition to the input data menu (Fig. 9.27), modify data menu (Fig. 9.28), main menu (Fig. 9.29), display menu (Fig. 9.30), and section menu (Fig. 9.31), this program has an additional menu called the *loading menu*, as shown in

PRESS 'ENTER' TO GO BACK TO DISPLAY MENU !   **Figure 9.42**

Fig. 9.42. This menu is used for displaying the girder loadings. From this menu the user can ask the program to display the dead load (Fig. 9.43), the live load (Fig. 9.44), load case number 1 ($1.4D_n$) (Fig. 9.45), or load case number 2 ($1.2D_n + 1.6L_n$) (Fig. 9.46). The program can also display the shear force and bending moment diagrams for each loading case, as shown in Figs. 9.47 through 9.50. These diagrams are for the plate girder designed in the interactive session of Sec. 9.15.2. The cross sections and elevation of this plate girder are shown in Figs. 9.51 to 9.53. At the time of writing this book, the LRFD specification for design of connections was not available. The design of welded connections in this program is based on the AISC specification [47].

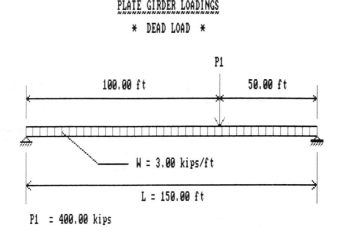

Figure 9.43

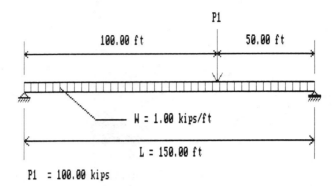

Figure 9.44

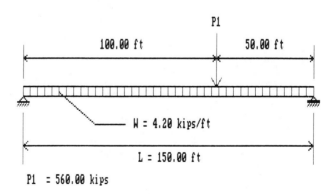

Figure 9.45

PLATE GIRDER LOADINGS

* LOAD CASE 2 *

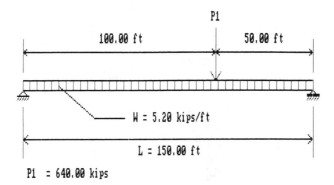

*** PRESS ANY KEY TO GO BACK TO DISPLAY MENU ***

Figure 9.46

SHEAR FORCE DIAGRAM

* LOAD CASE 1 : 1.4 D.L *

NOTE : VALUES OF SHEAR FORCES ARE IN kips.

* PRESS ANY KEY TO CONTINUE *

Figure 9.47

SHEAR FORCE DIAGRAM

* LOAD CASE 2 : 1.2 D.L + 1.6 L.L *

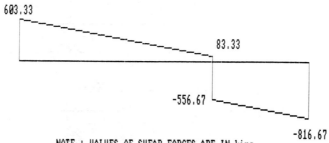

NOTE : VALUES OF SHEAR FORCES ARE IN kips.

* PRESS ANY KEY TO CONTINUE *

Figure 9.48

BENDING MOMENT DIAGRAM

* LOAD CASE 1 : 1.4 D.L *

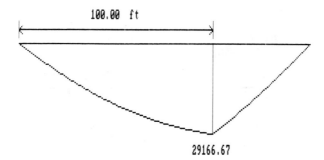

NOTE : VALUES OF BENDING MOMENTS ARE IN kips.ft

* PRESS ANY KEY TO CONTINUE *

Figure 9.49

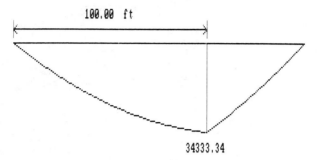

BENDING MOMENT DIAGRAM

* LOAD CASE 2 : 1.2 D.L + 1.6 L.L *

100.00 ft

34333.34

NOTE : VALUES OF BENDING MOMENTS ARE IN kips.ft

* PRESS ANY KEY TO CONTINUE *

Figure 9.50

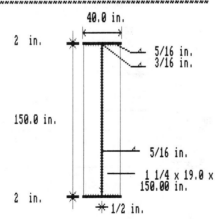

PLATE GIRDER SECTION AT CONCENTRATED LOAD # 1

40.0 in.

2 in.

5/16 in.
3/16 in.

150.0 in.

5/16 in.

1 1/4 x 19.0 x
150.00 in.

2 in.

1/2 in.

*** PRESS ANY KEY TO GO BACK TO SECTION MENU ***

Figure 9.51

PLATE GIRDER SECTION IN SEGMENT # 2

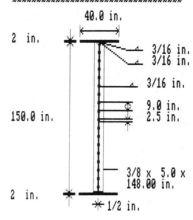

*** PRESS ANY KEY TO GO BACK TO SECTION MENU ***

Figure 9.52

ELEVATION OF THE PLATE GIRDER

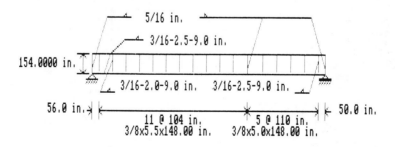

*** PRESS ANY KEY TO GO BACK TO DISPLAY MENU ***

Figure 9.53

9.15.2 Sample Interactive Session

***** PLATE GIRDER DESIGN *****

(((Based on LRFD Specification 1986)))

    This is a program written in Advanced BASIC that will design a simply
supported hybrid or non-hybrid steel plate girder interactively, based
on LRFD specification, 1986.
Sign convention : positive moment produces tension at the bottom fiber,
positive vertical shear acts clockwise on the transverse cross section.
A portion of a plate girder between two lateral supports is called a region and a portion of a plate girder between two concentrated loads or
between a reaction and a concentrated load is called a segment.
The segments and regions are numbered from left to right.
Single stiffeners or stiffeners furnished in pair are used.
Continuous or intermittent weld may be used for flange-to-web and intermediate stiffeners-to-web connections.

* PRESS ANY KEY TO CONTINUE *

*** INPUT DATA ***

* PROBLEM HEADING & DESCRIPTION *

ENTER PROBLEM HEADING ==> NON-HYBRID GIRDER BASED ON LRFD SPECIFICATION

ENTER PROBLEM DESCRIPTION ==> Fyw=Fyf=Fys=36 ksi

* CONTROL DATA *

ENTER NUMBER OF DESIGN ITERATIONS ==> 1

IS PLATE GIRDER HYBRID OR NON-HYBRID (H/NH) ? nh

IS THE PLATE GIRDER USE IN BUILDING OR BRIDGE (BD/BG) ? bd

ENTER LENGTH OF THE SPAN (ft) ==> 150

IS THE PLATE GIRDER FRAMED AT BOTH ENDS (Y/N) ? n

IS THE PLATE GIRDER USE FOR FLOOR OR ROOF (F/R) ? f

IS THE DEAD WEIGHT OF THE GIRDER AN ESTIMATED PART
OF THE DESIGN DEAD LOAD (Y/N) ? y

IS FLANGE-TO-WEB CONNECTION WELDED CONTINUOUSLY (Y/N) ? n

* LOAD DATA *

ENTER INTENSITY OF DISTRIBUTED DEAD-LOAD (kips/ft) ==> 3

ENTER INTENSITY OF DISTRIBUTED LIVE-LOAD (kips/ft) ==> 1

ENTER NUMBER OF CONCENTRATED LOADS ==> 1

ENTER DEAD-CONCENTRATED LOAD NO. 1 FROM LEFT SUPPORT (kips) ==> 400

✶ LATERAL SUPPORT ✶

DOES THE PLATE GIRDER HAVE TOTAL LATERAL SUPPORT (Y/N) ? n

ENTER NUMBER OF POINTS OF LATERAL SUPPORT ==> 4

ENTER DISTANCE OF LATERAL SUPPORT POINT NO. 1
(FROM LEFT SUPPORT IN ft) ==> 0

ENTER DISTANCE OF LATERAL SUPPORT POINT NO. 2
(FROM LEFT SUPPORT IN ft) ==> 50

ENTER DISTANCE OF LATERAL SUPPORT POINT NO. 3
(FROM LEFT SUPPORT IN ft) ==> 100

ENTER DISTANCE OF LATERAL SUPPORT POINT NO. 4
(FROM LEFT SUPPORT IN ft) ==> 150

✶ INTERMEDIATE STIFFENERS ✶

ARE INTERMEDIATE STIFFENERS PROVIDED (Y/N) ? y

SINGLE OR DOUBLE PLATE STIFFENERS (S/D) ? d

✶ YIELD/ULTIMATE STRESSES ✶

ENTER YIELD STRESS OF FLANGE AND WEB PLATES (kips/in/in) ==> 36

ENTER YIELD STRESS OF STIFFENER PLATES (kips/in/in) ==> 36

ENTER ULTIMATE STRESS OF WEB PLATE (kips/in/in) ==> 58

ENTER ULTIMATE STRESS OF STIFFENER PLATES (kips/in/in) ==> 58

ENTER TENSILE STRENGTH OF ELECTRODE (kips/in/in) ==> 70

✶ WIDTH/THICKNESS INCREMENT ✶

ENTER INCREMENT OF WIDTH FOR FLANGE-PLATE WIDTH (in.)
(e.g. 0.5 or 1 in.) ==> 1

ENTER INCREMENT OF THICKNESS FOR WEB AND FLANGE PLATES (in.)
(e.g. 1/16 = 0.0625 in.) ==> .0625

ENTER INCREMENT OF THICKNESS FOR STIFFENER PLATES (in.)
(e.g. 1/16 = 0.0625 in.) ==> .0625

✶✶✶ CALCULATION OF SHEAR FORCES ✶✶✶

✶ LOAD CASE 1 : 1.4 D.L ✶

LEFT SUPPORT REACTION : +501.67 kips

RIGHT SUPPORT REACTION : -688.33 kips

SHEAR FORCE TO THE LEFT OF POINT LOAD NO. 1 : +81.67 kips

SHEAR FORCE TO THE RIGHT OF POINT LOAD NO. 1 : -478.33 kips

✶ PRESS ANY KEY TO CONTINUE ✶

CALCULATION OF BENDING MOMENTS

LOAD CASE 1 : 1.4 D.L

BENDING MOMENT AT LOAD NO. 1   : +29166.67 kips.ft

MAXIMUM BENDING MOMENT   : +29166.67 kips.ft

LOCATION OF MAXIMUM BENDING MOMENT : 100.00 ft FROM LEFT SUPPORT

PRESS ANY KEY TO CONTINUE

CALCULATION OF SHEAR FORCES

LOAD CASE 2 : 1.2 D.L + 1.6 L.L

LEFT SUPPORT REACTION : +603.33 kips

RIGHT SUPPORT REACTION : -816.67 kips

SHEAR FORCE TO THE LEFT OF POINT LOAD NO. 1 : +83.33 kips

SHEAR FORCE TO THE RIGHT OF POINT LOAD NO. 1 : -556.67 kips

PRESS ANY KEY TO CONTINUE

CALCULATION OF BENDING MOMENTS

LOAD CASE 2 : 1.2 D.L + 1.6 L.L

BENDING MOMENT AT LOAD NO. 1   : +34333.34 kips.ft

MAXIMUM BENDING MOMENT   : +34333.34 kips.ft

LOCATION OF MAXIMUM BENDING MOMENT : 100.00 ft FROM LEFT SUPPORT

PRESS ANY KEY TO CONTINUE

DESIGN OF WEB PLATE

THE COMMON WEB DEPTH IS IN THE RANGE :

1/32 - 1/8 OF THE SPAN LENGTH

(56.25 - 225.00 in. for L = 150 ft.)

ENTER THE DEPTH OF THE WEB PLATE (in.) ==> 150

THICKNESS OF THE WEB : .5 in.

THIS DESIGN IS BASED ON THE MINIMUM WEB THICKNESS REQUIREMENT

DO YOU WANT TO INCREASE THE WEB THICKNESS BY ONE INCREMENT (Y/N) ? n

‡‡‡ DESIGN OF FLANGE PLATE ‡‡‡

WIDTH OF THE FLANGE   : 30 in.

THICKNESS OF THE FLANGE : 2.625 in.

AREA OF THE FLANGE   : 78.75 sq.in

DO YOU WANT ANY ALTERATION IN FLANGE DESIGN (Y/N) ? y

CHOOSE EITHER FLANGE WIDTH OR FLANGE THICKNESS (W/T) ==> t

ENTER FLANGE THICKNESS (in.) ==> 2

WIDTH OF THE FLANGE   : 40 in.

THICKNESS OF THE FLANGE : 2 in.

AREA OF THE FLANGE   : 80 sq.in

DO YOU WANT ANOTHER ALTERATION IN FLANGE DESIGN (Y/N) ? n

TENTATIVE SIZE OF PLATE GIRDER SECTION :

```
HEIGHT OF THE WEB PLATE     :  150  in.
WEB THICKNESS               :  .5  in.
WIDTH OF THE FLANGE PLATE   :  40  in.
FLANGE THICKNESS            :  2  in.
DEPTH OF THE GIRDER         :  154  in.
```

‡‡‡ SPACING OF INTERMEDIATE STIFFENER ‡‡‡

STIFFENER SPACING IN SEGMENT 1 : 104 in.

LOCATION OF FIRST INTERMEDIATE STIFFENER FROM LEFT SUPPORT : 56 in.

NUMBER OF SPACING   : 11

DO YOU WANT TO INCREASE THE WEB THICKNESS OR
REDUCE THE STIFFENER SPACING BY ONE INCREMENT (Y/N) ? n

STIFFENER SPACING IN SEGMENT 2 : 110 in.

LOCATION OF FIRST INTERMEDIATE STIFFENER FROM RIGHT SUPPORT : 50 in.

NUMBER OF SPACING   : 5

DO YOU WANT TO INCREASE THE WEB THICKNESS OR
REDUCE THE STIFFENER SPACING BY ONE INCREMENT (Y/N) ? n

‡‡‡ DESIGN OF INTERMEDIATE STIFFENER ‡‡‡

‡‡‡ STIFFENER IN SEGMENT 1 ‡‡‡

SIZE OF INTERMEDIATE STIFFENERS IN SEGMENT 1 :

WIDTH OF STIFFENERS   : 5.5 in.

THICKNESS OF STIFFENERS : .375 in.

AREA OF ONE PLATE   : 2.0625 sq.in

DO YOU WANT TO ALTER THE STIFFENER WIDTH (Y/N) ? n

SIZE OF STIFFENERS IN SEGMENT 1 : .375 x 5.5 x 148 in.

STIFFENER IN SEGMENT 2

SIZE OF INTERMEDIATE STIFFENERS IN SEGMENT 2 :

WIDTH OF STIFFENERS   : 5 in.

THICKNESS OF STIFFENERS : .375 in.

AREA OF ONE PLATE   : 1.875 sq.in

DO YOU WANT TO ALTER THE STIFFENER WIDTH (Y/N) ? n

SIZE OF STIFFENERS IN SEGMENT 2 : .375 x 5 x 148 in.

DESIGN OF BEARING STIFFENER

BEARING STIFFENERS MUST EXTEND APPROXIMATELY TO THE EDGES
OF THE FLANGE PLATES AND SHOULD BE PROVIDED IN PAIR !

SIZE OF BEARING STIFFENER AT SUPPORTS   : 19 x 1.25 x 150 in.

SIZE OF BEARING STIFFENER AT LOAD NO. 1 : 19 x 1.25 x 150 in.

DO YOU WANT TO ALTER THE WIDTH OF BEARING STIFFENERS (Y/N) ? n

DESIGN OF WELD CONNECTION

FLANGE TO WEB CONNECTION

WELD WITH MINIMUM SIZE : 5/16 in., IS SUFFICIENT.

DO YOU WANT TO INCREASE THE WELD SIZE BY ONE INCREMENT (Y/N) ? n

MINIMUM LENGTH OF FILLET WELD : 1.5 in.

MAXIMUM SPACING BETWEEN WELDS : 12 in.

CHOOSE EITHER LENGTH OF FILLET-WELD SEGMENTS OR SPACING (L/A) ==> a

ENTER SPACING BETWEEN FILLET WELDS (in.) ==> 12

FLANGE TO WEB CONNECTION :

        WELD SIZE   : 5/16 in.
        WELD LENGTH   : 5 in.
        WELD SPACING   : 12 in.

DO YOU WANT TO REDESIGN (Y/N) ? n

INTERMEDIATE STIFFENER-TO-WEB CONNECTION

DESIGN OF WELD SIZE IN SEGMENT 1 :

WELD WITH MINIMUM SIZE : 3/16 in., IS SUFFICIENT.

DO YOU WANT TO INCREASE THE WELD SIZE BY ONE INCREMENT (Y/N) ? n

MINIMUM LENGTH OF FILLET WELD : 1.5 in.

MAXIMUM SPACING BETWEEN WELDS : 9 in.

CHOOSE EITHER LENGTH OF FILLET-WELD SEGMENTS OR SPACING (L/A) ==> a

ENTER SPACING BETWEEN FILLET WELDS (in.) ==> 9

STIFFENERS-TO-WEB CONNECTION IN SEGMENT : 1

        WELD SIZE   : 3/16 in.
        WELD LENGTH   : 2 in.
        WELD SPACING   : 9 in.

DO YOU WANT TO REDESIGN (Y/N) ? n

DESIGN OF WELD SIZE IN SEGMENT 2 :

WELD WITH MINIMUM SIZE : 3/16 in., IS SUFFICIENT.

DO YOU WANT TO INCREASE THE WELD SIZE BY ONE INCREMENT (Y/N) ? n

MINIMUM LENGTH OF FILLET WELD : 1.5 in.

MAXIMUM SPACING BETWEEN WELDS : 9 in.

CHOOSE EITHER LENGTH OF FILLET-WELD SEGMENTS OR SPACING (L/A) ==> a

ENTER SPACING BETWEEN FILLET WELDS (in.) ==> 9

STIFFENERS-TO-WEB CONNECTION IN SEGMENT : 2

    WELD SIZE   : 3/16 in.
    WELD LENGTH   : 2.5 in.
    WELD SPACING   : 9 in.

DO YOU WANT TO REDESIGN (Y/N) ? n

           *** BEARING STIFFENERS TO WEB CONNECTION ***

WELD WITH MINIMUM SIZE : 5/16 in., IS SUFFICIENT.

DO YOU WANT TO INCREASE THE WELD SIZE BY ONE INCREMENT (Y/N) ? n

BEARING STIFFENERS TO WEB CONNECTION :

WELD SIZE : 5/16 in.   WELD CONTINUOUSLY !

          *** CALCULATION OF TOTAL WEIGHT ***

WEIGHT OF FLANGE AND WEB PLATES   : 119.95 kips

THE TOTAL WEIGHT OF THE PLATE GIRDER : 128.70 kips

                PRESS ANY KEY TO CONTINUE

   *** SUMMARY OF FINAL DESIGN OF PLATE GIRDER ***

            DESIGN NUMBER 1

   NON-HYBRID GIRDER BASED ON LRFD SPECIFICATION

DEPTH OF WEB PLATE   : 150 in.

THICKNESS OF WEB PLATE   : .5 in.

WIDTH OF FLANGE PLATE   : 40 in.

THICKNESS OF FLANGE PLATE   : 2 in.

DEPTH OF PLATE GIRDER   : 154 in.

            PRESS ANY KEY TO CONTINUE

LOCATION OF FIRST STIFFENER FROM LEFT SUPPORT   : 56 in.

SPACING OF INERMEDIATE STIFFENERS IN SEGMENT 1 : 104 in.

SIZE OF INTERMEDIATE STIFFENERS IN SEGMENT 1 :

```
          WIDTH OF STIFFENERS      :  5.5  in.
          THICKNESS OF STIFFENERS  :  .375 in.
          LENGTH OF STIFFENERS     :  148  in.

NUMBER OF INTERMEDIATE STIFFENERS IN SEGMENT  1  :  11

                    PRESS ANY KEY TO CONTINUE


LOCATION OF FIRST STIFFENER FROM RIGHT SUPPORT  :  50  in.

SPACING OF INERMEDIATE STIFFENERS IN SEGMENT 2 :  110  in.

SIZE OF INTERMEDIATE STIFFENERS IN SEGMENT  2  :

          WIDTH OF STIFFENERS      :  5  in.
          THICKNESS OF STIFFENERS  :  .375 in.
          LENGTH OF STIFFENERS     :  148  in.

NUMBER OF INTERMEDIATE STIFFENERS IN SEGMENT  2  :  5

                    PRESS ANY KEY TO CONTINUE


SIZE OF BEARING STIFFENERS AT SUPPORTS :

          WIDTH OF STIFFENER    :  19    in.
          THICKNESS OF STIFFENER :  1.25  in.
          LENGTH OF STIFFENER   :  150   in.

SIZE OF BEARING STIFFENER AT LOAD NO. 1  :

          WIDTH OF STIFFENER    :  19    in.
          THICKNESS OF STIFFENER :  1.25  in.
          LENGTH OF STIFFENER   :  150   in.

                    PRESS ANY KEY TO CONTINUE


FLANGE TO WEB CONNECTION :

          WELD SIZE    :  3/16 in.
          WELD LENGTH  :  2.5  in.
          WELD SPACING :  9    in.

INTERMEDIATE STIFFENERS-TO-WEB CONNECTION IN SEGMENT :  1

          WELD SIZE    :  3/16 in.
          WELD LENGTH  :  2    in.
          WELD SPACING :  9    in.


                    PRESS ANY KEY TO CONTINUE


INTERMEDIATE STIFFENERS-TO-WEB CONNECTION IN SEGMENT :  2

          WELD SIZE    :  3/16 in.
          WELD LENGTH  :  2.5  in.
          WELD SPACING :  9    in.

BEARING STIFFENERS TO WEB CONNECTION :

          WELD SIZE    :  5/16 in.


WEIGHT OF FLANGE AND WEB PLATES : 119.95 kips

TOTAL WEIGHT OF PLATE GIRDER    : 128.70 kips


                    PRESS ANY KEY TO CONTINUE
```

9.16 PROBLEMS

9.1 Design a simply supported welded hybrid plate girder subjected to a distributed load of intensity 3.5 K/ft including the girder weight and a concentrated load of 400 Kips at a distance of 50 ft from the left support, as shown in Fig. 9.54. The girder has lateral

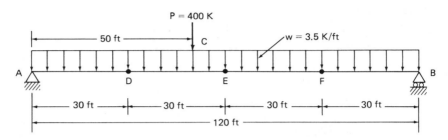

Figure 9.54

supports at the two ends and at points D, E, and F, as indicated on the figure. Double plates are used for intermediate stiffeners. Use A36 steel ($F_y = 36$ ksi) for the web plate and stiffeners and A572 ($F_y = 60$ ksi) for the flange plates. Use E70 electrode and shielded metal arc welding (SMAW) for connecting various plates.

9.2 A simply supported welded plate girder has a span of 200 ft and is subjected to a uniform load of 4 Kips/ft including its own weight. The girder is laterally supported at the supports at the midspan. Based on a preliminary design, a 180 in. $\times$ 0.75 in. plate is selected for the web, and a 40 in. $\times$ 1.5 in. plate is selected for each flange. Using A36 steel ($F_y = 36$ ksi) and considering the bending stresses in the flange only, check the adequacy of this preliminary design.

9.3 Design the intermittent web-to-flange fillet weld for the hybrid girder designed in Sec. 9.10 of the book, using submerged arc welding (SAW) and E70 electrode. Use the minimum weld size and minimum length for the intermittent fillet segments.

9.4 Solve Example 1 of Sec. 9.11, assuming that the plate girder consists of 2 WT12 $\times$ 52 with $F_y = 60$ ksi and a 100 in. $\times$ 0.50 in. plate with $F_y = 36$ ksi.

9.5 Design the bearing stiffeners in Example 1 of Sec. 9.11.

9.6 Design the intermediate stiffeners in Example 1 of Sec. 9.11, using
   a. single stiffeners
   b. double stiffeners

References

1. Abuyounes, S. and H. Adeli, "Optimization of Steel Plate Girders Via General Geometric Programming," *Journal of Structural Mechanics*, Vol. 14, no. 4 (1986), pp. 501–526.

2. Abuyounes, S. and H. Adeli, "Optimization of Hybrid Steel Plate Girders," *Computers and Structures*, Vol. 27, no. 2, 1987, pp. 241–248.

3. Adeli, H., "Microcomputer-Aided Instruction of Structural Steel Design," *Microcomputers in Civil Engineering*, Vol. 2, no. 1 (1987), pp. 75–82.

4. Adeli, H., "Microcomputers in Civil Engineering," in A. Kent and J. G. Williams, eds., *Encyclopedia of Microcomputers*, New York: Marcel Dekker, 1988 (in press).

5. Adeli, Ed., *Expert Systems in Construction and Structural Engineering*, London: Chapman and Hall, 1988.

6. Adeli, H. and K. V. Balasubramanyam, "Interactive Microcomputer-Aided Design of Circular Suspension Cable Roofs," *Journal of Computers and Structures*, Vol. 23, no. 6 (1986), pp. 837–844.

7. Adeli, H. and K. V. Balasubramanyam, "A Heuristic Approach for Interactive Analysis of Bridge Trusses Under Moving Loads," *Microcomputers in Civil Engineering*, Vol. 2, no. 1 (1987), pp. 1–18.

8. Adeli, H. and K. Chompooming, "Optimization of Multispan Plate Girders," to be published.

9. Adeli, H. and K. Chompooming, "Interactive Optimization of Nonprismatic Girders," to be published.

10. Adeli, H. and H. Chu, "Interactive Microcomputer-Aided Load and Resistance Factor Design of Steel Frames," *Journal of Computing in Civil Engineering*, ASCE, Vol. 2, no. 1, 1988.

11. Adeli, H. and H. Chyou, "Plastic Analysis of Irregular Frames on Microcomputers," *Journal of Computers and Structures*, Vol. 23, no. 2 (1986), pp. 233–240.

12. Adeli, H. and H. Chyou, "Microcomputer-Aided Optimal Plastic Design of Frames," *Journal of Computing in Civil Engineering*, ASCE, Vol. 1, no. 1 (1987), pp. 20–34.

13. Adeli, H. and J. Fiedorek, "Microcomputer-Aided Design and Drafting of Moment-Resisting Connections in Steel Buildings," *Microcomputers in Civil Engineering*, Vol. 1, no. 1 (1986), pp. 32–44.

14. Adeli, H. and J. Fiedorek, "A MICROCAD System for Design of Steel Connections—Program Structure and Graphic Algorithms," *Journal of Computers and Structures*, Vol. 24, no. 2 (1986), pp. 281–294.

15. Adeli, H. and J. Fiedorek, "A MICROCAD System for Design of Steel Connections—Applications," *Journal of Computers and Structures*, Vol. 24, no. 3 (1986), pp. 361–374.

16. Adeli, H. and J. Fiedorek, "Computer-Aided Design of Beam-Column Seated Angle Connections," *International Journal of Civil Engineering for Practicing and Design Engineers*, Vol. 5, no. 8 (1986), pp. 719–756.

17. Adeli, H. and J. Fiedorek, "Interactive Microcomputer-Aided Design of Shop-Welded and Field-Bolted Beam-Column Connections," *Computer-Aided Design*, Vol. 19, no. 3 (1987), pp. 115–121.

18. Adeli, H. and D. Hawkins, "A Graphics Preprocessor for Computer-Aided Design and Drafting of Frame Structures," *Microcomputers in Civil Engineering*, Vol. 1, no. 2 (1986), pp. 107–120.

19. Adeli, H. and O. Kamal, "Efficient Optimization of Space Trusses," *Computers and Structures*, Vol. 24, no. 3 (1986), pp. 501–511.

20. Adeli, H. and N. Mabrouk, "Optimum Plastic Design of Unbraced Frames of Irregular Configuration," *International Journal of Solids and Structures*, Vol. 22, no. 10 (1986), pp. 1117–1128.

21. Adeli, H. and Y. Paek, "Computer-Aided Design of Structures Using LISP," *Computers and Structures*, Vol. 22, no. 6 (1986), pp. 939–956.

22. Adeli, H. and K. Phan, "Interactive Design of Structures on Microcomputers," *International Journal of Civil Engineering for Practicing and Design Engineers*, Vol. 4, no. 5 (May 1985), pp. 413–437.

23. Adeli, H. and K. Phan, "Interactive Computer-Aided Design of Non-Hybrid and Hybrid Steel Plate Girders," *Journal of Computers and Structures*, Vol. 22, no. 3 (1986), pp. 267–289.

24. Artwick, B. A., *Applied Concepts in Microcomputer Graphics*. Englewood Cliffs, New Jersey: Prentice-Hall, Inc., 1984.

25. Boggs, R. A., *Advanced BASIC—Data Structures and File Techniques*, Reston, Virginia: Reston Publishing Company, 1984.

26. Chen, W. F. and E. M. Lui, "Column with End Restraint and Bending in Load and Resistance Design Factor," *Engineering Journal*, American Institute of Steel Construction, Vol. 22, no. 3 (1985), pp. 105–132.

27. Cooper, P. B., T. V. Galambos, and M. K. Ravindra, "LRFD Criteria for Plate Girders," *Journal of Structural Division*, ASCE, Vol. ST9 (1978).

28. Cowan, H. J. and F. Wilson, *Structural Systems*. New York: Van Nostrand Reinhold, 1981.

29. Fried, S. S., "Evaluating 8087 Performance On the IBM PC," *BYTE Guide to the IBM PC* (Fall 1984), pp. 197–208.

30. Gaggero, G. and H. B. Wilson, "Implementing Large Engineering Programs on Microcomputers," *Proceedings of the 1st National Conference on Microcomputers in Civil Engineering* (November 1-3, 1983), pp. 227-235.

31. Galambos, T. V., *Structural Members and Frames.* Englewood Cliffs, New Jersey: Prentice-Hall, Inc., 1968.

32. Hart, F., W. Henn, and H. Sontag, *Multi-Storey Buildings in Steel.* 2nd ed., London: Collins Professional and Technical Books, 1985.

33. Kittner, M. L. and B. Northcutt, *Basic BASIC—A Structured Approach.* Menlo Park, California: The Benjamin/Cummings Publishing Company, 1984.

34. Newell, M. E., R. G. Newell, and T. L. Sancha, "A New Approach to the Shaded Picture Problem," *Proceedings of the ACM National Conference* (1972).

35. Newman, W. M. and R. F. Sproull, *Principles of Interactive Computer Graphics*, 2nd ed. New York: McGraw Hill, 1979.

36. Perry, G., "Fast and BASIC," *PC World* (September 1986), pp. 213-217.

37. Salmon, C. G. and J. E. Johnson, *Steel Structures—Design and Behavior*, 2nd ed. New York: Harper & Row, 1980.

38. Salvadori, M. and R. Heller, *Structure in Architecture—The Building of Buildings*, 3rd ed. Englewood Cliffs, New Jersey: Prentice-Hall, Inc., 1986.

39. Teal, E. J., "Seismic Design Practice for Steel Buildings," *Engineering Journal, AISC* (fourth quarter, 1975).

40. Traister, R. J., *Music and Speech for the IBM PC.* Blue Ridge Summit, PA: Tab Books, 1983.

41. Wang, C. K., *Structural Analysis on Microcomputers*, New York: MacMillan Publishing Co., 1986.

42. Young, N. W. and R. O. Disque, "Design Aids for Single Plate Framing Connections," *Engineering Journal, AISC* (fourth quarter, 1981), pp. 129-148.

43. Yura, J. A., T. V. Galambos, and M. K. Ravindra, "The Bending Resistance of Steel Beams," *Journal of Structural Division, ASCE*, Vol. ST9 (1978).

44. *American National Standard Minimum Design Loads for Buildings and Other Structures.* American National Standards Institute, National Bureau of Standards, 1982.

45. *Building Code Requirements for Reinforced Concrete (ACI 318-83).* Detroit: American Concrete Institute, 1983.

46. *Engineering for Steel Construction.* Chicago: American Institute of Steel Construction, 1984.

47. *Manual of Steel Construction*, 8th ed. Chicago: American Institute of Steel Construction, 1980.

48. *Manual of Steel Construction—Load and Resistance Factor Design.* Chicago: American Institute of Steel Construction, 1986.

49. *Steel Bridges: The Best of Current Practice.* Chicago: American Institute of Steel Construction, 1985.

50. *Uniform Building Code.* Whittier, California: International Conference of Building Officials, 1985.

Index